Contributions to Statistics

For other titles published in this series, go to
www.springer.com/series/2912

Frédéric Ferraty
Editor

Recent Advances in Functional Data Analysis and Related Topics

Physica-Verlag
A Springer Company

Editor
Frédéric Ferraty
Mathematics Toulouse Institute
Toulouse University
Narbonne road 118
31062 Toulouse
France
frederic.ferraty@math.univ-toulouse.fr

ISSN 1431-1968
ISBN 978-3-7908-2735-4 e-ISBN 978-3-7908-2736-1
DOI 10.1007/978-3-7908-2736-1
Springer Heidelberg Dordrecht London New York

Library of Congress Control Number: 2011929779

Cover design: eStudio Calamar S.L.

Printed on acid-free paper

Physica-Verlag is a brand of Springer-Verlag Berlin Heidelberg
Springer-Verlag is part of Springer Science+Business Media (www.springer.com)

Preface

Nowaday, the progress of high-technologies allow us to handle increasingly large datasets. These massive datasets are usually called "high-dimensional data" . At the same time, different ways of introducing some continuum in the data appeared (use of sophisticated monitoring devices, function-based descriptors as the density function for instance, etc). Hence, the data can be considered as observations varying over a continuum defining a subcategory of high-dimensional data called functional data. Statistical methodologies dealing with functional data are called Functional Data Analysis (FDA), the "functional" word emphasizing the fact that the statistical method takes into accound the functional feature of the data. The failure of standard multivariate statistical analyses, the numerous fields of applications as well as the new theoretical challenges motivate an increasingly statistician community to develop new methodologies. The huge research activity around FDA and its related fields produces very fast progress. Then, it is necessary to propose regular snapshots about the most recent advances in this topic.

This is the main goal of the International Workshop on Functional and Operatorial Statistics (IWFOS'2011, Santander, Spain) which is the second edition of the first successful one (IWFOS'2008, Toulouse, France) initiated by the working group STAPH (Toulouse Mathematics Institute, France). This volume gathers peer-reviewed contributions authored by outstanding confirmed experts as well as young brillant researchers. The presentation of these contributions in a short (around 6 pages a contribution) and concise way makes the reading and use of this book very easy. As a by-product, the reader should find most of representative and significant recent advances in this field, mixing works oriented towards applications (with original datasets, computational issues, applications in numerous fields of Sciences - biometrics, chemometrics, economics, medicine, etc) with fundamental theoretical ones. This volume contents a wide scope of statistical topics: change point detection, clustering, conditional density/expectation/mode/quantiles/extreme quantiles, covariance operators, depth, forecasting, functional additive regression, functional extremality, functional linear regression, functional principal components analyses, functional single index model, functional varying coefficient models, generalized additive models, hilbertian processes, nonparametric models, noisy observations, quantiles in functions spaces, random fields, semi-functional models, statistical inference, structural tests, threshold-based procedures, time series, variable selection, wavelet-based smoothing, etc. These statistical advances deal with numerous kind of interesting datasets (functional data, high-dimensional data, longitudinal functional data, multidimensional curves, spatial functional data, sparse functional data, spatial-temporal data) and propose very attractive applications in various fields of Sciences: DNA minicircles, electoral behavior, electricity spot markets, electro-cardiogram records, gene expression, irradiance data (exploitation of solar energy), magnetic resonance spectroscopy data (neurocognitive impairment), material sciences, signature recognition, spectrometric curves (chemometrics), tractography data (multiple sclerosis), etc.

Clearly, this volume should be very attractive for a large audience, like academic researchers, graduate/PhD students as well as engineers using regularly recent statistical developments in his work.

At last, this volume is a by-product of the organization of IWFOS'2011 which is chaired by two other colleagues: Juan A. Cuesta-Albertos (Santander, Spain) and Wenceslao Gonzalez-Manteiga (Santiago de Compostela, Spain). Their trojan work as well as their permanent support and enthusiasm are warmly and gratefully thanked.

Toulouse, France
March 2011

Frédéric Ferraty
The Editor and co-Chair of IWFOS'2011

Acknowledgements

First of all, the vital material of this volume was provided by the contributors. Their outstanding expertise in this statistical area as well as their valuable contributions guarantee the high scientific level of this book and hence the scientific success of IWFOS'2011. All the contributors are warmly thanked.

This volume could not have existed without the precious and efficience help of the members of the IWFOS'2011 Scientific Committee named J. Antoch (Prague, Czech Rep.), E. del Barrio (Valladolid, Spain), G. Boente (Buenos Aires, Argentina), C. Crambes (Montpellier, France), A. Cuevas (Madrid, Spain), L. Delsol (Orléans, France), D. Politis (California, USA), M. Febrero-Bande (Santiago de Compostela, Spain), K. Gustafson (Colorado, USA), P. Hall (Melbourne, Ausralia), S. Marron (North Carolina, USA), P. Sarda (Toulouse, France), M. Valderrama (Granada, Spain), S. Viguier-Pla (Toulouse, France), Q. Yao (London, UK). Their helpful and careful involvement in the peer-reviewing process has contributed significantly to the high scientific level of this book; all of them are gratefully acknowledged.

Of course, this book is a by-product of IWFOS'2011 and its success is due to the fruitful collaboration of people from the University of Cantabria (Santander, Spain), the University of Santiago de Compostela (Spain) and the University of Toulouse (France). In particular, A. Boudou (Toulouse, France), A. Martínez Calvo (Santiago de Compostela, Spain), A. Nieto-Reyes (Santander, Spain), B. Pateiro-López (Santiago de Compostela), Gema R. Quintana-Portilla (Santander, Spain), Y. Romain (Toulouse, France) and P. Vieu (Toulouse, France), members of the Organizing Committee, have greatly contributed to the high quality of IWFOS'2011 and are gratefully thanked. It is worth noting that this scientific event is the opportunity to emphasize the links existing between these three universities and this is why these International Workshop is chaired by three people, one of each above-mentioned university. Clearly, this Workshop should strengthen the scientific collaborations between these three universities.

A special thank is addressed to the working group STAPH (http://www.math.univ-toulouse.fr/staph). Its intensive and dynamic research activities oriented towards functional and operatorial statistics with a special attention to Functional Data Analysis and High-Dimensional Data contributed to the development of numerous scientific collaborations with statisticians in the whole world. A first consequence was the creation, the organization and the management of the first edition of IWFOS (Toulouse, France, 2008). The success of IWFOS'2008 was certainly the necessary starting point allowing the emergence of IWFOS'2011. All its members and collaborators are warmly acknowledged.

The final thanks go to institutions/organizations which supported this Workshop via grants or administrative supports. In particular, the Chairs of IWFOS'2011 would like to express their grateful thanks to:

- the Departamento de Matemáticas, Estadística y Computación, the Facultad de Ciencias and the Vicerrectorado de Investigación y Transferencia del Conocimiento de la Universidad de Cantabria,
- the Programa Ingenio Mathematica, iMATH,
- the Acciones Complementarias del Ministerio Español de Ciencia e Innovación,
- the Toulouse Mathematics Institute,
- the IAP research network in statistics

March 2011

Juan A. Cuesta-Albertos
Frédéric Ferraty
Wenceslao Gonzalez-Manteiga
The co-Chairs of IWFOS'2011

Contents

Chapter 1
Penalized Spline Approaches for Functional Principal Component Logit Regression

A. Aguilera, M. C. Aguilera-Morillo, M. Escabias, M. Valderrama

Abstract The problem of multicollinearity associated with the estimation of a functional logit model can be solved by using as predictor variables a set of functional principal components. The functional parameter estimated by functional principal component logit regression is often unsmooth. To solve this problem we propose two penalized estimations of the functional logit model based on smoothing functional PCA using P-splines.

1.1 Introduction

The aim of the functional logit model is to predict a binary response variable from a functional predictor and also to interpret the relationship between the response and the predictor variables. To reduce the infinite dimension of the functional predictor and solve the multicollinearity problem associated to the estimation of the functional logit model, Escabias et al. (2004) proposed to use a reduced number of functional principal components (pc's) as predictor variables. A functional PLS based solution was also proposed by Escabias et al. (2006). The problem associated with these approaches is that in many cases the estimated functional parameter is not smooth and therefore difficult to interpret. Different penalized likelihood estimations with B-spline basis were proposed in the general context of functional generalized linear

Ana Aguilera
Department of Statistics and O. R. University of Granada, Spain, e-mail: aaguiler@ugr.es

Maria del Carmen Aguilera-Morillo
Department of Statistics and O. R. University of Granada, Spain, e-mail: caguilera@ugr.es

Manuel Escabias
Department of Statistics and O. R. University of Granada, Spain, e-mail: escabias@ugr.es

Mariano Valderrama
Department of Statistics and O. R. University of Granada, Spain, e-mail: valderra@ugr.es

models to solve this problem (Marx and Eilers, 1999; Cardot and Sarda, 2005). In this paper we introduce two different penalized estimation approaches based on smoothed functional principal component analysis (FPCA). On one hand, FPCA of P-spline approximation of sample curves is performed. On the other hand, a discrete P-spline penalty is included in the own formulation of FPCA.

1.2 Background

Let us consider a sample of functional observations $x_1(t), x_2(t), \ldots, x_n(t)$ of a fixed design functional variable and let $y_1, y_2, \ldots, y_n$ be a random sample of a binary response variable Y associated to them. That is, $y_i \in \{0,1\}, i = 1, \ldots, n$. The functional logistic regression model is given by

$$y_i = \pi_i + \varepsilon_i, \quad i = 1, \ldots, n,$$

where π_i is the expectation of Y given $x_i(t)$ modeled as

$$\pi_i = P[Y = 1 | \{x_i(t) : t \in T\}] = \frac{\exp\{\alpha + \int_T x_i(t)\beta(t)\,dt\}}{1 + \exp\{\alpha + \int_T x_i(t)\beta(t)\,dt\}}, \quad i = 1, \ldots, n,$$

α being a real parameter, $\beta(t)$ a parameter function, $\{\varepsilon_i : i = 1, \ldots, n\}$ independent errors with zero mean and T the support of the sample paths $x_i(t)$.

The logit transformations can be expressed as

$$l_i = \ln\left[\frac{\pi_i}{1 - \pi_i}\right] = \alpha + \int_T x_i(t)\beta(t)\,dt, \quad i = 1, \ldots, n. \tag{1.1}$$

A way to estimate the functional logit model is to consider that both, the sample curves and the parameter function, admit an expansion in terms of basis functions. Then, the functional logit model turns into a multiple logit model whose design matrix is the product between the matrix of basis coefficients of sample paths and the matrix of inner products between basis functions (Escabias et al., 2004). The estimation of this model is affected by multicollinearity due to the high correlation between the columns of the design matrix. In order to obtain a more accurate and smoother estimation of the functional parameter than the one provided by standard functional principal component logit regression (FPCLR), we present in this paper two penalized estimation approaches based on P-spline smoothing of functional PCA.

1.3 Penalized estimation of FPCLR

In general, the functional logit model can be rewritten in terms of functional principal components as

$$L = \alpha \mathbf{1} + \Gamma \gamma, \tag{1.2}$$

where $\Gamma = (\xi_{ij})_{n\times p}$ is a matrix of functional pc's of $x_1(t), \ldots, x_n(t)$ and γ is the vector of coefficients of the model.

By considering that the predictor sample curves admit the basis expansions $x_i(t) = \sum_{j=1}^{p} a_{ij}\phi_j(t)$, the functional parameter can be also expressed also in terms of the same basis, $\beta(t) = \sum_{k=1}^{p} \beta_k \phi_k(t)$, and the vector β of basis coefficients is given by $\hat{\beta} = F\hat{\gamma}$, where the way of computing F depends on the kind of FPCA used to obtain the pc's.

An accurate estimation of the parameter function can be obtained by considering only a set of optimal principal components as predictor variables. In this paper we select the optimal number of predictor pc's by using a leave-one-out cross validation method that maximizes the area under ROC curve computed by following the process outlined in Mason and Graham (2002). To obtain this area, observed and predicted values are required. In this case, we have considered y_i the ith observed value of the binary response and $\hat{y}_i^{(-i)}$ the ith predicted value obtained by deleting the ith observation of the design matrix in the iterative estimation process.

Let us consider that the sample curves are centered and belong to the space $L^2[T]$ with the usual inner product defined by $< f, g >= \int_T f(t)g(f)dt$. In the standard formulation of functional PCA, the jth principal component scores are given by

$$\xi_{ij} = \int_T x_i(t) f_j(t)\, dt, \; i = 1, \ldots, n, \tag{1.3}$$

where the weight function or factor loading f_j is obtained by solving

$$\begin{cases} Max_f \;\; Var\left[\int_T x_i(t) f(t)\, dt\right] \\ s.t. \; \|f\|^2 = 1 \; and \; \int f_\ell(t) f(t)\, dt = 0, \;\; \ell = 1, \ldots, j-1. \end{cases}$$

The weight functions f_j are the solutions to the eigenequation $Cf_j = \lambda_j f_j$, with $\lambda_j = Var[\xi_j]$ and C being the sample covariance operator defined by $Cf = \int c(.,t) f(t)\, dt$, in terms of the sample covariance function $c(s,t) = \frac{1}{n}\sum_{i=1}^{n} x_i(s) x_i(t)$.

In practice, functional PCA has to be estimated from discrete time observations of each sample curve $x_i(t)$ at a set of times $\{t_{i0}, t_{i1}, \ldots, t_{im_i} \in T, \; i = 1, \ldots, n\}$. The sample information is given by the vectors $x_i = (x_{i0}, \ldots, x_{im_i})'$, with x_{ik} the observed value for the ith sample path $x_i(t)$ at time t_{ik} $(k = 0, \ldots, m_i)$.

When the sample curves are smooth and observed with error, least squares approximation in terms of B-spline basis is an appropriate solution for the problem of reconstructing their true functional form. This way, the vector of basis coefficients of each sample curve that minimizes the least squares error is given by $\hat{a}_i = (\Phi_i'\Phi_i)^{-1}\Phi_i' x_i$, with $\Phi_i = (\phi_j(t_{ik}))_{m_i\times p}$ and $a_i = (a_{i1}, \ldots, a_{ip})'$.

Functional PCA is then equivalent to the multivariate PCA of $A\Psi^{\frac{1}{2}}$ matrix, $\Psi^{\frac{1}{2}}$ being the squared root of the matrix of the inner products between B-spline basis functions (Ocaña et al. 2007). Then, matrix F that provides the relation between the basis coefficients of the functional parameter and the parameters in terms of principal components is given by $F = \Psi_{p\times p}^{-\frac{1}{2}} G_{p\times n}$, where G is the matrix whose columns are the eigenvectors of the sample covariance matrix of $A\Psi^{1/2}$. This non smoothed FPCA estimation of functional logit models with B-spline basis was performed by Escabias et al. (2004).

1.3.1 Functional PCA via P-splines

Now, we propose a penalized estimation based on functional PCA of the P-spline approximation of the sample curves. The basis coefficients in terms of B-splines are computed by introducing a discrete penalty in the least squares criterion (Eilers and Marx, 1996), so that we have to minimize $(x_i - \Phi_i a_i)'(x_i - \Phi_i a_i) + \lambda a_i' P_d a_i$, where $P_d = \left(\triangle^d\right)' \triangle^d$ and $\triangle^d$ is the differencing matrix that gives the dth-order differences of a_i. The solution is then given by $\hat{a}_i = \left(\Phi_i'\Phi_i + \lambda P_d\right)^{-1} \Phi_i' x_i$, and the smoothed parameter is chosen by leave-one-out cross validation.

Then, we carry out the multivariate PCA of $A\Psi^{\frac{1}{2}}$ matrix as explained above. The difference between smoothed FPCA via P-splines and non smoothed FPCA is only the way of computing the basis coefficients (rows of matrix A), with or without penalization, respectively.

1.3.2 P-spline smoothing of functional PCA

Now we propose to obtain the principal components by maximizing a penalized sample variance that introduces a discrete penalty in the basis coefficients of principal component weights.

The jth principal component scores are defined as in equation (1.3) but now the weight functions f_j are obtained by solving

$$\begin{cases} Max_f \dfrac{var[\int x_i(t) f(t) dt]}{\|f\|^2 + \lambda PEN_d(f)} \\ s.t.\ \|f\|^2 = b'\Psi b = 1 \text{ and } b'\Psi b_l + b' P_d b_l = 0,\ \ell = 1, \ldots, j-1, \end{cases}$$

where $PEN_d(f) = b' P_d b$ is the discrete roughness penalty function, b being the vector of basis coefficients of the weight functions, $f(t) = \sum_{k=1}^{p} b_k \phi_k$, and λ the smoothing parameter estimated by leave-one-out cross validation.

Finally, this variance maximization problem is converted into an eigenvalue problem, so that, applying the Choleski factorization $LL' = \Psi + \lambda P_d$, P-spline smoothing of functional PCA is reduced to classic PCA of the matrix $A\Psi(L^{-1})'$. Then, the estimated vector β of basis coefficients of the functional parameter is given by $\hat{\beta} = F\hat{\gamma} = (L^{-1})' G\hat{\gamma}$, where G is the matrix of eigenvectors of the sample covariance matrix of $A\Psi(L^{-1})'$.

1.4 Simulation study

We are going to illustrate the good performance of the proposed penalty approaches following the simulation scheme developed in Ferraty and Vieu (2003) and Escabias et al. (2006). We simulated 1000 curves of two different classes of sample curves. For the first class we simulated 500 curves according to the random function $x(t) = uh_1(t) + (1-u)h_2(t) + \varepsilon(t)$, and another 500 curves were simulated for the second class according to the random function $x(t) = uh_1(t) + (1-u)h_3(t) + \varepsilon(t)$, u and $\varepsilon(t)$ being uniform and standard normal simulated random values, respectively, and $h_1(t) = \max\{6 - |t-11|, 0\}, h_2(t) = h_1(t-4), h_3(t) = h_1(t+4)$. The sample curves were simulated at 101 equally spaced points in the interval $[1,21]$.

As binary response variable, we considered $Y = 0$ for the curves of the first class and $Y = 1$ for the ones of the second class. After simulating the data, we performed least squares approximation of the curves, with and without penalization, in terms of the cubic B-spline functions defined on 30 equally spaced knots of the interval $[1,21]$.

	non smoothed FPCA	FPCA via P-splines	P-spline smoothed FPCA
Number pc's	3	2	3
ROC area	0.9986	0.9985	0.9988

Table 1.1: Area under the ROC curve for the test sample with the optimum models selected by cross validation with the three different FPCA approaches (non smoothed FPCA, FPCA via P-splines ($\lambda = 24.2$) and P-spline smoothed FPCA ($\lambda = 5$)).

In order to estimate the binary response Y from the functional predictor X we have estimated three different FPCLR models by using non smoothed FPCA and the two P-spline estimation approaches of FPCA proposed in this work. A training sample of 500 curves (250 of each class) was considered to fit the model and a test sample with the remaining 500 curves to evaluate the forecasting performance of the model. The pc's were included in the model by variability order and the optimum number of pc's selected by maximizing the cross validation estimation of the area under the ROC curve. In Table 1.1 we can see that P-spline smoothed FPCA estimation provides a slightly higher area and FPCA via P-splines requires fewer components.

Escabias et al. (2006) estimated the parameter function using different methods as functional PLS logistic regression and functional principal component logit regression, obtaining in both cases a non smooth estimation. In Figure 1.1 we can see that both penalized estimations of FPCA based on P-splines provide a smooth estimation of the functional parameter. This shows that using a smoothing estimation of FPCA is required in order to obtain a smooth estimation of the functional parameter that makes the interpretation easier. Although there are not significant differences between the estimation of the parameter function provided by FPCA via P-splines and P-spline smoothed FPCA, the second approach spends much more time in cross validation procedure so that, in practice, the estimation of FPCLR based on FPCA via P-splines is more efficient.

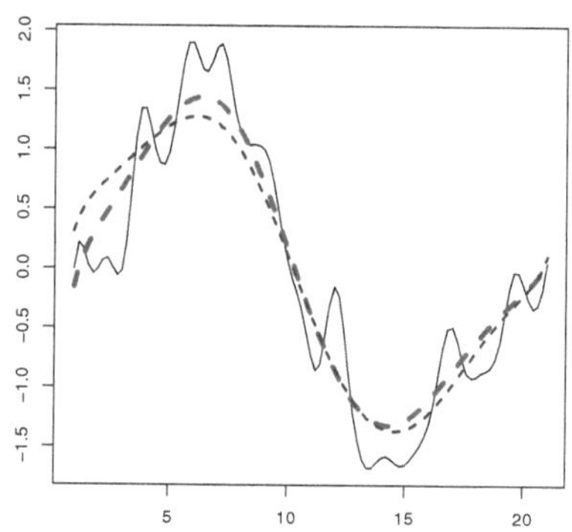

Fig. 1.1: Estimated parameter function with the three different considered FPCA estimations: non smoothed FPCA (black and continue line), FPCA via P-splines (red and long dashed line, $\lambda = 24.2$) and P-spline smoothed FPCA (blue and short dashed line, $\lambda = 5$)

Acknowledgements This research has been funded by project MTM2010-20502 from *Ministerio de Ciencia e Innovación, Spain.*

References

1. Eilers, P.H.C., Marx, B.D.: Flexible smoothing with B-splines and penalties. Stat. Sci. **11**(2), 89–121 (1996)
2. Cardot, H., Sarda, P.: Estimation in generalized linear models for functional data via penalized likelihood. J. Multivariate Anal. **92**(1), 24–41 (2005)

3. Escabias, M., Aguilera, A. M., Valderrama. M. J.: Principal component estimation of functional logistic regression: discussion of two different approaches. J. Nonparametr. Stat. **16**(3-4), 365–384 (2004)
4. Escabias, M., Aguilera, A. M., Valderrama. M. J.: Functional PLS logit regression model. Comput. Stat. Data An. **51**, 4891–4902 (2006)
5. Ferraty, F., Vieu, P.: Curves discrimination: a nonparametric functional approach. Comput. Stat. Data An. **44**, 161–173 (2003)
6. Marx, B.D., Eilers, P.H.C.: Generalized linear regression on sampled signals and curves. A P-spline approach. Technometrics **41**, 1–13 (1999)
7. Mason, S.J., Graham, N.E.: Areas beneath the relative operating characteristics (ROC) and relative operating levels (ROL) curves: Statistical significance and interpretation. Q. J. Roy. Meteor. Soc. **128**, 291–303 (2002)
8. Ocaña, F.A., Aguilera, A.M. and Escabias, M.: Computational considerations in functional principal component analysis. Computation. Stat. **22**(3), 449–465 (2007)

Chapter 2
Functional Prediction for the Residual Demand in Electricity Spot Markets

Germán Aneiros, Ricardo Cao, Juan M. Vilar-Fernández, Antonio Muñoz-San-Roque

Abstract The problem of residual demand prediction in electricity spot markets is considered in this paper. Hourly residual demand curves are predicted using nonparametric regression with functional explanatory and functional response variables. Semi-functional partial linear models are also used in this context. Forecasted values of wind energy as well as hourly price and demand are considered as linear predictors. Results from the electricity market of mainland Spain are reported. The new forecasting functional methods are compared with a naive approach.

2.1 Introduction

Nowadays, in many countries all over the world, the production and sale of electricity is traded under competitive rules in free markets. The agents involved in this market: system operators, market operators, regulatory agencies, producers, consumers and retailers have a great interest in the study of electricity load and price. Since electricity cannot be stored, the demand must be satisfied instantaneously and producers need to anticipate to future demands to avoid overproduction. Good forecasting of electricity demand is then very important from the system operator viewpoint. In the past, demand was predicted in centralized markets (see Gross and Galiana (1987)) but competition has opened a new field of study. On the other hand

Germán Aneiros
Universidade da Coruña, Spain, e-mail: ganeiros@udc.es

Ricardo Cao
Universidade da Coruña, Spain, e-mail: rcao@udc.es

Juan M. Vilar-Fernández
Universidade da Coruña, Spain, e-mail: eijvilar@udc.es

Antonio Muñoz-San-Roque
Universidad Pontificia de Comillas, Madrid, Spain, e-mail: antonio.munoz@iit.icai.upcomillas.es

prediction of residual demand of an agent is a valuable tool to establish good bidding strategies for the agent itself. Consequently, prediction of electricity residual demand is a significant problem in this sector.

Residual demand curves have been considered previously in the literature. In each hourly auction, the residual demand curve is defined as the difference of the combined effect of the demand at any possible price and the supply of the generation companies as a function of price. Consequently 24 hourly residual demand curves are obtained every day. These curves are useful tools to design optimal offers for companies operating in a day-ahead market (see Baillo et al. (2004) and Xu and Baldick (2007)). We focus on one day ahead forecasting of electricity residual demand curves. Therefore, for each day of the week, 24 curve forecasts need to be computed.

This paper proposes functional and semi-functional nonparametric and partial linear models to forecast electricity residual demand curves. Forecasted wind energy as well as forecasted hourly price and demand are incorporated as explanatory variables in the model. Nonparametric regression estimation under dependence is a useful tool for time series forecasting. Some relevant work in this field include Härdle and Vieu (1992), Hart (1996) and Härdle, Lütkepohl and Chen (1997). Other papers more specifically focused on prediction using nonparametric techniques are Carbon and Delecroix (1993), Matzner-Lober, Gannoun and De Gooijeret (1998) and Vilar-Fernández and Cao (2007). The literature on methods for time series prediction in the context of functional data is much more limited. The books by Bosq (2000) and Ferraty and Vieu (2006) are comprehensive references for linear and nonparametric functional data analysis, respectively. Faraway (1997) considered a linear model with functional response in a regression setup. Antoch et al. (2008) also used functional linear regression models to predict electricity consumption. Antoniadis, Paparoditis and Sapatinas (2006) proposed a functional wavelet-kernel approach for time series prediction and Antoniadis, Paparoditis and Sapatinas (2009) studied a method for smoothing parameter selection in this context. Aneiros-Pérez and Vieu (2008) have dealt with the problem of nonparametric time series prediction using a semi-functional partial linear model and Aneiros-Pérez, Cao and Vilar-Fernández (2010) used Nadaraya-Watson and local linear methods for functional explanatory variables and scalar response in time series prediction. Finally, Cardot, Dessertaine and Josserand (2010) use semi-parametric models for predicting electricity consumption and Vilar-Fernández, Cao and Aneiros (2010) use also semi-functional models with scalar response to predict next-day electricity demand and price.

The remaining of this paper is organized as follows. In Section 2, a mathematical description of the functional nonparametric model is given. The semi-functional partial linear model is presented in Section 3. Section 4 contains some information about the data and the empirical study concerning one-day ahead forecasting of electricity residual demand curves in Spain. The references are included at the final section of the paper.

2.2 Functional nonparametric model

The time series under study (residual demand curve) will be considered as a realization of a discrete time functional valued stochastic process, $\{\chi_t(p)\}_{t\in\mathbb{Z}}$, observed for $p \in [a,b)$. For a given hour, r, ($r \in \{1,\ldots,24\}$) of day t, the values of $\chi_t^{(r)}(p)$ indicate the energy that can be sold (positive values) or bought (negative values) at price p and the interval $[a,b)$ is the range for prices. We first concentrate on predicting the curve $\chi_{n+1}^{(r)}(p)$, after having observed a sample of values $\left\{\chi_i^{(r)}(p)\right\}_{i=1,2,\ldots,n}$. For simplicity the superindex r will be dropped off.

In the following we will assume that the sequence of functional valued random variables $\{\chi_t(p)\}_{t\in\mathbb{Z}}$ is Markovian. We may look at the problem of predicting the future curve $\chi_{n+1}(p)$ by computing nonparametric estimations, $\widehat{m}(\chi)$, of the autoregression function in the functional nonparametric (FNP) model

$$\chi_{i+1}(\bullet) = m(\chi_i) + \varepsilon_{i+1}(\bullet),\ i = 1,\ldots,n, \tag{2.1}$$

which states that the values of the residual demand at day $i+1$ is an unknown nonparametric function of the residual demand at the previous day plus some error term. These errors $\varepsilon_i(\bullet)$ are iid zero mean functional valued random variables. Thus, $\widehat{m}(\chi_n)$ gives a functional forecast for $\chi_{n+1}(\bullet)$.

In our context this approach consists on estimating the autoregression functional, m, using hourly residual demand curves and apply this estimated functional to the last observed day.

Whereas the Euclidean norm is a standard distance measure in finite dimensional spaces, the notion of semi-norm or semi-metric arises in this infinite-dimensional functional setup. Let us denote by $\mathscr{H} = \{f : C \to \mathbb{R}\}$ the space where the functional data live and by $d(\bullet,\bullet)$ a semi-metric associated with $\mathscr{H}$. Thus $(\mathscr{H}, d)$ is a semi-metric space (see Ferraty and Vieu (2006) for details).

A Nadaraya-Watson type estimator (see Nadaraya (1964) and Watson (1964)) for m in (2.1) is defined as follows

$$\widehat{m}_h^{FNP}(\chi) = \sum_{i=1}^{n-1} w_h(\chi,\chi_i)\chi_{i+1}(\bullet), \tag{2.2}$$

where the bandwidth $h > 0$ is a smoothing parameter,

$$w_h(\chi,\chi_i) = \frac{K(d(\chi,\chi_i)/h)}{\sum_{j=1}^{n} K(d(\chi,\chi_j)/h)}, \tag{2.3}$$

and the kernel function $K : [0,\infty) \to [0,\infty)$ is typically a probability density function chosen by the user.

The choice of the kernel function is of secondary importance. However, both the bandwidth and the semi-metric are relevant aspects for the good asymptotic and practical behavior of (2.2).

A key role of the semi-metric is that related to the so called "curse of dimensionality". From a practical point of view the "curse of dimensionality" can be explained as the sparsness of data in the observation region as the dimension of the data space grows. This problem is specially dramatic in the infinite-dimensional context of functional data. More specifically, Ferraty and Vieu (2006) have proven that it is possible to construct a semi-metric in such a way that the rate of convergence of the nonparametric estimator in the functional setting is similar to that of the finite-dimensional one. It is important to remark that we use a semi-metric rather than a metric. Indeed, the "curse of dimensionality" would appear if a metric were used instead of a semi-metric.

In functional data it is usual to consider semi-metrics based on semi-norms. Thus, Ferraty and Vieu (2006) recommend, for smooth functional data, to take as semi-norm the L_2 norm of some q-th derivative of the function. For the case of rough data curves, these authors suggest to construct a semi-norm based on the first q functional principal components of the data curves.

2.3 Semi-functional partial linear model

Very often there exist exogenous scalar variables that may be useful to improve the forecast. For the residual demand prediction this may be the case of the hourly wind energy in the market and the hourly price and demand. Although these values cannot be observed in advance, one-day ahead forecasts can be used to anticipate the values of these three explanatory variables. Previous experience also suggests that an additive linear effect of these variables on the values to forecast might occur. In such setups, it seems natural to generalize model (2.1) by incorporating a linear component. This gives the semi-functional partial linear (SFPL) model:

$$\chi_{i+1}(\bullet) = \mathbf{x}_{i+1}^T \beta(\bullet) + m(\chi_i) + \varepsilon_{i+1}(\bullet),\ i = 1, \ldots, n, \tag{2.4}$$

where $\mathbf{x}_i = (x_{i1}, \ldots, x_{ip})^T \in \mathbb{R}^p$ is a vector of exogenous scalar covariates and $\beta(\bullet) = (\beta_1(\bullet), \ldots, \beta_p(\bullet))^T$ is a vector of unknown functions to be estimated.

Now, based on the SFLP model, we may look at the problem of predicting $\chi_{n+1}(\bullet)$ by computing estimations $\widehat{\beta}$ and $\widehat{m}(\chi)$ of β and $m(\chi)$ in (2.4), respectively. Thus, $\mathbf{x}_{n+1}^T \widehat{\beta}(\bullet) + \widehat{m}(\chi_n)$ gives the forecast for $\chi_{n+1}(\bullet)$.

An estimator for $\beta(\bullet)$ based on kernel and ordinary least squares ideas was proposed in Aneiros-Pérez and Vieu (2006) in the setting of independent data. More specifically, recall the weights $w_h(\chi, \chi_i)$ defined in the previous subsection and denote $\widetilde{\mathbf{X}}_h = (\mathbf{I} - \mathbf{W}_h)\mathbf{X}$ and $\widetilde{\chi}_h = (\mathbf{I} - \mathbf{W}_h)\chi$, with $\mathbf{W}_h = (w_h(\chi_i, \chi_j))_{1 \le i,j \le n-1}$, $\mathbf{X} = (x_{ij})_{1 \le i \le n-1, 1 \le j \le p}$ and $\chi(\bullet) = (\chi_2(\bullet), \ldots, \chi_n(\bullet))^T$, the estimator for β is defined by

$$\widehat{\beta}_h(\bullet) = (\widetilde{\mathbf{X}}_h^T \widetilde{\mathbf{X}}_h)^{-1} \widetilde{\mathbf{X}}_h^T \widetilde{\chi}_h(\bullet). \tag{2.5}$$

It should be noted that $\widehat{\beta}_h$ is the ordinary least squares estimator obtained when one linearly links the vector of response variables $\widetilde{\chi}_h$ with the matrix of covariates $\widetilde{\mathbf{X}}_h$. It is worth mentioning that kernel estimation is used to obtain both $\widetilde{\chi}_h$ and $\widetilde{\mathbf{X}}_h$. Actually, both terms are computed as some nonparametric residuals.

Finally, nonparametric estimation is used to construct the estimator for $m(\chi)$ in (2.4)

$$\widehat{m}_h^{SFPL}(\chi) = \sum_{i=1}^{n-1} w_h(\chi,\chi_i)\left(\chi_{i+1}(\bullet) - \mathbf{x}_{i+1}^T\widehat{\beta}_h(\bullet)\right). \tag{2.6}$$

Other estimators for m in (2.1) (and therefore for β and $m(\chi)$ in (2.4)) could be obtained by means of wavelet-kernel approaches (see Antoniadis et al 2006) or local linear functional procedures (see Aneiros-Pérez, Cao and Vilar-Fernández 2010), among others.

2.4 Data description and empirical study

The data consists of the 24 hourly residual demand curves for all the days in years 2008 and 2009. One-day ahead forecasts for the hourly wind energy production and the hourly demand or price are also avaliable. Our aim is to predict the 24 hourly residual demand curves for all the days in eight different weeks along 2009. The learning sample considered for the whole forecasting process consists of 58 days (not necessarily consecutive). The whole sample is used to select the smoothing parameter and the semi-norm, while only the last 34 observations are used to build the predictor itself. The semi-norm used is the L_2 norm of the q-th derivative ($q = 0, 1, 2$) and q has been selected by minimizing some cross-validation criterion. This is also the criterion used to select the smoothing parameter h with a k-nearest neighbour approach.

Since working days and weekends have very different electricity demand patterns, four different scenarios are considered for prediction: (a) Sunday, (b) Monday, (c) Tuesday-Friday and (d) Saturday. The eight test samples were the eight weeks in February 8-21, May 3-16, August 2-15 and November 8-21, all in 2009. In scenarios (a), (b) and (d) the training sample consists of the hourly residual demand curve at the hour and day of the week to be predicted pertaining to the previous 58 weeks to the actual day. The training sample in scenario (c) uses the hourly demand curve for the 58 preceeding days in the range Tuesday-Friday within the current and the previous 15 weeks.

Several forecasting methods have been considered: (i) the naïve method (which just uses the hourly demand curve of previous day in the training sample), (ii) the functional nonparametric approach presented in Section 2, (iii) the semi-functional partial linear model, presented in Section 3, using the predicted demand as explanatory variable for the linear component, (iv) the semi-functional partial linear model using the predicted price as explanatory variable for the linear component, (v) the semi-functional partial linear model using the predicted wind energy as explanatory variable for the linear component, (vi) the semi-functional partial linear model using jointly the predicted demand, the predicted price and the predicted wind energy as explanatory linear variables.

Since the design of an optimal strategy for a production company is an inverse problem in terms of the residual demand curve, an alternative approach has been considered by just inverting all these curves. Inverse residual demand curves $\Upsilon_i(s) = \chi_i^{-1}(s)$ are considered and the previous methods have been applied to these new data.

Preliminary numerical results show the good behaviour of the functional nonparametric method and semi-functional partial linear model for residual demand forecasting. Final empirical results will be presented at IWFOS2011.

References

1. Aneiros-Pérez, G., Cao, R., Vilar-Fernández, J.M.: Functional methods for time series prediction: a nonparametric approach. To appear in *Journal of Forecasting* (2010)
2. Aneiros-Pérez, G., Vieu, P.: Semi-functional partial linear regression. Statist. Probab. Lett. **76**, 1102–1110 (2006)
3. Aneiros-Pérez, G., Vieu, P.: Nonparametric time series prediction: A semi-functional partial linear modeling. J. Multivariate Anal. **99**, 834–857 (2008)
4. Antoniadis, A., Paparoditis, E., Sapatinas, T.: A functional waveletkernel approach for time series prediction. J. Roy. Statist. Soc. Ser. B **68**, 837–857 (2006)
5. Antoniadis, A., Paparoditis, E., Sapatinas, T.: Bandwidth selection for functional time series prediction. Statist. Probab. Lett. **79**, 733–740 (2009)
6. Antoch, J., Prchal, L., De Rosa, M.R., Sarda, P. (2008). Functional linear regression with functional response: application to prediction of elecricity cosumption. In: Dabo-Niang, S., Ferraty, F. (eds.) Functional and Operatorial Statistics, pp. 23-29. Physica-Verlag, Heidelberg (2008)
7. Baillo, A., Ventosa, M., Rivier, M., Ramos, A.: Optimal Offering Strategies for Generation Companies Operating in Electricity Spot Markets. IEEE Transactions on Power Systems **19**, 745–753 (2004)
8. Bosq, D.: Linear Processes in Function Spaces: Theory and Applications. Lecture Notes in Statistics, 149, Springer (2000)
9. Carbon, M., Delecroix, M.: Nonparametric vs parametric forecasting in time series: a computational point of view. Applied Stochastic Models and Data Analysis **9**, 215–229 (1993)
10. Cardot, H., Dessertaine, A., Josserand E.: Semiparametric models with functional responses in a model assisted survey sampling setting. Presented at COMPSTAT 2010 (2010)
11. Faraway, J.: Regression analysis for a functional reponse. Technometrics **39**, 254–261 (1997)
12. Ferraty, F. and Vieu, P.: Nonparametric Functional Data Analysis. Series in Statistics, Springer, New York (2006)
13. Gross, G., Galiana, F.D.: Short-term load forecasting. Proc. IEEE **75**, 1558–1573 (1987)
14. Härdle, W., Lütkepohl, H., Chen, R.: A review of nonparametric time series analysis. International Statistical Review **65**, 49–72 (1997)
15. Härdle, W., Vieu, P.: Kernel regression smoothing of time series. J. Time Ser. Anal.**13**, 209–232 (1992)
16. Hart, J. D.: Some automated methods of smoothing time-dependent data. J. Nonparametr. Stat. **6**, 115–142 (1996)
17. Matzner-Lober, E., Gannoun, A., De Gooijer, J. G.: Nonparametric forecasting: a comparison of three kernel based methods. Commun. Stat.-Theor. M.**27**, 1593–1617 (1998)

18. Nadaraya, E. A.: On Estimating Regression. Theor. Probab. Appl. **9**, 141–142 (1964)
19. Vilar-Fernández, J.M., Cao, R.: Nonparametric forecasting in time series – A comparative study. Commun. Stat. Simulat. C. **36**, 311–334 (2007)
20. Vilar-Fernández, J.M., Cao, R., Aneiros-Pérez, G.: Forecasting next-day electricity demand and price using nonparametric functional methods. Preprint (2010)
21. Watson, G.S.: Smooth regression analysis. Sankhyā Ser. A **26**, 359–372 (1964)
22. Xu, L., and Baldick, R.: Transmission-constrained residual demand derivative in electricity markets. IEEE Transactions on Power Systems **22**, 1563–1573 (2007)

Chapter 3
Variable Selection in Semi-Functional Regression Models

Germán Aneiros, Frédéric Ferraty, Philippe Vieu

Abstract We deal with a regression model where a functional covariate enters in a nonparametric way, a divergent number of scalar covariates enter in a linear way and the corresponding vector of regression coefficients is sparse. A penalized-least-squares based procedure to simultaneously select variables and estimate regression coefficients is proposed, and some asymptotic results are obtained: rates of convengence and oracle property.

3.1 Introduction

Modeling the relationship between a response and a set of predictors is of main interest in order to predict values of the response given the predictors. The larger the number of predictors is, better fitted the model will be. But, if some predictors included in the model really do not influence the response, the model will not be good for predicting. Thus, in practice, it is needed some kind of methodology for selecting the significant covariates.

In a setting of linear regression with sparse regression coefficients, Tibshirani (1996) proposed the LASSO method, a version of Ordinary Least Squares (OLS) that constrains the sum of the absolute regression coefficients, and Efron et al. (2004) gave the LARS algorithm for model selection (a refinement of the LASSO method). Fan and Li (2001) proposed and studied the use of nonconcave penalized likelihood for variable selection and estimation of coefficients simultaneously. Fan and Peng (2004) generalized the paper of Fan and Li (2001) to the case where a

Germán Aneiros
Universidade da Coruña, Spain, e-mail: ganeiros@udc.es

Frédéric Ferraty
Institut de Mathématiques de Toulouse, France, e-mail: ferraty@math.univ-toulouse.fr

Philippe Vieu
Institut de Mathématiques de Toulouse, France, e-mail: vieu@math.univ-toulouse.fr

diverging number $p_n < n$ of parameters is considered, and they noted that the prize to pay is a slower rate of convergence ($(n/p_n)^{-1/2}$ instead of $n^{-1/2}$). Huang et al. (2008a) and Huang et al. (2008b) focused on particular classes of penalty functions (giving marginal bridge and adaptive LASSO estimators, respectively). Under a partial orthogonality condition on the covariance matrix, they obtained that their procedure can consistently identify the covariates with zero coefficients even when $p_n > n$.

Other authors dealt this topic in the setting where the regression function is the sum of a linear and a nonparametric component (that is, in Partial Linear Regression (PLR) models). Liang and Li (2009) considered a PLR model with fixed number p of covariates in the linear part, and measurement errors. In order to extend the procedure of Fan and Li (2001) to the new semi-parametric setting, they used local linear regression ideas. Ni et al. (2009) allowed a diverging number $p_n < n$ of parameters and studied a double-penalized least squares. These authors used spline smoothing to estimate the nonparametric part of the model, and penalized both the roughness of the nonparametric fit and the lack of parsimony. Xie and Huang (2009) used a penalized least squares function based on polynomial splines, and also considered the case of a diverging number $p_n < n$ of parameters. Their main contribution consists in building an estimator as a global minimum of a penalized least squares function (in general, the estimators proposed in the statistical literature are obtained as local minimum). The rate of convergence obtained by all these authors was the same as that obtained in pure linear models (i.e. $(n/p_n)^{-1/2}$).

In this paper we focus on a PLR model where the covariate that enters in a nonlinear way is of functional nature, such as a curve, an image, ... (see Aneiros-Pérez and Vieu (2006) for a first paper). In addition, the number of (scalar) covariates in the linear part is divergent, and the corresponding vector of regression coefficients is sparse. The topic we deal is that of variable selection and estimation of coefficients simultaneously. We extend to this new functional setting the methodology proposed when all the covariates are scalar, and we obtain rates of convengence and oracle property. Finally, in order to illustrate the practical interest of our procedure, a modest simulation study is reported. As far as we know, this is the first paper attacking (from a theoretical point of view) the problem of variable selection in a semi-functional PLR model.

3.2 The methodology

We are concerned with the semi-functional PLR model

$$Y_i = \mathbf{X}_i'\beta_0 + m(T_i) + \varepsilon_i, \forall i = 1, \dots, n, \tag{3.1}$$

where $\beta_0 = (\beta_{01}, ..., \beta_{0p_n})'$ is a vector of unknown sparse real parameters, m is an unknown smooth real function and ε_i are i.i.d. random errors satisfying

$$\mathbb{E}(\varepsilon_i \mid \mathbf{X}_i, T_i) = 0. \tag{3.2}$$

The covariates $\mathbf{X}_i = (X_{i1}, ..., X_{ip_n})'$ and T_i take values in $\mathbb{R}^{p_n}$ and some abstract semi-metric space $\mathscr{H}$, respectively.

The regression function in (3.1) has a parametric and a nonparametric component. Thus, we need to simultaneously use parametric and nonparametric techniques in order to construct good estimators. On the one hand, the nonparametric approach that we consider is that of kernel estimation. More specifically, a Nadaraya-Watson type estimator is constructed by using the weight function

$$w_{n,h}(t, T_i) = \frac{K(d(t, T_i)/h)}{\sum_{j=1}^{n} K(d(t, T_j)/h)}, \tag{3.3}$$

where $d(\cdot,\cdot)$ is the semi-metric associated to $\mathscr{H}$, $h > 0$ is a smoothing parameter and $K : \mathbb{R}^+ \to \mathbb{R}^+$ is a kernel function. On the other hand, the parametric procedure that we use is that of penalized least squares.

The steps to construct our estimator are, first, using kernel regression to transform the semi-parametric model (3.1) into a parametric model; then, apply to the transformed model the penalized-least-squared procedure in order to estimate β_0. To show this procedure clearer, let us denote $\mathbf{X} = (\mathbf{X}_1, ..., \mathbf{X}_n)'$, $\mathbf{Y} = (Y_1, ..., Y_n)'$ and, for any $(n \times q)$-matrix $\mathbf{A}$ $(q \geq 1)$, $\widetilde{\mathbf{A}}_h = (\mathbf{I} - \mathbf{W}_h)\mathbf{A}$, where $\mathbf{W}_h = \left(w_{n,h}(T_i, T_j)\right)_{i,j}$. Because

$$Y_i - \mathbb{E}(Y_i \mid T_i) = (\mathbf{X}_i - \mathbb{E}(\mathbf{X}_i \mid T_i))'\beta_0 + \varepsilon_i, \forall i = 1, \ldots, n$$

(see (3.1) and (3.2)), we consider the approximate model

$$\widetilde{\mathbf{Y}}_h \approx \widetilde{\mathbf{X}}_h'\beta_0 + \varepsilon,$$

where $\varepsilon = (\varepsilon_1, \ldots, \varepsilon_n)'$ (note that $\widetilde{\mathbf{Y}}_h$ and $\widetilde{\mathbf{X}}_h$ are formed by partial nonparametric residuals adjusting for T). Thus, in order to estimate β_0, we minimize the penalized least squares function

$$\mathscr{Q}(\beta) = \frac{1}{2}\left(\widetilde{\mathbf{Y}}_h - \widetilde{\mathbf{X}}_h\beta\right)'\left(\widetilde{\mathbf{Y}}_h - \widetilde{\mathbf{X}}_h\beta\right) + n\sum_{j=1}^{p_n} \mathscr{P}_{\lambda_{jn}}(|\beta_j|), \tag{3.4}$$

where $\mathscr{P}_{\lambda_{jn}}(\cdot)$ is a penalty function with a tuning parameter λ_{jn}. Once one has the Penalized Least Squares (PLS) estimator $\widehat{\beta}_0$, a natural estimator for $m(t)$ is

$$\widehat{m}(t) = \sum_{i=1}^{n} w_{n,h}(t, T_i)(Y_i - \mathbf{X}_i'\widehat{\beta}_0). \tag{3.5}$$

3.3 Asymptotic results

Under suitable conditions, we obtain a rate of convergence of $n^{-1/2}\log n$ for $\widehat{\beta}_0$, and we prove an oracle property; that is, with probability tending to 1, the estimator $\widehat{\beta}_0$ correctly identifies the null and non-null coefficients, and the corresponding estimator of the non-null coefficients is asymptotically normal with the same mean and covariance that it would have if the zero coefficients were known in advance. Thus, our approach gives sparse solutions and can be used as a methodology for variable selection and estimation of coefficients simultaneously in semi-functional PLR models: if the estimate of the parameter β_{0j} $(j=1,\ldots,p_n)$ is not equal to zero, then the corresponding covariate X_j is selected in the final model. In addition, for the nonparametric estimator $\widehat{m}(t)$, we obtain a uniform rate of convergence (on the compact set $\mathscr{C}$) of $h^{\alpha}+\sqrt{\psi_{\mathscr{C}}\left(n^{-1}\right)/\left(n\phi\left(h\right)\right)}$ (α denotes a constant coming from a Hölder condition, $\phi\left(\cdot\right)$ is the *small ball probability function* or *concentration function* and $\psi_{\mathscr{C}}\left(\varepsilon\right)$ denotes the ε-entropy of the set $\mathscr{C}$).

In summary, our main contributions are: (i) we extend the usual models to a functional setting, (ii) we improve the usual rate of convergence ($n^{-1/2}\log n$ instead of $n^{-1/2}p_n^{1/2}$) and (iii) we use weaker conditions on p_n than those in the statistical literature ($p_n^2 n^{-\log n}=o(1)$ instead of $p_n^2 n^{-k}=o(1)$).

3.4 A simulation study

A modest simulation study was designed in order to illustrate the practical behaviour of the proposed procedure.

The semi-functional PLR model

$$Y_i = X_{i1}\beta_{01}+X_{i2}\beta_{02}+\cdots+X_{ip_n}\beta_{0p_n}+m(T_i)+\varepsilon_i, \forall i=1,\ldots,n, \qquad (3.6)$$

was considered. The i.i.d. covariate vectors $\mathbf{X}_i=(X_{i1},\ldots,X_{ip_n})^T$ were normally distributed with mean zero and covariance matrix $\left(\rho^{|j-k|}\right)_{jk}$, and the functional covariates were $T_i(z)=a_i(z-0.5)^2+b_i$ $(z\in[0,1])$. Values $\rho=0$ and $\rho=0.5$ were considered, while a_i and b_i were i.i.d. according to a $U(0,1)$ and a $U(-0.5,0.5)$, respectively (these curves were discretized on the same grid of 100 equispaced points in $[0,1]$). The independent random errors ε_i were generated from a $N(0,\sigma_\varepsilon)$ distribution, where $\sigma_\varepsilon=0.1(\max_T m(T)-\min_T m(T))$. Finally, the unknown vector of parameters was

$$(\beta_{01},\ldots,\beta_{0p_n})'=(3,1.5,0,0,2,0,\ldots,0),$$

while the unknown function $m(\cdot)$ was

$$m(T_i)=\exp(-8f(T_i))-\exp(-12f(T_i)),$$

where

$$f(T_i) = sign(T_i'(1) - T_i'(0))\sqrt{3\int_0^1 (T_i'(z))^2 dz}.$$

$M = 50$ samples of sizes $n = 50$, 100 and 200 were drawn from model (3.6) and, for each of these values n, the size p_n of the vector of parameters was $p_{50} = 5$, $p_{100} = 7$ and $p_{200} = 10$, respectively. For each of the M replicates, we compute both the PLS and the OLS estimates, as well as the Oracle (OR) estimate (that is, the OLS estimate based on the true submodel). The smoothness of the curves T_i lead us to consider semi-metrics based on the L_2 norm of the q-th derivative of the curves. In addition, we considered $\lambda_j = \lambda sd(\widehat{\beta}_{0,j,OLS})$ and bandwidths h_k allowing to take into account k terms in (3.5). Values for the tuning parameters $\theta = (q, \lambda, k)$ were selected by means of the fivefold cross-validation method . The Epanechnikov kernel was used, while the penalty function was the SCAD penalty ($a = 3.7$).

Fig. 3.1 displays the M quadratic errors obtained for each combination considered (the quadratic error of $\widehat{\beta}$ for estimating β is defined as $(\widehat{\beta} - \beta)'(\widehat{\beta} - \beta)$). In addition, Table 3.1 reports the averages (on the M replicates) of both the number and the percentage (among the true null coefficients) of coefficients correctly set to zero (no coefficient was incorrectly set to zero).

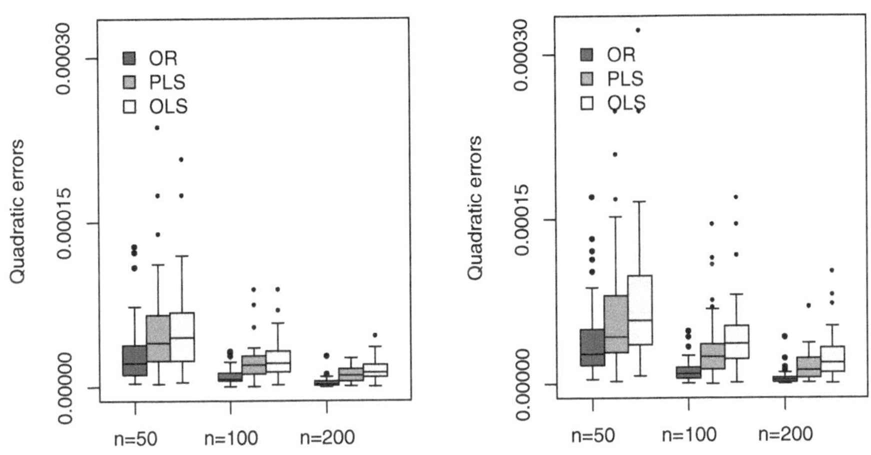

Fig. 3.1: Quadratic errors when $\rho = 0$ (left panel) and $\rho = 0.5$ (right panel).

Remark *Naturally, the results of any simulation study are valid only for the models considered, and in that sense should be interpreted. As expected, Fig. 3.1 suggests that the OR estimate performs the best, and the PLS is better than the OLS. Table 3.1 shows how, as the sample size increases, our procedure for selecting variables correctly detects a greater percentage of nonsignificant variables. In addition, it indicates that our procedure is not affected by the dependence structure in the vector of covariates.*

n	p_n	Zero coefficients			
		True	Correct		Incorrect
			$\rho = 0$	$\rho = 0.5$	
50	5	2	0.90 [45%]	0.96 [48%]	0
100	7	4	2.84 [71%]	2.72 [68%]	0
200	10	7	5.42 [77%]	5.44 [78%]	0

Table 3.1: Averages of both the number and the percentage of coefficients correctly and incorrectly set to zero.

Acknowledgements The research of G. Aneiros was partly supported by Grant number MTM2008-00166 from Ministerio de Ciencia e Innovacin (Spain). F. Ferraty and P. Vieu whish to thank all the participants of the working group STAPH on Functional Statistics in Toulouse for their numerous and interesting comments.

References

1. Aneiros-Perez, G., Vieu, P.: Semi-functional partial linear regression. Stat. Probabil. Lett. **76**, 1102–1110 (2006)
2. Efron, B., Hastie, T., Johnstone, I., Tibshirani, R.: Least angle regression. Ann. Stat. **32**, 407–499, (2004)
3. Fan, J., Li, R.: Variable selection via nonconcave penalized likelihood and its oracle properties. J. Am. Stat. Assoc. **96**, 1348–1360 (2001)
4. Fan, J., Peng, H.: Nonconcave penalized likelihood with a diverging number of parameters. Ann. Stat. **32**, 928–961 (2004)
5. Ferraty, F., Vieu, P.: Nonparametric Functional Data analysis. Springer, New York (2006)
6. Huang, J., Horowitz, J. L., Ma, S.: Asymptotic properties of bridge estimators in sparse high-dimensional regression models. Ann. Stat. **36**, 587–613 (2008a)
7. Huang, J., Ma, S., Zhang, C.-H.: Adaptive lasso for sparse high-dimensional regression models. Stat. Sinica **18**, 1606–1618 (2008b)
8. Liang, H., Li, L.: Variable selection for partially linear models with measurement errors. J. Am. Stat. Assoc. **104**, 234–248 (2009)
9. Ni, X., Zhang, H. H., Zhang, D.: Automatic model selection for partially linear models. J. Multivariate Anal. **100**, 2100–2111 (2009)
10. Tibshirani, R.: Regression shrinkage and selection via the Lasso. J. Roy. Stat. Soc. B **58**, 267–288 (1996)
11. Xie, H., Huang, J.: SCAD-penalized regression in high-dimensional partially linear models. Ann. Stat. **37**, 673–696 (2009)
12. Zou, H., Li, R.: One-step sparse estimates in nonconcave penalized likelihood models. Ann. Stat. **36**, 1509–1533 (2008)

Chapter 4
Power Analysis for Functional Change Point Detection

John A. D. Aston, Claudia Kirch

Abstract Change point detection in sequences of functional data is examined where the functional observations are dependent. The theoretical properties for tests for at most one change are derived with a special focus on power analysis. It is shown that the usual desirable properties of PCA to represent large amounts of the variation in a few components can actually be detrimental in the case of change point detection.

4.1 Introduction

This abstract is concerned with the power of detection of change-points, specifically at-most-one-change (AMOC) points in functional data. This work generalises remarks in Berkes et al (2009) with our results also extended to the case of weakly dependent functional data as defined in Hormann and Kokoszka (2010). The results show that a counter-intuitive effect occurs in the power analysis. Methods such as functional PCA rely on sparse representations of the data. However, in change point detection, if the data is generated from a process where the underlying system (without any change point) cannot be sparsely represented, then it can be easier to detect any change points present with a relatively small number of components. In contrast, data where the underlying system is very sparse may need large changes to be present before detection is possible.

The results in this abstract are for the AMOC model which is given by

$$X_i(t) = Y_i(t) + \mu_1(t)1_{\{i \leq \theta n\}} + \mu_2(t)1_{\{\theta n < i \leq n\}}, \tag{4.1}$$

John A. D. Aston
University of Warwick, UK, e-mail: j.a.d.aston@warwick.ac.uk

Claudia Kirch
Karlsruhe Institute of Technology, Germany, e-mail: claudia.kirch@kit.edu

where the mean functions before and after the change $\boldsymbol{\mu}_j = \mu_j(\cdot)$ as well as the functional time series $\{Y_i(\cdot) : 1 \le i \le n\}$ are elements of $L^2(\mathscr{Z})$, that are $(a.s.)$ continuous, $0 < \theta \le 1$ describes the position of the change, $EY_i(t) = 0$. $\boldsymbol{\mu}_1$, $\boldsymbol{\mu}_2$ as well as θ are unknown.

4.2 Testing for a change

We are interested in testing the null hypothesis of no change in the mean

$$H_0 : EX_i(\cdot) = \mu_1(\cdot), \quad i = 1,\ldots,n,$$

versus the AMOC alternative

$$\begin{aligned} H_1^{(A)} : \qquad & EX_i(\cdot) = \mu_1(\cdot), \quad i = 1,\ldots,\lfloor \theta n \rfloor, \\ & EX_i(\cdot) = \mu_2(\cdot) \neq \mu_1(\cdot), \quad i = \lfloor n\theta \rfloor + 1,\ldots,n, \quad 0 < \theta < n \end{aligned}$$

Note that the null hypothesis corresponds to the case where $\theta = 1$.

The idea is to use a projection into a lower dimensional space and use standard change-point statistics for the projected data. Berkes et al. propose, for instance, to use the space spanned by the first d principal components, where frequently this subspace is not known but needs to be estimated from the data.

Denote by $\widehat{\eta}_{i,l}$ the estimated scores, i.e. the projection coefficients of a d-dimensional estimated orthonormal system $\widehat{v}_l$, $l = 1,\ldots,d$. To elaborate

$$\widehat{\eta}_{i,l} = \langle X_i, \widehat{v}_l \rangle = \int X_i(t)\widehat{v}_l(t)\,dt, \quad i = 1,\ldots,n, \quad l = 1,\ldots,d.$$

Since $\{\widehat{v}_l\}$ forms an orthonormal system $\widehat{\eta}_{i,1},\ldots,\widehat{\eta}_{i,d}$ are uncorrelated for any fixed i. Furthermore $\widehat{\boldsymbol{\eta}}_i = (\widehat{\eta}_{i,1},\ldots,\widehat{\eta}_{i,d})^T$ is a d-dimensional time series exhibiting the same type of change as the functional sequence $\{X_i(\cdot) : 1 \le i \le n\}$ if the change is not orthogonal to the subspace spanned by $\widehat{v}_1(\cdot),\ldots,\widehat{v}_d(\cdot)$. To see this, let

$$\check{\eta}_{i,l} = \langle Y_i, \widehat{v}_l \rangle = \int Y_i(t)\widehat{v}_l(t)\,dt. \tag{4.2}$$

Then it holds

$$\widehat{\eta}_{i,l} = \check{\eta}_{i,l} + 1_{\{i \le \theta n\}} \int \mu_1(t)\widehat{v}_l(t)\,dt + 1_{\{i > \theta n\}} \int \mu_2(t)\widehat{v}_l(t)\,dt \tag{4.3}$$

in case of AMOC change.

Furthermore let $\widehat{\Sigma}$ be a consistent estimator for Σ, the long run covariance matrix of the d-dimensional time series, and $B_l(\cdot)$, $l = 1,\ldots,d$, be independent standard Brownian bridges. It can be proved that the following statistic is suitable to detect AMOC-change alternatives. Under H_0 it holds:

$$T_n^{(A1)} := \frac{1}{n^2}\widehat{\Upsilon}'\Sigma^{-1}\widehat{\Upsilon} \longrightarrow \sum_{1\leq l\leq d}\int_0^1 B_l^2(x)\,dx$$

where the lth element of Υ is estimated as $\widehat{\Upsilon}_l = \sum_{k=1}^{n}\left(\sum_{1\leq i\leq k}\widehat{\eta}_{i,l} - \frac{k}{n}\sum_{i=1}^{n}\widehat{\eta}_{i,l}\right)$.

4.3 Asymptotic Power Analysis

From above, it is easy to prove that the tests have asymptotic power one and the estimators are consistent if the change $\Delta(t) = \mu_1(t) - \mu_2(t)$ is not orthogonal to the contaminated projection subspace defined from taking the empirical covariance function and using it to define an orthonormal eigen-system. The contaminated covariance function depends directly on both the change point and the change itself as well as on the underlying true covariance function $c(s,t)$. The following theorem allows a characterisation of detectable changes in terms of the non-contaminated projection subspace and even more importantly shows that the change has a tendency to influence the contaminated projection subspace in such a way that it becomes detectable.

Theorem

1. (a) Let $\Delta(t) = \mu_1(t) - \mu_2(t)$, then

$$\int \Delta(t)v_l(t)\,dt \neq 0 \text{ for some } l = 1,\ldots,d$$
$$\Rightarrow \int \Delta(t)w_l(t)\,dt \neq 0 \text{ for some } l = 1,\ldots,d,$$

   where $v_l(t)$, $l = 1,\ldots,d$, are eigenfunctions of the uncontaminated covariance and $w_k(t)$ are those of the contaminated covariance. This shows, that any change that is not orthogonal to the non-contaminated subspace is detectable.
2. (b) Let $\Delta_D(t) = D\Delta(t)$, $\int \Delta^2(t)\,dt \neq 0$. Then, there exists $D_0 > 0$ such that

$$\int \Delta_D(t)w_{1,D}(t)\,dt \neq 0$$

   for all $|D| \geq D_0$, where $w_{1,D}$ is the eigenfunction belonging to the largest eigenvalue of the contaminated covariance kernel. This shows, that any large enough change is detectable.

The theorem part a) shows that we will be able to detect at least all changes that are not orthogonal to the non-contaminated subspace spanned by the first d principle components. Part b) shows that frequently changes can be detected even if they are orthogonal to the non-contaminated covariance. The reason is that large mean changes lead to a larger variability of the empirical covariance function and thus the contaminated covariance function $k(t,s) = c(t,s) + \theta(1-\theta)\Delta(t)\Delta(s)$ in the components that are not orthogonal to the change, while not changing the variability in the components that are orthogonal. In the following example such a change in the subspace takes place: Let $\{b_j : j \geq 1\}$ be an orthonormal basis of the continuous functions on $\mathscr{Z}$. Furthermore X, Y are i.i.d. $N(0,1)$, and $Y(t) = 2Xb_1(t) + Yb_2(t)$. Obviously $c(t,s)$ has the eigenvalues 4 with eigenfunction b_1 as well as the eigenvalue 1 with eigenfunction b_2 in addition to the eigenvalue 0. Let $\theta = 1/2$ and consider $\Delta(t) = 4b_2(t)$ which for $d = 1$ is obviously orthogonal to b_1, but it is easy to see that the eigenvalues of $k(t,s)$, are now 5 corresponding to b_2 and 4 corresponding to b_1 in addition to the eigenvalue 0. This shows that the mean change is no longer orthogonal to the space spanned by the eigenfunction corresponding to the largest eigenvalue, which is the one spanned by b_2.

An immediate corollary to the Theorem also gives rise to a surprising fact for multivariate data. PCA is well known to work poorly as a representation of the data when the covariance matrix of the multivariate observations is close to a multiple of the identity matrix. In fact, the scree plot will be linear in nature in the case when the covariance is an exact multiple of the identity. This implies that there is no effective sparse representation of the data. However, by the theorem above, this situation for the uncontaminated covariance is optimal for detecting a change point. Only choosing a single principal component from the contaminated covariance will guarantee the power of detection is asymptotically one (as by the theorem, the change will cause the first component to be non-orthogonal to the change with largest eigenvalue). Thus PCA based change point detection (for either epidemic or AMOC) works best when PCA itself works worst for the uncontaminated system regardless of the direction of the change.

This fact also translates over to functional data, but by the nature that the eigenvalues are square summable, the degenerate case will not occur. However, situations where the eigenvalues decay very rapidly in the uncontaminated case will require bigger changes in the mean to occurring in directions orthogonal to eigenfunctions associated with the large eigenvalues and naturally situations with more slowly decreasing eigenvalues will require smaller changes, to achieve asymptotic power one of detection with a small number of chosen basis functions.

References

1. Berkes, I., Gabrys, R., Horvath, L., Kokoszka, P.: Detecting changes in the mean of functional observations. J. R. Stat. Soc. Ser. B **71**, 927–946 (2009)
2. Hörmann, S., Kokoszka, P.: Weakly dependent functional data. Ann. Stat. **38**, 1845–1884 (2010)

Chapter 5
Robust Nonparametric Estimation for Functional Spatial Regression

Mohammed K. Attouch, Abdelkader Gheriballah, Ali Laksaci

Abstract This contribution deals with robust nonparametric regression analysis when the regressors are functional random fields. More precisely, we propose a family of robust nonparametric estimators for nonparametric functional spatial regression based on the kernel method. The main results of this work are the establishment of the almost complete convergence rate of these estimators.

5.1 Introduction

The statistical problems involved in the modelization of spatial data have received an increasing interest in the literature. The infatuation for this topic is linked with many fields of applications in which the data are collected in the spatial order. The nonparametric treatment of such data is relatively recent. The first results have been obtained by Tran (1990). For relevant works on the nonparametric modelization of spatial data, see Biau and Cadre (2004), Carbon *et al.* (2007), Li *et al.* (2009) or Gheriballah *et al.* (2010). In this work, we are interested in the nonparametric spatial regression, when the covariates are of functional nature, by using a robust approach.

Currently, the progress of informatics tools and the modern technology permits the recovery of increasingly bulky data which are recorded densely over time. They are typically treated as curve or functional data. This presents the advantage to give a framework which fits better to the functional nature of the observations. For an

Mohammed K. Attouch
Université de Sidi Bel Abbès, Algeria, e-mail: attou_kadi@yahoo.fr

Abdelkader Gheriballah
Université de Sidi Bel Abbès, Algeria, e-mail: gheribaek@yahoo.fr

Ali Laksaci
Université de Sidi Bel Abbès, Algeria, e-mail: alilak@yahoo.fr

overview on functional data analysis, we refer the reader to the monographs of Ramsay and Silverman (2005), Bosq (2000) for parametric models and Ferraty and Vieu (2006) for the nonparametric area . In this nonparametric context, the robust estimation of the regression function is an interesting problem in statistical inference. It is used as an alternative approach to classical methods, in particular when the data are affected by the presence of outliers. There is an extensive literature on robust estimation (see, for instance Huber (1964), Robinson (1984), Collomb and Härdle (1986), Fan *et al.* (1994) for previous results and Boente *et al.* (2009) for recent advances and references). The first results concerning the nonparametric robust estimation in functional statistic were obtained by Azzedine *et al.* (2008). They studied the almost complete convergence of robust estimators based on a kernel method, considering independent observations. Crambes *et al.* (2008) stated the convergence in L^q norm in both cases (i.i.d and strong mixing). While the asymptotic normality of these estimators is proved by Attouch *et al.* (2010).
The main aim of this contribution is to extend the results of Collomb and Härdle (1986) and Gheriballah *et al.* (2010) in the real case to the functional spatial processes. In our knowledge, this work is the first contribution on nonparametric robust regression for functional spatial variables. Specifically, we investigate almost complete convergence of the kernel estimator of the robust regression function. The interest comes mainly from the fact that an important field of application of functional statistical methods relates to the analysis of continuously indexed spatial processes.

5.2 The model

Consider $Z_{\mathbf{i}} = (X_{\mathbf{i}}, Y_{\mathbf{i}})$, $\mathbf{i} \in \mathbb{N}^N$ be a $\mathscr{F} \times \mathrm{IR}$-valued measurable strictly stationary spatial process, defined on a probability space $(\Omega, \mathscr{A}, \mathrm{IP})$, where $\mathscr{F}$ is a semi-metric space, d denoting the semi-metric. We assume that the process under study $(Z_{\mathbf{i}})$ is observed over a rectangular domain $\mathscr{I}_{\mathbf{n}} = \{\mathbf{i} = (i_1, ..., i_N) \in \mathbb{N}^N,\ 1 \le i_k \le n_k, k = 1, ..., N\}$, $\mathbf{n} = (n_1, ..., n_N) \in \mathbb{N}^N$. A point $\mathbf{i}$ will be referred to as a *site*. We will write $\mathbf{n} \to \infty$ if $\min\{n_k\} \to \infty$ and $|\frac{n_j}{n_k}| < C$ for a constant C such that $0 < C < \infty$ for all j, k such that $1 \le j, k \le N$. For $\mathbf{n} = (n_1, ..., n_N) \in \mathbb{N}^N$, we set $\widehat{\mathbf{n}} = n_1 \times \cdots \times n_N$. The nonparametric model studied in this paper, denoted by θ_x, is implicitly defined, for all $x \in \mathscr{F}$, as a zero with respect to (w.r.t.) $t \in \mathrm{IR}$ of the equation

$$\Psi(x,t) := \mathrm{IE}\left[\psi(Y_{\mathbf{i}}, t) \mid X_{\mathbf{i}} = x\right] = 0.$$

where ψ is a real-valued Borel function satisfying some regularity conditions to be stated below. In what follows, we suppose that, for all $x \in \mathscr{F}$, θ_x exists and is unique (see, for instance, Boente and Fraiman (1989)).

For all $(x,t) \in \mathscr{F} \times \mathrm{IR}$, we propose a nonparametric estimator of $\Psi(x,t)$ given by

$$\widehat{\Psi}(x,t) := \frac{\sum_{\mathbf{i} \in \mathscr{I}_{\mathbf{n}}} K(h^{-1} d(x, X_{\mathbf{i}})) \psi(Y_{\mathbf{i}}, t)}{\sum_{\mathbf{i} \in \mathscr{I}_{\mathbf{n}}} K(h^{-1} d(x, X_{\mathbf{i}}))},$$

where K is a kernel and $h = h_{\mathbf{n}}$ is a sequence of positive real numbers. A natural estimator $\widehat{\theta_x}$ of θ_x is a zero w.r.t. t of the equation

$$\widehat{\Psi}(x,t) = 0.$$

In this work, we will assume that the random field $(Z_{\mathbf{i}}, \mathbf{i} \in \mathbb{N}^N)$ satisfies the following mixing condition:

$$\begin{cases} \text{There exists a function } \varphi(t) \downarrow 0 \text{ as } t \to \infty, \text{ such that} \\ \forall E,\, E^{'} \text{ subsets of } \mathbb{N}^N \text{ with finite cardinals} \\ \alpha\left(\mathscr{B}(E), \mathscr{B}\left(E^{'}\right)\right) = \sup_{B \in \mathscr{B}(E), C \in \mathscr{B}\left(E^{'}\right)} \left|\mathrm{IP}(B \cap C) - \mathrm{IP}(B)\,\mathrm{IP}(C)\right| \\ \leq s\left(\mathrm{Card}(E), \mathrm{Card}\left(E^{'}\right)\right) \varphi\left(\mathrm{dist}\left(E, E^{'}\right)\right), \end{cases} \tag{5.1}$$

where $\mathscr{B}(E)$(*resp.* $\mathscr{B}\left(E^{'}\right)$) denotes the Borel σ-field generated by $(Z_{\mathbf{i}},\, \mathbf{i} \in E)$ (*resp.* $\left(Z_{\mathbf{i}},\, \mathbf{i} \in E^{'}\right)$), $\mathrm{Card}(E)$ (*resp.* $\mathrm{Card}\left(E^{'}\right)$) the cardinality of E (*resp.* $E^{'}$), $\mathrm{dist}\left(E, E^{'}\right)$ the Euclidean distance between E and $E^{'}$ and $s : \mathbb{N}^2 \to \mathrm{IR}^+$ is a symmetric positive function nondecreasing in each variable such that

$$s(n,m) \leq C \min(n,m), \quad \forall n, m \in \mathbb{N}. \tag{5.2}$$

We also assume that the process satisfies the following mixing condition:

$$\sum_{i=1}^{\infty} i^{\delta} \varphi(i) < \infty, \; \delta > 4N. \tag{5.3}$$

The conditions (5.2) and (5.3) measure the spatial dependence of the process. These conditions are used in Tran (1990). They are satisfied by many spatial models (see Guyon (1987) for some examples).

5.3 Main results

From now on, x stand for a fixed point in $\mathscr{F}$, we assume that the $Z_{\mathbf{i}}$'s have the same distribution with (X,Y) and all along the paper, when no confusion is possible, we denote by C and/or C' any generic positive constant. For $r > 0$, let $B(x,r) := \{x' \in \mathscr{F} / \; d(x',x) < r\}$. Moreover, for all $\mathbf{i} \in \mathscr{I}_{\mathbf{n}}$, we put $K_{\mathbf{i}}(x) = K(h^{-1} d(x, X_{\mathbf{i}}))$ and we set

$$\widehat{\Psi}(x,t) = \frac{\widehat{\Psi}_N(x,t)}{\widehat{\Psi}_D(x)}$$

with

$$\widehat{\Psi}_D(x) = \frac{1}{\widehat{\mathbf{n}}\mathbb{E}[K_{\mathbf{1}}(x)]}\sum_{\mathbf{i}\in\mathscr{I}_{\mathbf{n}}} K_{\mathbf{i}}(x) \text{ and } \widehat{\Psi}_N(x,t) = \frac{1}{\widehat{\mathbf{n}}\mathbb{E}[K_{\mathbf{1}}(x)]}\sum_{\mathbf{i}\in\mathscr{I}_{\mathbf{n}}} K_{\mathbf{i}}(x)\psi(Y_{\mathbf{i}},t).$$

where $\mathbf{1}$ is the site of components fixed to 1.
In order to derive the almost complete convergence (*a. co.*) of the kernel estimate $\widehat{\theta}_x$ of θ_x, some conditions are necessary. Recall that a sequence Z_n is said to converge *a. co.* to Z if and only if, for any $\varepsilon > 0$, $\sum_n \mathbb{P}(|Z_n - Z| > \varepsilon) < \infty$.

(H1) $\forall r > 0, \mathbb{P}(X \in B(x,r)) =: \phi_x(r) > 0$. Moreover, $\phi_x(r) \longrightarrow 0$ as $r \longrightarrow 0$.
(H2) $\forall \mathbf{i} \neq \mathbf{j}$,

$$0 < \sup_{\mathbf{i}\neq\mathbf{j}} \mathbb{P}\left[(X_{\mathbf{i}},X_{\mathbf{j}}) \in B(x,h)\times B(x,h)\right] \leq C(\phi_x(h))^{(a+1)/a}, \text{ for some } 1 < a < \delta N^{-1}.$$

(H3) ψ is bounded function, strictly monotone and continuously differentiable function, w.r.t. the second component, and its derivative $\frac{\partial \psi(y,t)}{\partial t}$ is bounded and continuous at θ_x uniformly in y.
(H4) The function $\Psi(\cdot,\cdot)$ satisfies Hölder's condition w.r.t. the first one, that is: there exist strictly positives constants b_1 and δ_0 such that:

$$\forall x_1,x_2 \in \mathscr{N}_x, \quad \forall t \in [\theta_x - \delta_0, \theta_x + \delta_0], \mid \Psi(x_1,t) - \Psi(x_2,t) \mid \leq C d^{b_1}(x_1,x_2)$$

where $\mathscr{N}_x$ is a fixed neighborhood of x.
(H5) The function $\Gamma(\cdot,\cdot) := \mathbb{E}\left[\psi'_x(Y,\cdot)|X = \cdot\right]$ satisfies Hölder's condition w.r.t. the first one, that is: there exists a strictly positive constant b_2 such that:

$$\forall x_1,x_2 \in \mathscr{N}_x, \forall t \in [\theta_x - \delta_0, \theta_x + \delta_0], |\Gamma(x_1,t) - \Gamma(x_2,t)| \leq C' d^{b_2}(x_1,x_2).$$

(H6) K is a function with support $[0,1]$ such that $C\mathbb{I}_{(0,1)}(\cdot) \leq K(\cdot) \leq C'\mathbb{I}_{(0,1)}(\cdot)$.
(H7) There exists $\eta_0 > 0$, such that, $C\widehat{\mathbf{n}}^{\frac{4N-\delta}{\delta}+\eta_0} \leq \phi_x(h)$.

Remarks on the assumptions. Our conditions are very standard in this context. Indeed, the conditions (H1) is the same as those used by Ferraty *et al.* (2006). Noting that, the function $\phi_x(.)$ defined in this assumption can be explicited for several continuous processes (see Ferraty *et al.* (2006). The local dependence (H2) allows to get the same convergence rate as in the i.i.d. case (see Azzedine *et al.* (2008). These hypotheses could be weakened, but the convergence rate would be perturbed by the presence of covariance terms. Condition (H3) controls the robustness properties of our model. We point out that the boundedness hypotheses over ψ can be dropped by using the truncation method as in Laïb and Ould-Saïd (2000). But it is well known that the boundedness of the score function is an fundamental constraint of the robustness properties of the M-estimators. Conditions (H4) and (H5) are regularity conditions which characterize the functional space of our model and are needed to evaluate the bias term in the asymptotic properties. Assumptions (H6) and (H7) are standard technical conditions in nonparametric estimation. They are imposed for the sake of simplicity and brevity of the proofs.

The following result ensures almost complete consistency of the kernel robust regression function when the observations $(X_{\mathbf{i}}, Y_{\mathbf{i}})$ satisfy (5.1), (5.2) and (5.3), in the previous Section.

Theorem 5.1. *Assume that (H1)-(H7) are satisfied and if $\Gamma(x, \theta_x) \neq 0$, then $\widehat{\theta}_x$ exists and is unique a.s. for all sufficiently large $\widehat{\mathbf{n}}$, and we have*

$$\widehat{\theta}_x - \theta_x = O\left(h^{b_1}\right) + O\left(\sqrt{\frac{\log \widehat{\mathbf{n}}}{\widehat{\mathbf{n}}\,\phi_x(h)}}\right) \quad a.co. \quad as \quad \mathbf{n} \to \infty$$

References

1. Attouch, M., Laksaci, A., Ould Saïd, E.: Asymptotic normality of a robust estimator of the regression function for functional time series data. J. Korean Stat. Soc. **39**, 489–500 (2010)
2. Azzedine, N., Laksaci, A., Ould Saïd, E.: On the robust nonparametric regression estimation for functional regressor. Stat. Probab. Lett. **78**, 3216–3221 (2008)
3. Biau, G., Cadre, B.: Nonparametric Spatial Prediction. Stat. Infer. Stoch. Proc. **7**, 327–349 (2004)
4. Boente, G., Fraiman, R.: Nonparametric regression estimation. J. Multivariate Anal. **29**, 180–198 (1989)
5. Boente, G., Gonzalez-Manteiga, W., Gonzalez, A.: Robust nonparametric estimation with missing data. J. Stat. Plan. Infer. **139**, 571–592 (2009)
6. Bosq, D.: Linear processes in function spaces. Theory and Application. Lectures Notes in Statistics, **149**, Springer Verlag, New-York (2000)
7. Carbon, M., Francq, C., Tran, L.T.: Kernel regression estimation for random fields. J. Stat. Plan. Infer. **137**, 778–798 (2007)
8. Collomb, G., Härdle, W.: Strong uniform convergence rates in robust nonparametric time series analysis and prediction: Kernel regression estimation from dependent observations. Stoch. Proc. Appl. **23**, 77–89 (1986)
9. Crambes, C, Delsol, L., Laksaci, A.: Robust nonparametric estimation for functional data. J. Nonparametr. Stat. **20**, 573–598 (2008)
10. Gheriballah, A., Laksaci, A., Rouane, R.: Robust nonparametric estimation for spatial regression. J. Stat. Plan. Infer. **140**, 1656–1670 (2010)
11. Guyon, X.: Estimation d'un champ par pseudo-vraisemblance conditionnelle: Etude asymptotique et application au cas Markovien. Proceedings of the Sixth Franco-Belgian Meeting of Statisticia (1987)
12. Fan, J., Hu, T.C., Truong, Y.K.: Robust nonparametric function estimation. *Scand. J. Stat.* **21**, 433–446 (1994)
13. Ferraty, F., Vieu, P.: Nonparametric functional data analysis: Theory and Practice . Springer, New York (2006)
14. Huber, P.J.: Robust estimation of a location parameter. Ann. Math. Stat. **35**, 73–101 (1964)
15. Laïb, N., Ould-Saïd, E.: A robust nonparametric estimation of the autoregression function under an ergodic hypothesis. Canad. J. Stat. **28**, 817–828 (2000)
16. Li, J., Tran, L.T.: Nonparametric estimation of conditional expectation. J. Stat. Plan. Infer. **139**, 164–175 (2009)
17. Ramsay, J. O., Silverman, B. W.: Functional data analysis (Second Edition). Springer, New York (2005)
18. Robinson, R.: Robust Nonparametric Autoregression. Lecture Notes in Statistics, **26**, Springer Verlag, New York (1984)
19. Tran, L.T.: Kernel density estimation on random fields. J. Multivariate Anal. **34**, 37–53 (1990)

Chapter 6
Sequential Stability Procedures for Functional Data Setups

Alexander Aue, Siegfried Hörmann, Lajos Horváth, Marie Hušková

Abstract The talk concerns sequential procedures detection of changes in linear relationship $Y_k(t) = \int_0^1 \Psi_k(t,s)X_k(s)ds + \varepsilon_k(t)$, $1 \leq k < \infty$, between random functions Y_k and X_k on $[0,1]$, where errors $\{\varepsilon_k\}$ are curves on $[0,1]$, and $\{\Psi_k\}$ are operators. Test procedures for testing the constancy of the operators Ψ_k's (i.e., $\Psi_1 = \Psi_2 = \ldots$) against a change point alternative when a training sample is available is proposed and studied. The procedure utilizes the functional principal component analysis. Limit behavior of the developed test procedures are investigated.

6.1 Introduction

We assume that the explanatory variables $X_k(t)$ and the response variables $Y_k(t)$ are connected via the linear relation

$$Y_k(t) = \int_0^1 \Psi_k(t,s)X_k(s)ds + \varepsilon_k(t), \quad 1 \leq k < \infty, \tag{6.1}$$

where $Y_k(t), X_k(t)$ and $\varepsilon_k(t)$ are random functions on $[0,1]$. The considered setup is sequential with a training sample of size m with no change (i.e., Ψ_k does not depend on $k \leq m$) is available.

Alexander Aue
University of California, Davis, USA, e-mail: alex.aue@gmail.com

Siegfried Hörmann
Université Libre de Bruxelles, Belgium, e-mail: shormann@ulb.ac.be

Lajos Horváth
University of Utah, Salt Lake City, USA, e-mail: horvath@math.utah.edu

Marie Hušková
Charles University of Prague, Czech Republic, e-mail: huskova@karlin.mff.cuni.cz

We are interested in testing if the relations in (6.1) hold with the same Ψ's, i.e. we want to check if

$$H_0: \quad \Psi_1 = \Psi_2 = \Psi_3 = \ldots \tag{6.2}$$

against the alternative that the Ψ's have changed at an unknown time during the observation period. More precisely, the following alternative is considered:

$$H_A: \text{ there is } k^* \geq 1 \text{ such that } \Psi_1 = \Psi_2 = \ldots = \Psi_m = \ldots = \Psi_{m+k^*-1} \neq \Psi_{m+k^*} = \ldots, \tag{6.3}$$

k^* is unknown.

There is a number of practical situations where such problems occur. In econometrics or finance, for example, $Y_k(t)$ and $X_k(t)$ represent the selling prices of two stocks during day k or the exchange rates between two currencies, see Cyree et al. (2004). Another example is the connection between bid and ask curves investigated by Elazović (2009).

So far the sequential setup formulated above has been considered for finite dimensional situations, typically for a change in linear models or time series, e.g., Chu et al (1996), Horváth et al (2004), Aue et al (2006), Berkes et al (2004). Their procedures are typically based on functionals of partial sums of various residuals. Here we propose a sequential procedure that suits for functional data model. The procedures are described by the stopping rule

$$\eta_m = \inf\{k \geq 1 : Q(m,k) \geq c\, q_\gamma^2(k/m)\}, \tag{6.4}$$

with $\inf\emptyset := +\infty$, where $Q(m,k)$'s are statistics (detectors) based on the observations $(Y_1, X_1) \ldots, (Y_{m+k}, X_{m+k})$, $k = 1, 2, \ldots$, the function $q(t), t \in (0, \infty)$, is a (critical) boundary function, and the constant $c = c(\alpha)$ is chosen such that, under H_0, for $\alpha \in (0,1)$ (fixed),

$$\lim_{m\to\infty} P\big(\eta_m < \infty\big) \;=\; \alpha, \tag{6.5}$$

and, under H_A,

$$\lim_{m\to\infty} P\big(\eta_m < \infty\big) \;=\; 1. \tag{6.6}$$

In other words, we require the asymptotic level α and a consistent test. Alternatively, the procedure can be described as follows: we stop and reject the null hypothesis as soon as at first time $Q(m,k) \geq c\, q_\gamma^2(k/m)$, we continue otherwise. The fundamental problem is a suitable choice of the sequence $\{Q(m,k), k \geq 1\}$. This will be discussed in the next section.

6.2 Test procedures

The scalar product and the norm of functions in $L_2([0,1])$ are denoted by $< f, g >= \int_0^1 f(t)g(t)dt$ and $\| f \| = (< f, g >)^{1/2}$, respectively.

Since by the assumption the observations $\{Y_k(t), X_k(t), t \in [0,1]\}$ are functions, i.e. infinite dimensional, a method which uses the projections of the observations into finite dimensional spaces is proposed. So a functional version of the principle component analysis is employed. The projections into a finite dimensional space should explain a large percentage of the randomness in the observations. The sequence $\{X_k(\cdot), \varepsilon_k(\cdot)\}$ is allowed to be dependent. It is required:

Assumption A.1 $\{(X_k(\cdot), \varepsilon_k(\cdot)), -\infty < k < \infty\}$ is a stationary and ergodic sequence satisfying $EX_k(t) = E\varepsilon_k(t) = 0 \forall t \in [0,1]$,

$$\int_0^1 E|X_k(t)|^{4+\kappa} dt < \infty \quad \text{and} \quad \int_0^1 E|\varepsilon_n(t)|^{4+\kappa} dt < \infty \tag{6.7}$$

for some $\kappa > 0$.

Under this assumption

$$C(t,s) = EX_k(t)X_k(s), \quad D(t,s) = EY_k(t)Y_k(s), \quad t,s \in [0,1]$$

exist and do not depend on k. Since $C(t,s)$ is a positive semi-definite function, the eigenvalues $\lambda_1 \geq \lambda_1 \geq \dots$ are non-negative. The corresponding eigenfunctions are denoted by $v_1, v_2, \dots$, they can be assumed orthonormal. We project the X_k's into the subspace spanned by $\{v_i, 1 \leq i \leq p\}$. Choosing p appropriately, the projections can explain a large percentage of randomness in the X_k's. Since $C(t,s)$ and therefore $\{\lambda_i, 1 \leq i \leq p\}$ and $\{v_i, 1 \leq i \leq p\}$ are unknown, we need to estimate them from the observations. Since the training sample is stable, we use the estimator

$$\widehat{C}_m(t,s) = \frac{1}{m} \sum_{k=1}^{m} X_k(t)X_k(s)$$

Let $\widehat{\lambda}_{1,m} \geq \widehat{\lambda}_{2,m} \geq \dots \geq \widehat{\lambda}_{p,m}$ denote the p largest eigenvalues of $\widehat{C}_m$ and $\widehat{v}_{1,m}, \dots, \widehat{v}_{p,m}$ be the corresponding eigenfunctions of $\widehat{C}_m$. It is assumed that $\{\widehat{v}_{i,m}, 1 \leq i \leq m\}$ is an orthonormal system.

Similarly for Y_k we introduce

$$D(t,s) = EY_k(t)Y_k(s).$$

and denote by $\tau_1 \geq \tau_2 \geq \dots$ the eigenvalues and by $w_1, w_2, \dots$ the corresponding eigenfunctions. Using the training sample we estimate $D(t,s)$ it by

$$\widehat{D}_m(t,s) = \frac{1}{m} \sum_{1 \leq k \leq m} Y_k(t)Y_k(s).$$

The eigenvalues and the corresponding eigenfunctions of $\widehat{D}_m$ are denoted by $\widehat{\tau}_{1,m} \geq \widehat{\tau}_{2,m} \geq \dots$ and $\widehat{w}_{1,m}, \widehat{w}_{2,m}, \dots$, respectively. It is assumed that $\widehat{w}_{1,m}, \widehat{w}_{2,m}, \dots$ are orthonormal functions. We also assume that

Assumption A.2

$$||\Psi||^2 = \int_0^1 \int_0^1 \psi^2(t,s)dtds < \infty$$

holds, where Ψ denotes the common value of the Ψ_k's under H_0 and $\psi(t,s)$ denotes a kernel function in $L^2([0,1]^2)$.

Since $\{w_i(t)v_j(s), (t,s) \in [0,1]^2 1 \le i,j < \infty\}$ is an orthonormal basis of $L_2([0,1]^2)$ we have that

$$\psi(t,s) = \sum_{1 \le i,j < \infty} \psi_{i,j} w_i(t) v_j(s)$$

with some $\psi_{i,j}$'s satisfying $\sum_{1 \le i,j < \infty} \psi_{i,j}^2 < \infty$.
Therefore projecting Y_n into the subspace spanned by $\widehat{w}_{1,m}, \dots, \widehat{w}_{q,m}$ and projecting X_n into the subspace spanned by $\widehat{v}_{1,m}, \dots, \widehat{v}_{p,m}$ under H_0 we get from (6.1)

$$< Y_i, \widehat{w}_{j,m} > = \sum_{s=1}^{p} \beta_{js} < X_i, \widehat{v}_{s,m} > + \Delta_{ij}, \quad i = 1,2,\dots, \quad j = 1,\dots q \tag{6.8}$$

where

$$\beta_{js} = \widehat{d}_{jm} \psi_{js} \widehat{c}_{sm}, \quad j = 1,\dots,q, \quad s = 1,\dots,p$$

with $\widehat{d}_{jm}$ and $\widehat{c}_{sm}$ being random signs such that $\widehat{d}_{jm} w_j$ are close $\widehat{w}_{j,m}$ and $\widehat{c}_{sm} v_s$ are close $\widehat{w}_{s,m}$ in certain sense. Also Δ_{ij}'s play formally the role of the error terms, they include not only $< \varepsilon_i, \widehat{w}_{jm} >$ but also other terms in order the equations (6.1) and (6.8) are in accordance.
Next we rewrite the relations (6.8) differently. Let

$$\boldsymbol{\beta} = vec(\beta_{js}, \quad j = 1,\dots,q, \quad s = 1,\dots,p),$$

$$\widehat{\boldsymbol{Y}}_i = (< Y_i, \widehat{w}_{1,m} >, \dots, < Y_i, \widehat{w}_{q,m} >)^T,$$

$$\widehat{\boldsymbol{\Delta}}_i = (< \Delta_i, \widehat{w}_{1,m} >, \dots, < \Delta_i, \widehat{w}_{q,m} >)^T,$$

$$\widehat{\boldsymbol{Y}}_{n,N}^T = (\widehat{\boldsymbol{Y}}_n^T, \dots, \widehat{\boldsymbol{Y}}_N^T), \quad \widehat{\boldsymbol{\Delta}}_{n,N}^T = (\widehat{\boldsymbol{\Delta}}_n^T, \dots, \widehat{\boldsymbol{\Delta}}_N^T).$$

Now the equations in (6.8) for the variables $(Y_i, X_i), n < i \le N$ can be rewritten as

$$\widehat{\boldsymbol{Y}}_{n,N} = \widehat{\boldsymbol{Z}}_{n,N} \boldsymbol{\beta} + \widehat{\boldsymbol{\Delta}}_{n,N},$$

where

$$\widehat{\boldsymbol{Z}}_{n,N}^T = (\widehat{\boldsymbol{Z}}_n^T, \dots, \widehat{\boldsymbol{Z}}_N^T)$$

with $\widehat{\boldsymbol{Z}}_i = \widehat{\boldsymbol{x}}_i \otimes \boldsymbol{I}_p$, $\widehat{\boldsymbol{x}}_i = (< X_i, \widehat{v}_{1,m} >, \dots, < X_i, \widehat{v}_{p,m} >)^T$, $\boldsymbol{I}_p$ stands for the $p \times p$ identity matrix and $\otimes$ denotes the Kronecker product. The least squares estimator is given by

$$\widehat{\boldsymbol{\beta}}_{n,N} = \left(\widehat{\boldsymbol{Z}}_{n,N}^T \widehat{\boldsymbol{Z}}_{n,N}\right)^{-1} \widehat{\boldsymbol{Z}}_{n,N}^T \widehat{\boldsymbol{Y}}_{n,N}.$$

Now the detector is defined as

$$Q(m,k) = (\widehat{\boldsymbol{\beta}}_{m,m+k} - \widehat{\boldsymbol{\beta}}_{0,m})^T \widehat{\boldsymbol{V}}_m \widehat{\boldsymbol{\Sigma}}_m^{-1} \widehat{\boldsymbol{V}}_m (\widehat{\boldsymbol{\beta}}_{m,m+k} - \widehat{\boldsymbol{\beta}}_{0,m}), \quad k \geq 1$$

where $\widehat{\boldsymbol{V}}_m = \widehat{\boldsymbol{Z}}_{0,m} \widehat{\boldsymbol{Z}}_{0,m}^T / m$ and $\widehat{\boldsymbol{\Sigma}}_m$ is a suitable standardization matrix based on on the training data. Particularly, it is assumed that, as $m \to \infty$,

$$|\widehat{\boldsymbol{\Sigma}}_m - (\widehat{\boldsymbol{d}}_m \otimes \widehat{\boldsymbol{c}}_m)\boldsymbol{\Sigma}| = o_P(1), \tag{6.9}$$

where $\widehat{\boldsymbol{d}}_m = vec(\widehat{d}_{1,m}, \ldots, \widehat{d}_{q,m})$, $\widehat{\boldsymbol{c}}_m$ is defined analogously, $\otimes$ is the Kronecker product of matrices. Here $\boldsymbol{\Sigma}$ is the asymptotic variance matrix of $\widehat{\boldsymbol{\Delta}}_i$.

Clearly, due to the definition of β_{ij} the above statistics are sensitive w.r.t. a change in ψ_{ij}, $1 \leq i \leq q$, $1 \leq j \leq p$. The procedure is not sensitive w.r.t. a change in ψ_{ij} if either $i > q$ or/and $j > p$. Limit properties are stated in the following section.

6.3 Asymptotic properties

We still need some assumptions on the dependence structure:

Assumption A.3 There are functionals a and b such that

$$X_n = a(\gamma_n, \gamma_{n-1}, \ldots) \quad \text{and} \quad \varepsilon_n = b(\delta_n, \delta_{n-1}, \ldots),$$

where $\{\gamma_k(t), -\infty < k < \infty\}$ and $\{\delta_k(t), -\infty < k < \infty\}$ are i.i.d. sequences of random elements with values in some measurable spaces.

The assumption states that both X_n and ε_n are Hilbert space valued Bernoulli shifts. We consider only weakly dependent random processes in this paper which is formulated as

Assumption A.4 There are C_0 and $A > 2$ such that

$$\left(E|\parallel X_n - X_n^{(k)} \parallel|^{4+\kappa}\right)^{1/(4+\kappa)} + \left(E|\parallel \varepsilon_n - \varepsilon_n^{(k)} \parallel|^{4+\kappa}\right)^{1/(4+\kappa)} \leq C_0 k^{-A}, \quad 1 \leq k < \infty \tag{6.10}$$

with

$$X_n^{(k)} = a(\gamma_n, \gamma_{n-1}, \ldots, \gamma_{n-k+1}, \gamma_{n-k}^{(k)}, \gamma_{n-k-1}^{(k)}, \ldots)$$

where $\{\gamma_\ell^{(k)}, -\infty < k, \ell < \infty\}$ are i.i.d. copies of γ_0., $\{\varepsilon_\ell^{(k)}, -\infty < k, \ell < \infty\}$ are defined accordingly.

Note that these assumptions means that (X_n, ε_n) can be approximated with the k dependent sequences $(X_n^{(k)}, \varepsilon_n^{(k)})$, $-\infty < n < \infty$ and this approximation is improving with the rate k^{-A} as k increases.

The following requirement is standard in functional data analysis (cf. Bosq (2000)):

Assumption 2.5

$$\lambda_1 > \lambda_2 > \ldots > \lambda_p > \lambda_{p+1}, \qquad \tau_1 > \tau_2 > \ldots > \tau_q > \tau_{q+1}.$$

The last set of conditions are on the boundary function g:

Assumption A.6 (i) $g(t)$ is continuous on $[0,1]$
(ii) $\inf_{\varepsilon \le t \le 1} g(t) > 0$ for every $0 < \varepsilon < 1$
(iii) there are $C_0 > 0$ and $0 \le \gamma < 1/2$ such that $C_0 x^\gamma \le g(x)$ for all $0 < x \le 1$.

Now we are ready to state the main result of this paper.

Theorem If H_0, if assumptions A.1, – A.6 and (6.9) hold, then

$$\lim_{m\to\infty} P\left\{Q(m,k) > c\frac{m}{k^2}\left(1+\frac{k}{m}\right)g^2\left(\frac{k}{k+m}\right) \text{ for some} k \ge 1\right\} = P\left\{\sup_{0<\le t\le 1} \frac{\Gamma(t)}{g^2(t)} > c\right\}$$

with

$$\{\Gamma(t), 0 \le t \le 1\} \stackrel{\mathcal{D}}{=} \{\sum_{\ell=1}^{pq} W_\ell^2, 0 \le t \le 1\},$$

where $\{W_\ell(t), 1 \le \ell \le pq\}$ are independent standard Brownian motions.

The above theorem provides reasonable approximation for the constant c in the stopping rule. Under mild conditions the tests are consistent.
In the talk some further results, discussion and some examples will be presented. The above results are part of the paper in preparation by Aue et al (2011).

Acknowledgements Research partially supported by NSF grant DMS 0905400 and by GAČR 2010/09/J006, MSM 0021620839.

References

1. Aue, A., Hörmann, S. Horváth, L., Hušková, M.: Sequential stability test for functional linear models. In preparation (2011)
2. Aue, A., Horváth, L., Hušková, M., Kokoszka, P.: Change-point monitoring in linear models. Econometrics J. **9**, 373–403 (2006)
3. Berkes, I., Gombay, E., Horváth, L., Kokoszka, P.: Sequential change-point detection in GARCH(p,q) models. Economet. Theor. **20**, 1140–1167 (2004)
4. Berkes, I., Hörmann, S., Schauer, J. (2009+). Split invariance principles for stationary sequences. To appear in Ann. Probab. (2009+)
5. Chu, C.-S., Stinchcombe, M., White, H.: Monitoring structural change. Econometrica **64**, 1045–1065 (1996)
6. Cyree, K. K., Griffiths, M. D. and Winters, D. B.: An empirical examination of intraday volatility in euro-dollar rates. The Quaterly Review of Economics and Finance **44**, 44–57 (2004)
7. Elazović, S.: Functional modelling of volatility in the Swedish limit order book. Comput. Statist. Data Anal. **53**, 2107–2118 (2009)
8. Hörmann, S., Horváth, L., Reeder, R.: Functional volatility sequences. Preprint, University of Utah, Salt Lake City, Utah, USA (2009+)

9. Horváth, L., Hušková, M., Kokoszka, P.: Testing the stability of the functional autoregressive process. J. Multivariate Anal. **101**, 352–367 (2010)
10. Horváth, L., Hušková, M., Kokoszka, P., Steinebach, J.: Monitoring changes in linear models. J. Stat. Plan. Infer. **126**, 225–251 (2004)

Chapter 7
On the Effect of Noisy Observations of the Regressor in a Functional Linear Model

Mareike Bereswill, Jan Johannes

Abstract We consider the estimation of the slope function in functional linear regression, where a scalar response Y is modeled in dependence of a random function X, when Y and only a panel $Z_1,\dots,Z_L$ of noisy observations of X are observable. Assuming an iid. sample of $(Y,Z_1,\dots,Z_L)$ we derive in terms of both, the sample size and the panel size, a lower bound of a maximal weigthed risk over certain ellipsoids of slope functions. We prove that a thresholded projection estimator can attain the lower bound up to a constant.

7.1 Introduction

A common problem in a diverse range of disciplines is the investigation of the dependence of a real random variable Y on the variation of an explanatory random function X (see for instance Ramsay and Silverman [2005] and Ferraty and Vieu [2006]). We assume that X takes its values in an infinite dimensional separable Hilbert space $\mathbb{H}$ which is endowed with an inner product $\langle\cdot,\cdot\rangle$ and its associated norm $\|\cdot\|$. In functional linear regression the dependence of the response Y on the regressor X is then modeled by

$$Y = \langle\beta,X\rangle + \sigma\varepsilon, \quad \sigma > 0, \tag{7.1a}$$

where $\beta \in \mathbb{H}$ is unknown and the error ε has mean zero and variance one. In this paper we suppose that we have only access to Y and a panel of noisy observations of X,

$$Z_\ell = X + \varsigma\,\Xi_\ell, \quad \varsigma \geqslant 0, \quad \ell = 1,\dots,L, \tag{7.1b}$$

Mareike Bereswill
University of Heidelberg, Germany, e-mail: mareike.bereswill@web.de

Jan Johannes
Université Catholique de Louvain, Belgium, e-mail: jan.johannes@uclouvain.be

where $\Xi_1,\dots,\Xi_L$ are measurement errors. One objective is then the non-parametric estimation of the slope function β based on an iid. sample of $(Y,Z_1,\dots,Z_L)$.

In recent years the non-parametric estimation of the slope function β from a sample of (Y,X) has been of growing interest in the literature (c.f. Cardot et al. [1999], Marx and Eilers [1999], Bosq [2000] or Cardot et al. [2007]). In this paper we follow an approach based on dimension reduction and thresholding techniques, which has been proposed by Cardot and Johannes [2010] and borrows ideas from the inverse problems community (c.f. Efromovich and Koltchinskii [2001] and Hoffmann and Rei [2008]).

The objective of this paper is to establish a minimax theory for the non-parametric estimation of β in terms of both, the size L of the panel $Z_1,\dots,Z_L$ of noisy measurements of X and the size n of the sample of $(Y,Z_1,\dots,Z_L)$. In order to make things more formal let us reconsider model (1a) - (1b). Given an orthonormal basis $\{\psi_j\}_{j\geqslant 1}$ in $\mathbb{H}$ (not necessarily corresponding to the eigenfunctions of Γ) we assume real valued random variables $\xi_{j,\ell} := \langle \Xi_\ell, \psi_j\rangle$ and observable blurred versions of the coefficient $\langle X, \psi_j\rangle$ of X,

$$Z_{j,\ell} := \langle X, \psi_j\rangle + \varsigma\, \xi_{j,\ell}, \quad \ell = 1,\dots,L \text{ and } j \in \mathbb{N}. \tag{7.2}$$

The motivating example for our abstract framework consists in irregular and sparse repeated measures of a contaminated trajectory of a random function $X \in L^2[0,1]$ (c.f. Yao et al. [2005] and references therein). To be more precise, suppose that there are L uniformly-distributed and independent random measurement times $U_1,\dots,U_L$ for X. Let $V_\ell = X(U_\ell) + \eta_\ell$ denote the observation of the random trajectory X at a random time U_ℓ contaminated with measurement error η_ℓ, $1 \leqslant \ell \leqslant L$. The errors η_ℓ are assumed to be iid. with mean zero and finite variance. If the random function X, the random times $\{U_\ell\}$ and the errors $\{\eta_\ell\}$ are independent, then, it is easily seen that for each $\ell = 1,\dots,L$ and $j \in \mathbb{N}$ the observable quantity $Z_{j,\ell} := V_\ell \psi_j(U_\ell)$ is just a blurred version of the coefficient $\langle X, \psi_j\rangle$ corrupted by an uncorrelated additive measurement error $V_\ell \psi_j(U_\ell) - \langle X, \psi_j\rangle$. Moreover, those errors are uncorrelated for all $j \in \mathbb{N}$ and different values of ℓ. It is interesting to note that recently Crambes et al. [2009] prove minimax-optimality of a spline based estimator in the situation of deterministic measurement times. However, the obtained optimal rates are the same as for a known regressor X since the authors suppose the deterministic design to be sufficiently dense. In contrast to this result we seek a minimax theory covering also sparse measurements. In particular, it enables us to quantify the minimal panel size in order to recover the minimal rate for a known X.

In Section 2 we introduce our basic assumptions and recall the minimax theory derived in Cardot and Johannes [2010] for estimating β non-parametrically given an iid. sample of (Y,X). Assuming an iid. sample of size n of $(Y,Z_1,\dots,Z_L)$ we derive in Section 3 a lower bound in terms of both, n and L, for a maximal weighted risk. We propose an estimator based on dimension reduction and thresholding techniques that can attain the lower bound up to a constant. All proofs can be found in Bereswill and Johannes [2010].

7.2 Background to the methodology

For sake of simplicity we assume that the measurement errors ε and $\{\xi_{j,\ell}\}_{j\in\mathbb{N},1\leqslant\ell\leqslant L}$ are independent and standard normally distributed, i.e, $\Xi_1,\dots,\Xi_L$ are independent Gaussian white noises in $\mathbb{H}$. Furthermore, we suppose that the regressor X is Gaussian with mean zero and a finite second moment, i.e., $\mathbb{E}\|X\|^2 < \infty$, as well as independent of all measurement errors. Taking the expectation after multiplying both sides in (1a) by X we obtain $g := \mathbb{E}[YX] = \mathbb{E}[\langle\beta, X\rangle X] =: \Gamma\beta$, where g belongs to $\mathbb{H}$ and Γ denotes the covariance operator associated with the random function X. In what follows we always assume that there exists in $\mathbb{H}$ a unique solution of the equation $g = \Gamma\beta$, i.e., that g belongs to the range of the strictly positive Γ (c.f. Cardot et al. [2003]). It is well-known that the obtainable accuracy of any estimator of β can essentially be determined by the regularity conditions imposed on both, the slope parameter β and the covariance operator Γ. We formalize now these conditions, which are characterized in this paper by different weighted norms in $\mathbb{H}$ with respect to the pre-specified basis $\{\psi_j\}_{j\geqslant}$.

Given a positive sequence of weights $w := (w_j)_{j\geqslant 1}$ we define the weighted norm $\|f\|_w^2 := \sum_{j\geqslant 1} w_j |\langle f, \psi_j\rangle|^2$, $f \in \mathbb{H}$, the completion $\mathscr{F}_w$ of $\mathbb{H}$ with respect to $\|\cdot\|_w$ and the ellipsoid $\mathscr{F}_w^c := \{f \in \mathscr{F}_w : \|f\|_w^2 \leqslant c\}$ with radius $c > 0$. Here and subsequently, given strictly positive sequences of weights $\gamma := (\gamma_j)_{j\geqslant 1}$ and $\omega := (\omega_j)_{j\geqslant 1}$ we shall measure the performance of any estimator $\widehat{\beta}$ by its maximal $\mathscr{F}_\omega$-risk over the ellipsoid $\mathscr{F}_\gamma^\rho$ with radius $\rho > 0$, that is $\sup_{\beta\in\mathscr{F}_\gamma^\rho} \mathbb{E}\|\widehat{\beta} - \beta\|_\omega^2$. This general framework allows us with appropriate choices of the basis $\{\psi_j\}_{j\geqslant}$ and the weight sequence ω to cover the estimation not only of the slope function itself (c.f. Hall and Horowitz [2007]) but also of its derivatives as well as the optimal estimation with respect to the mean squared prediction error (c.f. Crambes et al. [2009]). For a more detailed discussion, we refer to Cardot and Johannes [2010]. Furthermore, as usual in the context of ill-posed inverse problems, we link the mapping properties of the covariance operator Γ and the regularity conditions on β. Denote by $\mathscr{N}$ the set of all strictly positive nuclear operators defined on $\mathbb{H}$. Given a strictly positive sequence of weights $\lambda := (\lambda_j)_{j\geqslant 1}$ and a constant $d \geqslant 1$ define the subset $\mathscr{N}_\lambda^d := \{\Gamma \in \mathscr{N} : \|f\|_\lambda^2/d^2 \leqslant \|\Gamma f\|^2 \leqslant d^2\|f\|_\lambda^2, \forall f \in \mathbb{H}\}$ of $\mathscr{N}$. Notice that $\langle\Gamma\psi_j, \psi_j\rangle \geqslant d^{-1}\lambda_j^{1/2}$ for all $\Gamma \in \mathscr{N}_\lambda^d$, and hence the sequence $(\lambda_j^{1/2})_{j\geqslant 1}$ is necessarily summable. All the results in this paper are derived with respect to the three sequences ω, γ and λ. We do not specify these sequences, but impose from now on the following minimal regularity conditions.

ASSUMPTION (A.1) *Let $\omega := (\omega_j)_{j\geqslant 1}$, $\gamma := (\gamma_j)_{j\geqslant 1}$ and $\lambda := (\lambda_j)_{j\geqslant 1}$ be strictly positive sequences of weights with $\gamma_1 = 1$, $\omega_1 = 1$ and $\lambda_1 = 1$ such that γ and $(\gamma_j/\omega_j)_{j\geqslant 1}$ are non decreasing, λ and $(\lambda_j/\omega_j)_{j\geqslant 1}$ are non increasing with $\Lambda := \sum_{j=1}^\infty \lambda_j^{1/2} < \infty$.*

Given a sample size $n \geqslant 1$ and sequences ω, γ and λ satisfying Assumption A.1 define

$$m_n^* := m_n^*(\gamma,\omega,\lambda) := \arg\min_{m\geqslant 1}\left\{\max\left(\frac{\omega_m}{\gamma_m}, \sum_{j=1}^{m}\frac{\omega_j}{n\sqrt{\lambda_j}}\right)\right\} \text{ and}$$

$$\delta_n^* := \delta_n^*(\gamma,\omega,\lambda) := \max\left(\frac{\omega_{m_n^*}}{\gamma_{m_n^*}}, \sum_{j=1}^{m_n^*}\frac{\omega_j}{n\sqrt{\lambda_j}}\right). \quad (7.3)$$

If in addition $\triangle := \inf_{n\geqslant 1}\{(\delta_n^*)^{-1}\min(\omega_{m_n^*}\gamma_{m_n^*}^{-1}, \sum_{j=1}^{m_n^*}\omega_j(n\sqrt{\lambda_j})^{-1})\} > 0$, then there exists $C > 0$ depending on σ^2, ρ, d, $\triangle$ only such that (c.f. Cardot and Johannes [2010]),

$$\inf_{\breve\beta}\inf_{\Gamma\in\mathcal{N}_\lambda^d}\sup_{\beta\in\mathcal{F}_\gamma^\rho}\left\{\mathbb{E}\|\breve\beta-\beta\|_\omega^2\right\} \geqslant C\,\delta_n^* \quad \text{for all } n\geqslant 1.$$

Assuming an iid. sample $\{(Y^{(i)},X^{(i)})\}$ of size n of (Y,X), it is natural to consider the estimators $\widetilde{g} := \frac{1}{n}\sum_{i=1}^n Y^{(i)}X^{(i)}$ and $\widetilde{\Gamma} := \frac{1}{n}\sum_{i=1}^n\langle\cdot,X^{(i)}\rangle X^{(i)}$ for g and Γ respectively. Given $m\geqslant 1$, we denote by $[\widetilde{\Gamma}]_{\underline{m}}$ the $m\times m$ matrix with generic elements $[\widetilde{\Gamma}]_{j,\ell} := \langle\widetilde{\Gamma}\psi_\ell,\psi_j\rangle = n^{-1}\sum_{i=1}^n\langle X^{(i)},\psi_\ell\rangle\langle X^{(i)},\psi_j\rangle$, and by $[\widetilde{g}]_{\underline{m}}$ the m vector with elements $[\widetilde{g}]_\ell := \langle\widetilde{g},\psi_\ell\rangle = n^{-1}\sum_{i=1}^n Y^{(i)}\langle X^{(i)},\psi_\ell\rangle$, $1\leqslant j,\ell\leqslant m$. Obviously, if $[\widetilde{\Gamma}]_{\underline{m}}$ is non singular then $[\widetilde{\Gamma}]_{\underline{m}}^{-1}[\widetilde{g}]_{\underline{m}}$ is a least squares estimator of the vector $[\beta]_{\underline{m}}$ with elements $\langle\beta,\psi_\ell\rangle, 1\leqslant\ell\leqslant m$. The estimator of β consists now in thresholding this projection estimator, that is,

$$\widetilde{\beta}_m := \sum_{j=1}^m \widetilde{[\beta]}_j\psi_j \quad\text{with}\quad \widetilde{[\beta]}_{\underline{m}} := \begin{cases}[\widetilde{\Gamma}]_{\underline{m}}^{-1}[\widetilde{g}]_{\underline{m}}, & \text{if } [\widetilde{\Gamma}]_{\underline{m}} \text{ is non-singular}\\ & \text{and } \|[\widetilde{\Gamma}]_{\underline{m}}^{-1}\|\leqslant n,\\ 0, & \text{otherwise.}\end{cases} \quad (7.4)$$

Under Assumption A.1 and $\sup_{m\geqslant 1} m^4\lambda_m/\gamma_m < \infty$ it is shown in Cardot and Johannes [2010] that there exists $C>0$ depending on σ^2,ρ,d,Λ only such that

$$\sup_{\Gamma\in\mathcal{N}_\lambda^d}\sup_{\beta\in\mathcal{F}_\gamma^\rho}\left\{\mathbb{E}\|\widetilde{\beta}_{m_n^*}-\beta\|_\omega^2\right\}\leqslant C\,\delta_n^*,$$

where the dimension parameter m_n^* is given in (4).

Examples of rates. We compute in this section the minimal rate δ_n^* for two standard configurations for γ, ω, and λ. In both examples, we take $\omega_j = j^{2s}$, $s\in\mathbb{R}$, for $j\geqslant 1$. Here and subsequently, we write $a_n\lesssim b_n$ if there exists $C>0$ such that $a_n\leqslant C b_n$ for all $n\in\mathbb{N}$ and $a_n\sim b_n$ when $a_n\lesssim b_n$ and $b_n\lesssim a_n$ simultaneously.

(*p-p*) For $j\geqslant 1$ let $\gamma_j = j^{2p}$, $p>0$, and $\lambda_j = j^{-2a}$, $a>1$, then Assumption A.1 holds, if $-a<s<p$. It is easily seen that $m_n^*\sim n^{1/(2p+a+1)}$ if $2s+a>-1$, $m_n^*\sim n^{1/[2(p-s)]}$ if $2s+a<-1$ and $m_n^*\sim(n/\log(n))^{1/[2(p-s)]}$ if $a+2s=-1$. The minimal rate δ_n^* attained by the estimator is $\max(n^{-(2p-2s)/(a+2p+1)}, n^{-1})$, if $2s+a\neq -1$ (and $\log(n)/n$ if $2s+a=-1$). Since an increasing value of a leads to a slower minimal rate, it is called degree of ill-posedness (c.f. Natterer [1984]). Moreover, the case $0 \leqslant s < p$ can be interpreted as the L^2-risk of an estimator of the s-th derivative of β. On the other hand $s = -a/2$ corresponds to the mean-prediction error (c.f. Cardot and Johannes [2010]).

(p-e) For $j \geqslant 1$ let $\gamma_j = j^{2p}$, $p > 0$, and $\lambda_j = \exp(-j^{2a})$, $a > 0$, where Assumption A.1 holds, if $p > s$. Then $m_n^* \sim (\log n - \frac{2p+(2a-1)_+}{2a}\log(\log n))^{1/(2a)}$ with $(q)_+ := \max(q,0)$. Thereby, $(\log n)^{-(p-s)/a}$ is the minimal rate attained by the estimator.

7.3 The effect of noisy observations of the regressor

In order to formulate the lower bound below let us define for all $n, L \geqslant 1$ and $\varsigma \geqslant 0$

$$m^*_{n,L,\varsigma} := m^*_{n,L,\varsigma}(\gamma,\omega,\lambda) := \arg\min_{m\geqslant 1}\left\{\max\left(\frac{\omega_m}{\gamma_m}, \sum_{j=1}^m \frac{\omega_j}{n\sqrt{\lambda_j}}, \sum_{j=1}^m \frac{\varsigma^2\,\omega_j}{Ln\lambda_j}\right)\right\} \text{ and}$$

$$\delta^*_{n,L,\varsigma} := \delta^*_{n,L,\varsigma}(\gamma,\omega,\lambda) := \max\left(\frac{\omega_{m^*_{n,L,\varsigma}}}{\gamma_{m^*_{n,L,\varsigma}}}, \sum_{j=1}^{m^*_{n,L,\varsigma}} \frac{\omega_j}{n\sqrt{\lambda_j}}, \sum_{j=1}^{m^*_{n,L,\varsigma}} \frac{\varsigma^2\,\omega_j}{Ln\lambda_j}\right). \quad (7.5)$$

The lower bound given below needs the following assumption.

ASSUMPTION (A.2) *Let ω, γ and λ be sequences such that*

$$0 < \triangle := \inf_{L,n\geqslant 1}\left\{(\delta^*_{n,L,\varsigma})^{-1}\min\left(\frac{\omega_{m^*_{n,L,\varsigma}}}{\gamma_{m^*_{n,L,\varsigma}}}, \sum_{j=1}^{m^*_{n,L,\varsigma}} \frac{\omega_j}{n\sqrt{\lambda_j}}, \sum_{j=1}^{m^*_{n,L,\varsigma}} \frac{\varsigma^2\,\omega_j}{Ln\lambda_j}\right)\right\} \leqslant 1.$$

THEOREM (Lower bound) *If the sequences ω, γ and λ satisfy Assumptions A.1 - A.2, then there exists $C > 0$ depending on $\sigma^2, \varsigma^2, \rho, d$, and $\triangle$ only such that*

$$\inf_{\breve\beta}\ \inf_{\Gamma\in\mathcal{N}_\lambda^d}\ \sup_{\beta\in\mathcal{F}_\gamma^\rho}\left\{\mathbb{E}\|\breve\beta-\beta\|_\omega^2\right\} \geqslant C\,\delta^*_{n,L,\varsigma} \quad \textit{for all } n,L\geqslant 1.$$

Observe that the lower rate $\delta^*_{n,L,\varsigma}$ is never faster than the lower rate δ^*_n for known X defined in (3). Clearly, we recover δ^*_n for all $L \geqslant 1$ in case $\varsigma = 0$. On the other hand given an iid. sample $\{(Y^{(i)}, Z_1^{(i)}, \dots, Z_L^{(i)})\}$ of size n of $(Y, Z_1, \dots, Z_L)$ we define estimators for the elements $[g]_j := \langle g, \psi_j\rangle$ and $[\Gamma]_{k,j} := \langle \Gamma\psi_k, \psi_j\rangle$, $k, j \geqslant 1$, respectively as follows

$$\widehat{[g]}_j := \frac{1}{n}\sum_{i=1}^n Y^i \frac{1}{L}\sum_{\ell=1}^L Z^{(i)}_{j,\ell}, \quad\text{and}\quad \widehat{[\Gamma]}_{k,j} := \frac{1}{n}\sum_{i=1}^n \frac{1}{L(L-1)}\sum_{\substack{\ell_1,\ell_2=1\\ \ell_1\neq\ell_2}}^L Z^{(i)}_{j,\ell_1}Z^{(i)}_{k,\ell_2}. \quad (7.6)$$

We replace in definition (4) then the unknown matrix $[\widetilde\Gamma]_{\underline{m}}$ and vector $[\widetilde g]_{\underline{m}}$ respectively by the matrix $\widehat{[\Gamma]}_{\underline{m}}$ with elements $\widehat{[\Gamma]}_{k,j}$ and the vector $\widehat{[g]}_{\underline{m}}$ with elements $\widehat{[g]}_j$, that is,

$$\widehat{\beta}_m := \sum_{j=1}^{m} [\widehat{\beta}]_j \psi_j \quad \text{with} \quad [\widehat{\beta}]_{\underline{m}} := \begin{cases} [\widehat{\Gamma}]_{\underline{m}}^{-1} [\widehat{g}]_{\underline{m}}, & \text{if } [\widehat{\Gamma}]_{\underline{m}} \text{ is non-singular} \\ & \text{and } \|[\widehat{\Gamma}]_{\underline{m}}^{-1}\| \leqslant n, \\ 0, & \text{otherwise.} \end{cases} \tag{7.7}$$

The next theorem establishes the minimax-optimality of the estimator $\widehat{\beta}_m$ provided the dimension parameter m is chosen appropriate, i.e $m := m^*_{n,L,\varsigma}$ given in (5).

THEOREM (Upper bound) *If Assumptions A.1 - A.2 and* $\sup_{m\geqslant 1} m^4 \lambda_m \gamma_m^{-1} < \infty$ *are satisfied, then there exists* $C > 0$ *depending on* $\sigma^2, \varsigma^2, \rho, d, \Lambda$ *only such that*

$$\sup_{\Gamma \in \mathscr{N}_\lambda^d} \sup_{\beta \in \mathscr{F}_\gamma^\rho} \left\{ \mathbb{E} \| \widehat{\beta}_{m^*_{n,L,\varsigma}} - \beta \|_\omega^2 \right\} \leqslant C\, \delta^*_{n,L,\varsigma} \quad \textit{for all } n \geqslant 1, L \geqslant 2 \textit{ and } \varsigma \geqslant 0.$$

Examples of rates (continued). Suppose first that the panel size $L \geqslant 2$ is constant and $\varsigma > 0$. In example (*p-p*) if $2s + 2a + 1 > 0$ it is easily seen that $m^*_{n,L,\varsigma} \sim n^{1/(2p+2a+1)}$ and the minimal rate attained by the estimator is $\delta^*_{n,L,\varsigma} \sim n^{-(2p-2s)/(2a+2p+1)}$. Let us compare this rate with the minimal rates in case of a functional linear model (FLM) with known regressor and in case of an indirect regression model (IRM) given by the covariance operator Γ and Gaussian white noise $\dot{W}$, i.e., $g_n = \Gamma\beta + n^{-1/2}\dot{W}$ (c.f. Hoffmann and Rei [2008]). The minimal rate in the FLM with known X is $n^{-2(p-s)/(a+2p+1)}$, while $n^{-2(p-s)/(2a+2p+1)}$ is the minimal rate in the IRM. We see that in a FLM with known X the covariance operator Γ has the *degree of ill-posedness a* while it has in a FLM with noisy observations of X and in the IRM a *degree of ill-posedness* $2a$. In other words only in a FLM with known regressor we do not face the complexity of an inversion of Γ but only of its square root $\Gamma^{1/2}$. The same remark holds true in the example (*p-e*), but the minimal rate is the same in all three cases due to the fact that for $\lambda_j \sim \exp(-r|j|^{2a})$ the dependence of the minimal rate on the value r is hidden in the constant. However, it is rather surprising that in this situation a panel of size $L = 2$ is sufficient to recover the minimal but logarithmic rate when X is known. In contrast, in example (*p-p*) the minimal rate for known X can only be attained in the presence of noise in the regressor if the panel size satisfies $L_n^{-1} = O(n^{-a/(a+2p+1)})$ as the sample size n increases, since $\delta^*_{n,L,\varsigma} \sim \max(n^{-(2p-2s)/(a+2p+1)}, (L_n n)^{-(2p-2s)/(2a+2p+1)})$.

Acknowledgements This work was supported by the IAP research network no. P6/03 of the Belgian Government (Belgian Science Policy).

References

1. Bereswill, M., Johannes, J.: On the effect of noisy observations of the regressor in a functional linear model. Technical report, Université catholique de Louvain (2010)
2. Bosq. D.: Linear Processes in Function Spaces. Lecture Notes in Statistics, 149, Springer-Verlag (2000)
3. Cardot, H., Johannes, J.: Thresholding projection estimators in functional linear models. J. Multivariate Anal. **101** (2), 395–408 (2010)
4. Cardot, H., Ferraty, F., Sarda, P.: Functional linear model. Stat. Probabil. Lett. **45**, 11–22 (1999)
5. Cardot, H., Ferraty, F., Sarda, P.: Spline estimators for the functional linear model. Stat. Sinica **13** 571–591 (2003)
6. Cardot, H., Ferraty, F., Mas, A., Sarda, P.: Clt in functional linear regression models. Probab. Theor. Rel. **138**, 325–361 (2007)
7. Crambes, C., Kneip, A., Sarda, P.: Smoothing splines estimators for functional linear regression. Ann. Stat. **37** (1), 35–72 (2009)
8. Efromovich, S., Koltchinskii, V.: On inverse problems with unknown operators. IEEE T. Inform. Theory **47** (7), 2876–2894 (2001)
9. Ferraty, F., Vieu, P.: Nonparametric Functional Data Analysis: Practice and Theory. Springer, New York (2006)
10. Hall, P., Horowitz, J. L.: Methodology and convergence rates for functional linear regression. Ann. Stat. **35** (1), 70–91 (2007)
11. Hoffmann, M., Reiß, M.: Nonlinear estimation for linear inverse problems with error in the operator. Ann. Stat. **36** (1), 310–336 (2008)
12. Marx, B. D., Eilers, P. H.: Generalized linear regression on sampled signals and curves: a p-spline approach. Technometrics **41**, 1–13 (1999)
13. Müller, H.-G., Stadtmüller, U.: Generalized functional linear models. Ann. Stat. **33** (2), 774–805 (2005)
14. Natterer, F.: Error bounds for Tikhonov regularization in Hilbert scales. Appl. Anal. **18**, 29–37 (1984)
15. Ramsay, J., Silverman, B. Functional Data Analysis (Second Edition). Springer, New York (2005)
16. Yao, F., Müller, H.-G., Wang, J.-L.: Functional linear regression analysis for longitudinal data. Ann. Stat. **33** (6), 2873–2903 (2005)

Chapter 8
Testing the Equality of Covariance Operators

Graciela Boente, Daniela Rodriguez, Mariela Sued

Abstract In many situations, when dealing with several populations, equality of the covariance operators is assumed. In this work, we will study a hypothesis test to validate this assumption.

8.1 Introduction

Functional data analysis provides modern analytical tools for data that are recoded as images or as a continuous phenomenon over a period of time. Because of the intrinsic nature of these data, they can be viewed as realizations of random functions often assumed to be in $L^2(I)$, with I a real interval or a finite dimensional Euclidean set.

On the other hand, when working with more than one population, as in the finite dimensional case, a common assumption is to assume the equality of covariance operators. In the case of finite-dimensional data, test for equality of covariance matrices have been extensively studied, see for example Seber (1984), even when the sample size is smaller than the size of the variables see Ledoit and Wolf (2002) and Schott (2007). Ferraty et.al. (2007) have proposed tests for comparison of groups of curves based on the comparison of covariances. The hypothesis tested are that of equality, proportionality, and others based on the spectral decomposition of these covariances.

In the functional setting, we will study a proposal for testing the hypothesis that the covariance operators of $k-$populations of random objects are equal. If we have

Graciela Boente
Universidad de Buenos Aires and CONICET, Argentina e-mail: gboente@dm.uba.ar

Daniela Rodriguez
Universidad de Buenos Aires and CONICET, Argentina e-mail: drodrig@dm.uba.ar

Mariela Sued
Universidad de Buenos Aires and CONICET, Argentina e-mail: msued@dm.uba.ar

two populations where $\boldsymbol{\Gamma}_1$ and $\boldsymbol{\Gamma}_2$ are their covariance operators, we can consider consistent estimators $\widehat{\Gamma_2}$ and $\widehat{\Gamma_2}$ of both operators, such as those given by Dauxois, Pousse, and Romain (1982). It is clear that under the null hypothesis the difference in the estimates of the operators of both populations should be small. The idea is to build a test based on the norm of the difference between the estimates of the operators and then, generalize this approach to the $k-$populations case. We will obtain the asymptotic distribution of the test statistics under the null hypothesis. Also, we will study bootstrap procedures and their validation.

8.2 Notation and preliminaries

Let $X_{i,1}(t),\cdots,X_{i,n_i}(t) \in L^2(\mathscr{I})$ for $i=1,\ldots,k$ be independent observations from k independent samples of smooth random functions with mean $\mu_i(t)$, without loss of generality, we will assume that $\mathscr{I}=[0,1]$. Denote by γ_i and Γ_i the covariance function and operator, respectively, related to each population. To be more precise, we are assuming that $\{X_{i,1}(t) : t \in [0,1]\}$ are k stochastic processes defined in $(\Omega,\mathscr{A},P)$ with continuous trajectories, mean μ_i and finite second moment, i.e., $E\left(X_{i,1}(t)\right)=\mu_i(t)$ and $E\left(X_{i,1}^2(t)\right)<\infty$ for $t\in[0,1]$. We will denote by

$$\gamma_i(t,s)=E\left(\left(X_{i,1}(t)-\mu_i(t)\right)\left(X_{i,1}(s)-\mu_i(s)\right)\right)$$

their covariance functions, which is just the functional version of the variance–covariance matrix in the classical multivariate analysis. As in the finite–dimensional case, each covariance function has an associated linear operator $\Gamma_i : L^2[0,1]\to L^2[0,1]$ defined as $(\Gamma_i u)(t)=\int_0^1 \gamma_i(t,s)u(s)ds$, for all $u\in L^2[0,1]$. Throughout this paper, we will assume that the covariance operators satisfy

$$\int_0^1\int_0^1 \gamma_i^2(t,s)dtds<\infty. \tag{8.1}$$

Cauchy-Schwartz inequality implies that $|\Gamma_i u|^2 \le \|\gamma_i\|^2|u|^2$, where $|u|$ stands for the usual norm in the space $L^2[0,1]$, while $\|\gamma\|$ denotes the norm in the space $\mathscr{F}=L^2\left([0,1]\times[0,1]\right)$. Therefore, Γ_i is a self–adjoint continuous linear operator.

An natural way to estimate the covariance operators $\widehat{\Gamma}_i$ for $i=1,\ldots,k$ is to consider the empirical covariance operator given by

$$\widehat{\Gamma_i}=\frac{1}{n_i}\sum_{j=1}^{n_i}\left(X_{i,j}-\overline{X_i}\right)\otimes\left(X_{i,j}-\overline{X_i}\right),$$

where $\overline{X_i}(t)=1/n_i\sum_{j=1}^{n_i}X_{i,j}(t)$. Dauxois, Pousse and Romain (1982) proved that $\sqrt{n_i}\left(\widehat{\Gamma}_i-\Gamma_i\right)$ converges in distribution to a zero mean gaussian random element, $\boldsymbol{U}_i$, on $\mathscr{F}$ with covariance operator $\boldsymbol{\Upsilon}_i$ given by

$$\boldsymbol{\Upsilon}_i = E\left((X_{i,1}\otimes X_{i,1})\tilde{\otimes}(X_{i,1}\otimes X_{i,1})\right) - E\left(X_{i,1}\otimes X_{i,1}\right)\tilde{\otimes}E\left(X_{i,1}\otimes X_{i,1}\right). \qquad (8.2)$$

Smooth estimators $\widehat{\Gamma}_i^s$ of the covariance operators was studied in Boente and Fraiman (2000) and they proved that the smooth estimators have the same asymptotic distribution that the empirical version, under mild conditions. The smoothed version is defined as

$$\widehat{\gamma}_i^s(t,s) = \sum_{j=1}^{n_i} \left(X_{i,j,h}(t) - \overline{X}_{i,h}(t)\right)\left(X_{i,j,h}(s) - \overline{X}_{i,h}(s)\right)/n_i\,,$$

where $X_{i,j,h}(t) = \int K_h(t-x)X_{i,j}$ are the smoothed trajectories and $K_h(\cdot) = h^{-1}K(\cdot/h)$ is a nonnegative kernel function and h a smoothing parameter.

8.3 Hypothesis Test

In this Section, we study the problem of testing the hypothesis

$$H_0 : \Gamma_1 = \Gamma_2 \quad \text{against} \quad H_1 : \Gamma_1 \neq \Gamma_2\,. \qquad (8.3)$$

A natural approach is to consider the empirical covariance operators of each population $\widehat{\Gamma}_i$ and construct a statistic based on the difference between the estimators corresponding to the covariance operator at each population.

The following result allows to construct a test for the hypothesis (8.3) of equality of covariance operators when we consider two populations.

Theorem 3.1. *Let $\widehat{\Gamma}_i$ be an estimator of the $i-$th population covariance operator and assume that $E(\|X_{i,1}\|^4) < \infty$ for $i = 1,2$. Denote by $\{\theta\}_{i\geq 1}$ the sequence of eigenvalues associated to the operator $\frac{1}{\tau_1}\boldsymbol{\Upsilon}_1 + \frac{1}{\tau_2}\boldsymbol{\Upsilon}_2$, where $\boldsymbol{\Upsilon}_i$ are the covariance operator associated for the asymptotic distribution of $\widehat{\Gamma}_i$ and $n_i/N \to \tau_i$, for $i = 1,2$. Assume that $\sum_{i\geq 1}\theta_i < \infty$. Then,*

$$T_n = N\|(\widehat{\Gamma}_1 - \Gamma_1) - (\widehat{\Gamma}_2 - \Gamma_2)\|^2 \xrightarrow{\mathscr{D}} \sum_{i\geq 1}\theta_i Z_i^2\,, \qquad (8.4)$$

where Z_i are i.i.d. standard normal distributions and $N = n_1 + n_2$.

The previous results motivate the use of the bootstrap methods, due the fact that the asymptotic distribution obtained in (8.4) depends on the unknown eigenvalues θ_i. We will consider a bootstrap calibration for the distribution of the test that can be described as follows,

Step 1 Given a sample $X_{i,1}(t),\cdots,X_{i,n_i}(t)$ we estimate $\widehat{\boldsymbol{\Upsilon}} = \frac{n_1+n_2}{n_1}\widehat{\boldsymbol{\Upsilon}}_1 + \frac{n_1+n_2}{n_2}\widehat{\boldsymbol{\Upsilon}}_2$, where $\widehat{\boldsymbol{\Upsilon}}_i$ are consistent estimators of $\boldsymbol{\Upsilon}_i$ for $i = 1,2$.

Step 2 For $i = 1, \ldots k_n$ denote by $\widehat{\theta}_i$ the positive eigenvalues of $\widehat{\boldsymbol{\Upsilon}}$.
Step 3 Generate $Z_1^*, \ldots, Z_{k_n}^*$ random variables i.i.d. according to a standar normal distribution. Let $T_n^* = \sum_{j=1}^{k_n} \widehat{\theta}_j Z_j^{*2}$.
Step 4 Repeat **Step 3** Nboot times, to get Nboot values of T_{ni}^* for $1 \le i \le Nboot$.

The $(1-\alpha)-$quantile of the asymptotic distribution of T_n can be approximated by the $(1-\alpha)-$quantile of the empirical distribution of T_{ni}^* for $1 \le i \le Nboot$. The p-value can be estimated by $\widehat{p} = \frac{s}{Nboot}$ where s is the number of T_{ni}^* which are larger or equal than the observed value of T_n.

Remark 3.2. *Note that this procedure depends only on the asymptotic distribution of $\widehat{\Gamma}_i$. If we consider any other asymptotically normally estimator of Γ_i for example the smoothed estimators Γ_i^s, the results may be adapted to this new setting.*

The following theorem entails the validity of the bootstrap method. It is important to note that the following theorem entails that, under H_0 the bootstrap distribution of T_n converges to the asymptotic null distribution of T_n which ensures that the asymptotic significance level of the test based on the bootstrap critical value is indeed α.

Theorem 3.3. *Let k_n such that $k_n/\sqrt{n} \to 0$ and $\tilde{X}_n = (X_{1,1}, \cdots, X_{1,n_1}, X_{2,1}, \cdots, X_{2,n_2})$. Consider $F_{T_n^*|\tilde{X}_n}(\cdot) = P(T_n^* \le \cdot \,|\tilde{X}_n)$. Then, under the same hypothesis of Theorem 3.1, we get that*

$$\rho_{\mathrm{K}}(F_{T_n^*|\tilde{X}_n}, F_T) \xrightarrow{p} 0\,, \tag{8.5}$$

where F_T denotes the distribution function of $T = \sum_{i\ge1} \theta_i Z_i^2$, with Z_i are i.i.d. standard normal distributions and ρ_{K} is the Kolmogorov distance between distribution functions.

8.4 Generalization to k-populations

In this Section, we consider tests for the equality of the covariance operators of k populations. That is, if $\boldsymbol{\Gamma}_i$ denotes the covariance operator of the ith population, we wish to test the null hypothesis

$$H_0 : \boldsymbol{\Gamma}_1 = \cdots = \boldsymbol{\Gamma}_k \quad \text{against} \quad H_1 : \exists\, i \ne j \text{ such that } \boldsymbol{\Gamma}_i \ne \boldsymbol{\Gamma}_j \tag{8.6}$$

Let $N = n_1 + \cdots + n_k$ and assume that $n_i/N \to \tau_i$. A natural generalization of the proposal given in Section 3 is to consider the following statistic test

$$T_{k,n} = N \sum_{j=2}^{k} \|\widehat{\Gamma}_j - \widehat{\Gamma}_1\|^2,$$

where $\widehat{\Gamma}_i$ are the empirical covariance operators of ith population. The following result states the asymptotic distributions under the null hypothesis of $T_{k,n}$.

Theorem 4.1. *Let $\widehat{\Gamma}_i$ be an estimator of the covariance operator of the ith population such that $\sqrt{N}(\widehat{\Gamma}_i - \Gamma_i) \xrightarrow{\mathscr{D}} U_i$, where U_i is zero mean gaussian random element of $\mathscr{F}$ with covariance operator $\frac{1}{\tau_i}\boldsymbol{\Upsilon}_i$.*
Denote by θ_i the sequence of eigenvalues associated to the operator $\boldsymbol{\Upsilon}_W$ given by $\boldsymbol{\Upsilon}_W(y_1,\ldots,y_{k-1}) = \left(\frac{1}{\tau_2}\boldsymbol{\Upsilon}_2(y_1),\ldots,\frac{1}{\tau_k}\boldsymbol{\Upsilon}_k(y_{k-1})\right) + \frac{1}{\tau_1}\boldsymbol{\Upsilon}_1(\sum_{i=1}^{k-1} y_i)$. If $\sum_{i\geq 1}\theta_i < \infty$, we have

$$T_{k,n} = N\sum_{j=2}^{k} \|\widehat{\Gamma}_j - \widehat{\Gamma}_1\|^2 \xrightarrow{\mathscr{D}} \sum_{i\geq 1}\theta_i Z_i^2$$

where Z_i are i.i.d standard normal distribution.

As in Section 3, a bootstrap procedure can be considered. In order to estimate θ_j for $j \geq 1$, we can consider estimators of the operators $\boldsymbol{\Upsilon}_i$ for $1 \leq i \leq k$ and thus estimate $\boldsymbol{\Upsilon}_W$. Therefore, if $\widehat{\theta}_i$ are the positive eigenvalues of $\widehat{\boldsymbol{\Upsilon}}_W$, a bootstrap procedure follows as in Steps 3 and 4.

References

1. Boente, G., Fraiman, R.: Kernel-based functional principal components. Statist. Probab. Lett. **48**, 335–345 (2000)
2. Dauxois, J., Pousse, A., Romain, Y.: Asymptotic theory for the principal component analysis of a vector random function: Some applications to statistical inference. J. Multivariate Anal. **12**, 136–154 (1982)
3. Ferraty, F., View, P., Viguier-Pla, S.: Factor-based comparison of groups of curves. Comput. Stat. Data An. **51**, 4903–4910 (2007)
4. Ledoit, O., Wolf, M.: Some hypothesis tests for the covariance matrix when the dimension is large compared to the sample size. Ann. Stat. **30** (4), 1081–1102 (2002)
5. Schott, J.: A test for the equality of covariance matrices when the dimension is large relative to the sample sizes. Comput. Stat. Data An. **51** (12), 6535–6542 (2007).
6. Seber, G.: Multivariate Observations. John Wiley and Sons (1984)

Chapter 9
Modeling and Forecasting Monotone Curves by FDA

Paula R. Bouzas, Nuria Ruiz-Fuentes

Abstract A new estimation method and forecasting of monotone sample curves is performed from observations in a finite set of time points without a previous transformation of the original data. Monotone spline cubic interpolation is proposed for the reconstruction of the sample curves. Then, the interpolation basis is adapted to apply FPCA and forecasting is done by means of principal components prediction.

9.1 Introduction

Functional data analysis (FDA) deals with the modeling of sample curves. Ramsay and Silverman (1997) is a basic review of some techniques of FDA as functional principal components analysis (FPCA) or functional linear models. Ramsay and Silverman (2002) presents interesting applications of FDA to real data. Valderrama et al. (2000) presents several ways of approximation of FPCA and reviews models of linear prediction by principal components in order to forecast a stochastic process in terms of its past.

The aim of this work is to apply techniques of FDA to the sample paths of a stochastic process which are monotone. The usual way to treat this type of sample paths is to transform them to unconstrained curves and work with their transformed values. This paper propose to work with the original data in the following way. In practice, the sample paths of a stochastic process can be observed only in a finite set of time points so it is needed to reconstruct their functional form. We propose to use the cubic monotone interpolation of Fritsh and Carlson (1980) to reconstruct the sample paths. Then, the interpolation basis is adapted in order to apply FPCA.

Paula R. Bouzas
University of Granada, Spain, e-mail: paula@ugr.es

Nuria Ruiz-Fuentes
University of Jaén, Spain, e-mail: nfuentes@ujaen.es

Having derived the stochastic estimation, the forecasting of a new sample path can be achieved by prediction with principal components (PCP). Finally, the modeling and forecasting is applied to the real data of the growth of girls up to 18 years.

9.2 Functional reconstruction of monotone sample paths

Let $\{X(t); t \in [T_0, T_1]\}$ be a second order stochastic process, continuous in quadratic mean and monotonous sample paths. Let us consider a sample of n realizations of it observed in a finite set of instants of time $t_0 = T_0, \ldots, t_p = T_1$, which will be denoted by $\{X_w(t) : t \in [t_0, t_p], w = 1, \ldots, n\}$.

Firstly, the functional form of each sample path must be reconstructed estimating them in a finite space generated by a functions basis. In our case, the interpolation data are $(t_0, X_\omega(t_0)), \ldots, (t_p, X_\omega(t_p))$. In order to preserve the monotonicity, the first derivative has to be nonnegative for the nondecreasing case and nonpositive for the opposite case. The derivatives in the observed points, denoted by $d_{\omega 0}, \ldots, d_{\omega p}$, are calculated as proposed by Fritsch and Carlson (1980). The same authors propose the following monotone piecewise polynomial interpolation:

$$IX_{\omega j}(t) = X_\omega(t_j)H_1(t) + X_\omega(t_{j+1})H_2(t) + d_{\omega j}H_3(t) + d_{\omega j+1}H_4(t),$$

for $t \in [t_j, t_{j+1}]$, $j = 0, \ldots, p-1$, where $d_{\omega j} = \left.\frac{dIX_{\omega j}(t)}{dt}\right|_{t=t_j}$, $d_{\omega j+1} = \left.\frac{dIX_{\omega j}(t)}{dt}\right|_{t=t_{j+1}}$ and $H_s(t)$ are the usual Hermite functions for the interval $[t_j, t_{j+1}]$. We have chosen this type of interpolation because it joints the flexibility of cubic spline with the monotonicity preservation.

In order to use the interpolation in FPCA, it should be expressed in the whole observation interval in terms of a basis. After some manipulations it can be written as

$$IX_\omega(t) = \sum_{j=0}^{p} X_\omega(t_j)\Phi_j(t) + \sum_{j=0}^{p} d_{\omega j}\Psi_j(t), \quad t \in [t_0, t_p],\ \omega = 1, \ldots, k \tag{9.1}$$

where the functions are

$$\begin{aligned}
\Phi_j(t) &= \begin{cases} \phi(\frac{t-t_{j-1}}{h_{j-1}}),\ t \in [t_{j-1}, t_j] \\ \phi(\frac{t_{j+1}-t}{h_j}),\ t \in [t_j, t_{j+1}] \end{cases},\ j \neq 0, p \\
\Psi_j(t) &= \begin{cases} h_{j-1}\psi(\frac{t-t_{j-1}}{h_{j-1}}),\ t \in [t_{j-1}, t_j] \\ -h_j\psi(\frac{t_{j+1}-t}{h_j}),\ t \in [t_j, t_{j+1}] \end{cases}, \qquad j \neq 0, p \\
\Phi_0(t) &= \phi(\frac{t_1 - t}{h_0}),\ \Psi_0(t) = -h_0\psi(\frac{t_1 - t}{h_0}), \qquad t \in [t_0, t_1] \\
\Phi_p(t) &= \phi(\frac{t - t_{p-1}}{h_{p-1}}),\ \Psi_p(t) = h_{p-1}\psi(\frac{t - t_{p-1}}{h_{p-1}}), \qquad t \in [t_{p-1}, t_p]
\end{aligned} \tag{9.2}$$

with $h_j = t_{j+1} - t_j$, $\phi(x) = 3x^2 - 2x^3$ and $\psi(x) = x^3 - x^2$. Then, the functions of (9.2) form the Lagrange basis of cubic splines of dimension $2(p+1)$ (see Bouzas et al., 2006). Equation (9.1) for all the sample paths can be written jointly as

$$IX(t) = A\,B(t), \quad t \in [t_0, t_p]$$

where the matrices are defined as

$$IX(t) = (IX_1(t), \ldots, IX_n(t))^T; \qquad B(t) = (\Phi_0(t), \ldots, \Phi_p(t), \Psi_0(t), \ldots, \Psi_p(t))^T$$

$$A = \begin{pmatrix} X_1(t_0) & \ldots & X_1(t_p) & d_{10} & \ldots & d_{1p} \\ \vdots & \vdots & \vdots & \vdots & \vdots & \vdots \\ X_n(t_0) & \ldots & X_n(t_p) & d_{n0} & \ldots & d_{np} \end{pmatrix}$$

where ^{T} denotes the transpose matrix. But in order to unify the notation, let us rewrite the basis and the coefficients matrix as

$$B(t) = \left(B_1(t), \ldots, B_{p+1}(t), B_{p+2}(t), \ldots, B_{2(p+1)}(t)\right)^T$$
$$A = (a_{\omega l})_{\omega=1,\ldots,k\,;\, l=1,\ldots,2(p+1)}$$

so, the interpolation polynomial of equation (9.1) becomes

$$IX_\omega(t) = \sum_{l=1}^{2(p+1)} a_{\omega l} B_l(t); \quad t \in [t_0, t_p],\ \omega = 1, \ldots, n.$$

9.3 Modeling and forecasting

The stochastic structure of the process $X(t)$ with monotone sample paths is derived applying the usual methodology of FPCA with the proper basis found in Section 2. In this case, it is specially interesting because the basis dimension has been increased due to the coefficients of the monotone interpolation.

Considering the centered interpolated process

$$\overline{IX}(t) = IX(t) - \mu_{IX}(t) = \left(A - \bar{A}\right) B(t)$$

where $\bar{A} = (\overline{a}_{\omega l})$ with elements $\overline{a}_{\omega l} = \frac{1}{n}\sum_{\omega=1}^{n} a_{\omega l}$ $(l = 1, \ldots, 2(p+1),\ \omega = 1, \ldots, n)$, and $\mu_{IX}(t) = \bar{A}\,B(t)$. Let us denote by $\mathbb{P}$ the matrix whose elements are the usual inner products of the basis functions given by $\langle B_i, B_j \rangle_u = \int_{t_0}^{t_p} B_i(t) B_j(t)\, dt$. Then, the FPCA of $\overline{IX}_\omega(t)$ in the space generated by the basis $B(t)$ with respect to the usual metric in $L^2[t_0, t_p]$ is equivalent to the multivariant PCA of the matrix $\left(A - \bar{A}\right)\mathbb{P}^{1/2}$ with respect to the usual one in $\mathbb{R}^{2(p+1)}$.

Once the eigenvectors, g_j, of the covariance matrix of $\left(A - \bar{A}\right)\mathbb{P}^{1/2}$ are obtained, the sample paths of $\overline{IX}(t)$ are represented in terms of their principal components as

$$\overline{IX}_{\omega}(t) = \sum_{j=1}^{2(p+1)} \zeta_{\omega j} f_j(t), \quad \omega = 1,\ldots,n \tag{9.3}$$

were $f_j(t)$ are the eigenfunctions of the sample covariance of $X(t)$ given by $f_j(t) = \sum_{l=1}^{2(p+1)} f_{lj} B_l(t)$ where the vector of coefficients $f_j = \mathbb{P}^{-1/2} g_j$, and the principal components are obtained as generalized linear combinations of the sample paths of the interpolated process

$$\zeta_{\omega j} = \int_{t_0}^{t_p} \overline{IX}_{\omega}(t) f_j(t)\, dt = \left(A_{\omega} - \bar{A}_{\omega}\right) \mathbb{P}^{1/2} g_j$$

where $\left(A_{\omega} - \bar{A}_{\omega}\right)$ is the ω-*th* row of $\left(A - \bar{A}\right)$.

Finally, the stochastic estimation is the orthogonal representation which minimizes the mean squared error after truncating expression (9.3)

$$X^q(t) = \mu_{IX}(t) + \sum_{j=1}^{q} \zeta_j f_j(t).$$

Then, the dimension $2(p+1)$ is reduced to q, so that an amount of variability as closed to 1 as wanted is reached and given by

$$\frac{\sum_{j=1}^{q} \lambda_j}{\sum_{j=1}^{2(p+1)} \lambda_j},$$

where λ_j is the variance of the j-*th* principal component, ζ_j, given by the j-*th* eigenvalue of the covariance matrix of $\left(A - \bar{A}\right) \mathbb{P}^{1/2}$ associated to the j-*th* eigenvalue g_j.

Prediction by means of principal components of a stochastic process gives a continuous prediction of the process in a future time interval from discrete observations of the process in the past which was introduced by Aguilera et al. (1997). Having known the evolution of an stochastic process $\{X(t); t \in [T_0, T_1]\}$, PCP models estimate it in a future interval $\{X(t); t \in [T_1, T_2]\}$ using FPCA. The process must be of second order, continuous in quadratic mean and squared integrable sample paths in their corresponding intervals. If the available data are several sample paths of $X(t)$, the PCP model has to be estimated (see also Aguilera et al. (1999) and Valderrama et al. (2000) for a deeper study).

Firstly, FPCA of the process in both intervals is carried out

$$\begin{aligned} X^{q_1}(t) &= \mu_{IX}^1(t) + \sum_{j=1}^{q_1} \xi_j f_j(t); \qquad t \in [t_0 = T_0, T_1] \\ X^{q_2}(s) &= \mu_{IX}^2(s) + \sum_{j=1}^{q_2} \eta_j g_j(s); \qquad s \in (T_1, T_2) \end{aligned} \tag{9.4}$$

Secondly, the principal components of the past that predict the principal components of the future are selected by means of having significantly high correlation.

Let us denote by $\tilde{\eta}_j^{p_j} = \sum_{i=1}^{p_j} b_i^j \xi_i$ the estimator of η_j, $j = 1, \ldots, q_2$ in terms of the p_j principal components ξ_j. Therefore, we can rewrite (9.4) so that

$$X^{q_2}(s) = \mu_{IX}^2(s) + \sum_{j=1}^{q_2} \left(\sum_{i=1}^{p_j} b_i^j \xi_i \right) g_j(s); \qquad s \in (T_1, T_2) \tag{9.5}$$

This is the estimated stochastic structure of $X(t)$ in a future time interval from its knowledge in the past. The selected PCP model contents those pairs of future-past principal components with significant linear correlation, which are included in order of magnitude of the proportion of future variance explained by a PCP model only including the pair, until the relative high proportion of future variance explained is achieved.

Finally, the evolution of any other new sample path of the process observed in the past is predicted in the future interval using the FPCA in the future with the principal components predicted by the past ones using equation (9.5).

9.4 Application to real data

In order to illustrate the method explained in the previous sections, we have chosen a known example of real data, the heights of 54 girls measured at a set of 31 ages unequally spaced along their first 18 years, which have been analyzed by Ramsay et al. (2009). The data was organized in two groups, the first one contains 50 sample paths to model the process and the other 4 are kept apart to forecast.

Modeling theory of Section 2 has been applied to the data and it was found out that 4 principal components explain 98.77% of the total variability. Figure 1 illustrates it in two sample paths. Forecasting theory of Section 2 has been applied in order illustrate the forecasting method. The past interval has been chosen $[0, 7]$ so the future one is $(7, 18]$. Figure 2 shows two examples. The MSE of the predictions has become 0.5451.

9.5 Conclusions

This paper proposes a methodology for modeling monotone curves from the original data by means of fitting cubic splines that preserve the monotonicity. The results are similar to those of Ramsay et al. (2009) but this present modeling is more direct and has much less computational cost.

Acknowledgements This work was partially supported by projects MTM2010-20502 of Dirección General de Investigación y Gestión del Plan Nacional I+D+I and grants FQM-307 and FQM-246 of Consejería de Innovación de la Junta de Andalucía, both in Spain.

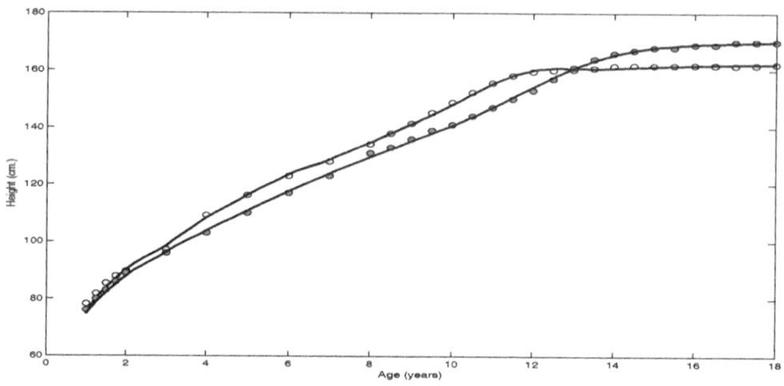

Fig. 9.1: Monotone curve modeling (solid line) to the observed heights of two girls.

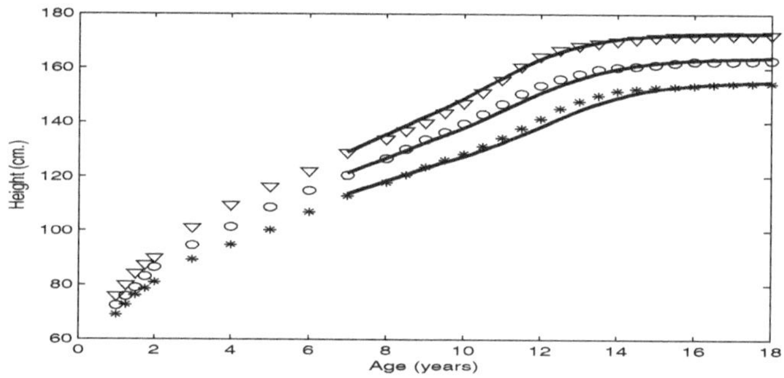

Fig. 9.2: Forecasting (solid line) the heights of three girls.

References

1. Aguilera, A. M., Ocaña, F. A., Valderrama, M. J.: An approximated principal component prediction model for continuous-time stochastic processes. Appl. Stoch. Model D. A. **13**, 61–72 (1997)
2. Aguilera, A. M., Ocaña, F. A., Valderrama, M. J.: Forecasting time series by functional PCA. Discussion of several weighted approaches. Computation. Stat. **14**, 443–467 (1999)
3. Bouzas, P. R., Valderrama, M. J., Aguilera, A. M., Ruiz-Fuentes, N.: Modelling the mean of a doubly stochastic Poisson process by functional data analysis. Comput. Statist. Data Anal. **50**, 2655–2667 (2006)
4. Fritsch, F. N., Carlson, R. E.: Monotone piecewise cubic interpolation. SIAM J. Numer. Anal. **17**, 238–246 (1980)
5. Ramsay, J. O., Silverman, B. M.: Functional Data Analysis. Springer-Verlag, New York (1997)
6. Ramsay, J. O., Silverman, B. M.: Applied Functional Data Analysis. Springer-Verlag, New York (2002)
7. Ramsay, J. O., Hooker, G., Graves, S.: Analysis with R and MatLab. Springer, New York (2009)

8. Valderrama, J. M., Aguilera, A. M., Ocaña, F. A.: Predicción dinámica mediante análisis de datos funcionales. Ed. Hespérides-La Muralla, Madrid 2000)

Chapter 10
Wavelet-Based Minimum Contrast Estimation of Linear Gaussian Random Fields

Rosa M. Crujeiras, María-Dolores Ruiz-Medina

Abstract Weak consistency of the wavelet periodogram is established for a class of linear Gaussian random fields, considering a Haar type isotropic wavelet basis. A minimum contrast estimation procedure is introduced and the weak consistency of the estimator is derived, following the methodology introduced by Ruiz-Medina and Crujeiras (2011).

10.1 Introduction

Consider the class of $d-$dimensional linear Gaussian random fields (RFs)

$$X(z) = \int_D a(\beta, z, y)\varepsilon(y)dy = \mathscr{A}^{\beta}(\varepsilon)(z),$$

given in terms of the kernel $a(\beta,\cdot,\cdot)$, with $\beta \in \Lambda$, being Λ a compact subset of $\mathbb{R}_+$, which defines the integral operator $\mathscr{A}^{\beta}$, where ε denotes a Gaussian white noise on $D \subseteq \mathbb{R}^d$, i.e., a generalized zero-mean Gaussian RF satisfying $\mathbb{E}[\varepsilon(\phi)\varepsilon(\psi)] = \langle \phi, \psi \rangle_{L^2(D)}$, with $\varepsilon(\psi) = \int_D \psi(x)\varepsilon(x)dx, \quad \forall \psi, \phi \in L^2(D)$. Here, $a(\beta,\cdot,\cdot)$ is a semi-parametric kernel satisfying the following condition:

C1. When $\|x-y\| \longrightarrow 0$, the following asymptotic behavior holds, for a certain positive constant C:

$$\frac{a(\beta,x,y)}{\|x-y\|^{\beta-d/2}} \longrightarrow C.$$

Rosa M. Crujeiras
University of Santiago de Compostela, Spain, e-mail: rosa.crujeiras@usc.es

María-Dolores Ruiz-Medina
University of Granada, Spain, e-mail: mruiz@ugr.es

Remark. If the reproducing kernel Hilbert space (RKHS) of X is isomorphic to a fractional Sobolev space of order β, with $\beta \in (d/2, d)$, the class of RFs considered is included in the one studied in Ruiz-Medina and Crujeiras (2011), which was previously introduced in Ruiz-Medina, *et al.* (2003) in a fractional generalized framework.

In this work, we propose a minimum contrast estimator for β in the class of RFs given by condition C1. The wavelet-based estimation procedure is similar to the one proposed by Ruiz-Medina and Crujeiras (2011). Weak consistency of the wavelet periodogram, based on Hermite expansion, and weak consistency of the minimum contrast estimator are derived. This paper is organized as follows. In Section 2, the wavelet scenario is specified. Asymptotic properties in the scale for the two-dimensional wavelet transform of the kernel and the covariance function are also obtained. Consistency of the wavelet periodogram is studied in Section 3. From these results, the minimum contrast estimator proposed in Section 4 is proved to be consistent. Final comments are given in Section 5, with discussion on possible extensions.

10.2 Wavelet generalized RFs

For simplicity and without loss of generality, assume that the compact domain D where the wavelet functions are defined, is of the form $D = [-M, M]^d$, for a certain positive constant M. In dimension d, the continuous discrete wavelet transform is defined in terms of the basic wavelet functions ψ^i, $i = 1, \ldots, 2^d - 1$. For each resolution level $j \in \mathbb{Z}$, and for every $x \in D$

$$\boldsymbol{\Psi}_{j,b}(x) = \sum_{i=1}^{2^d-1} \psi^i_{j,b}(x) = 2^{jd/2} \sum_{i=1}^{2^d-1} \psi^i(2^j x - b) = 2^{jd/2} \boldsymbol{\Psi}(2^j x - b),\ b \in L_j,$$

is the $d-$dimensional wavelet function translated at the center b. Domain L_j can be defined as $L_j = [0, 2^j]^d$, becoming the $d-$dimensional space $\mathbb{R}^d_+$ as $j \longrightarrow \infty$. Note that, since the asymptotic results derived in this paper hold for an increasing resolution level in the wavelet domain, this approach corresponds to a fixed domain asymptotic setting in the spatial domain D. Denoting by β_0 the true parameter characterizing the kernel, the continuous discrete wavelet RF, for $b \in L_j$ and $j \in \mathbb{Z}$, is given by:

$$X(\boldsymbol{\Psi}_{j,b}) = \int_{D \times D} \boldsymbol{\Psi}_{j,b}(x) a(\beta_0, x, y) \varepsilon(y) dy dx = \int_D \mathscr{A}^{\beta_0}(\boldsymbol{\Psi}_{j,b})(y) \varepsilon(y) dy$$

and the two-dimensional wavelet transform of the kernel $a(\beta_0, \cdot, \cdot)$ is computed as:

$$\mathscr{A}^{\beta_0}(\boldsymbol{\Psi}_{j,u})(\boldsymbol{\Psi}_{j,b}) = \int_{D \times D} \boldsymbol{\Psi}_{j,b}(x) a(\beta_0, x, y) \boldsymbol{\Psi}_{j,u}(y) dx dy, \quad b, u \in L_j, \quad j \in \mathbb{Z}.$$

In order to derive Lemma 1, the following condition is also assumed.
C2. The wavelet basis selected is an isotropic wavelet basis with the *mother wavelet* function $\boldsymbol{\Psi}$ satisfying:

$$\boldsymbol{\Psi}(\|z\|) = \begin{cases} 1, & 0 \leq \|z\| < 1/2, \\ -1, & 1/2 \leq \|z\| < 1, \\ 0, & \text{otherwise.} \end{cases}$$

Lemma 1. *(i) Under condition C1, as $j \longrightarrow \infty$, the following asymptotic approximation holds:*

$$\mathscr{A}^{\beta_0}(\boldsymbol{\Psi}_{j,u})(\boldsymbol{\Psi}_{j,b}) \simeq [2^{-j}]^{\beta_0+d/2} C_{\boldsymbol{\Psi}}^{\mathscr{A}}(\beta_0, b, u, j),\ b \in L_j,$$

where $\lim_{j\longrightarrow\infty} C_{\boldsymbol{\Psi}}^{\mathscr{A}}(\beta_0, b, u, j) = C_{\boldsymbol{\Psi}}^{\mathscr{A}}(\beta_0, b-u, \infty) = \int_{\mathbb{R}^d} \|h\|^{\beta_0-d/2} \gamma_{\boldsymbol{\Psi}}(h-(b-u))dh,$
with $\gamma_{\boldsymbol{\Psi}}(h-(b-u)) = \int_{\mathbb{R}^d} \boldsymbol{\Psi}(z)\boldsymbol{\Psi}(z-h+(b-u))dz.$
(ii) Moreover, under condition C2, for $\|b-u\| \gg 1$,

$$C_{\boldsymbol{\Psi}}^{\mathscr{A}}(\beta_0, b, u, \infty) = \int_{\mathbb{R}^d} \|z\|^{\beta_0-d/2} \left[\int_{\mathbb{R}^d} \boldsymbol{\Psi}(x)\boldsymbol{\Psi}(x-z+(b-u))dx\right] dz \sim \mathscr{O}\left(\|b-u\|^{\beta_0-5d/2}\right). \tag{10.1}$$

Otherwise, $C_{\boldsymbol{\Psi}}^{\mathscr{A}}(\beta_0, b, u, \infty) = \dfrac{1}{(\beta_0+d/2)(\beta_0+3d/2)} \xi_v^{\mathscr{A}}(b-u)$, *and*

$$\begin{aligned} \xi_v^{\mathscr{A}}(b-u) &= \|b-u-v\|^{\beta_0+3d/2} - 4(\|b-u-v/2\|)^{\beta_0+3d/2} \\ &\quad + 6\|b-u\|^{\beta_0+3d/2} - 4\|b-u+v/2\|^{\beta_0+3d/2} + \|b-u+v\|^{\beta_0+3d/2} \end{aligned} \tag{10.2}$$

for $v \in \mathbb{R}_+^d$ *such that* $\|v\| = 1$. *In particular,*

$$\lim_{j\to\infty} C_{\boldsymbol{\Psi}}^{\mathscr{A}}(\beta_0, b, j) = C_{\boldsymbol{\Psi}}^{\mathscr{A}}(\beta_0, \infty) = \widetilde{L}_d^{\mathscr{A}}(\beta_0) = \frac{2(1-2^{-(\beta_0+3d/2-2)})}{(\beta_0+d/2)(\beta_0+3d/2)}.$$

Proof of Lemma 1. Under condition C1, as $j \longrightarrow \infty$,

$$\begin{aligned} \mathscr{A}^{\beta_0}(\boldsymbol{\Psi}_{j,u})(\boldsymbol{\Psi}_{j,b}) &\simeq \int_{D\times D} 2^{jd}\boldsymbol{\Psi}(2^j x - b)\boldsymbol{\Psi}(2^j y - u)\|x-y\|^{\beta_0-d/2} dxdy \\ &= 2^{-j(\beta_0+d/2)} \int_{D_j\times D_j} \boldsymbol{\Psi}(z-b)\boldsymbol{\Psi}(v-u)\|z-v\|^{\beta_0-d/2} dzdv \\ &= 2^{-j(\beta_0+d/2)} C_{\boldsymbol{\Psi}}^{\mathscr{A}}(\beta_0, b, u, j), \end{aligned}$$

where $D_j = 2^j D = [-2^j M, 2^j M]^d$, and by direct computation, we obtain

$$\lim_{j\longrightarrow\infty} C_{\boldsymbol{\Psi}}^{\mathscr{A}}(\beta_0, b, u, j) = C_{\boldsymbol{\Psi}}^{\mathscr{A}}(\beta_0, b-u, \infty) = \int_{\mathbb{R}^d} \|h\|^{\beta_0-d/2} \left[\int_{\mathbb{R}^d} \boldsymbol{\Psi}(z)\boldsymbol{\Psi}(z-h+(b-u))dz\right] dh$$

(ii) From the fourth-order expansion of (10.2) in $(1/\|b-u\|)$, equation (10.1) is obtained. Note that equation (10.2) is derived by direct computation of $C_{\boldsymbol{\Psi}}^{\mathscr{A}}(\beta_0, b-u, \infty)$ under condition C2 (see, for example, Ruiz-Medina and Crujeiras (2011) for RFs with RKHS isomorphic to a fractional Sobolev space).

Corollary 1. *Under conditions C1 and C2, as $j \longrightarrow \infty$, the two-dimensional wavelet transform of the covariance function of X can be asymptotically approximated by*

$$B_X^{\beta_0}(\boldsymbol{\Psi}_{j,b}, \boldsymbol{\Psi}_{j,u}) = \int_{D\times D} B_X^{\beta_0}(x,y)\boldsymbol{\Psi}_{j,b}(x)\boldsymbol{\Psi}_{j,u}(y)dxdy \simeq [2^{-j}]^{(2\beta_0+d)} C_{\boldsymbol{\Psi}}^{B_X}(\beta_0, b, u, j),$$

for $b, u \in L_j$, where $C_{\boldsymbol{\Psi}}^{B_X}(\beta_0, b, u, j) = \int_{L_j} C_{\boldsymbol{\Psi}}^{\mathscr{A}}(\beta_0, b, v, j) C_{\boldsymbol{\Psi}}^{\mathscr{A}}(\beta_0, v, u, j)dv$ and

$$\lim_{j\longrightarrow\infty} C_{\boldsymbol{\Psi}}^{B_X}(\beta_0, b, u, j) = C_{\boldsymbol{\Psi}}^{B_X}(\beta_0, b-u, \infty) = \langle \boldsymbol{\Psi}_{0,b}, \boldsymbol{\Psi}_{0,u} \rangle_{[\mathscr{H}_{W_{\beta_0-d/2}}]^*},$$

with $\mathscr{H}_{W_{\beta_0-d/2}}$ denoting the RKHS of the fractional Brownian motion $W_{\beta_0-d/2}$. Moreover, for $\|b-u\| >> 1$,

$$C_{\boldsymbol{\Psi}}^{B_X}(\beta_0, b, u, \infty) \simeq \mathscr{O}\left(\|b-u\|^{2\beta_0-d}\right).$$

Otherwise, $C_{\boldsymbol{\Psi}}^{B_X}(\beta_0, b, u, \infty) = \dfrac{1}{(2\beta_0+d)(2\beta_0+2d)} \xi_v^{B_X}(b-u)$, *with*

$$\begin{aligned}\xi_v^{B_X}(b-u) = {} & \|b-u-v\|^{2\beta_0+2d} - 4(\|b-u-v/2\|)^{2\beta_0+2d} \\ & + 6\|b-u\|^{2\beta_0+2d} - 4\|b-u+v/2\|^{2\beta_0+2d} + \|b-u+v\|^{2\beta_0+2d},\end{aligned}$$

for $v \in \mathbb{R}_+^d$ such that $\|v\| = 1$. In particular,

$$\lim_{j\to\infty} C_{\boldsymbol{\Psi}}^{B_X}(\beta_0, b, j) = C_{\boldsymbol{\Psi}}^{B_X}(\beta_0, \infty) = \widetilde{L}_d^{B_X}(\beta_0) = \frac{1-2^{-(2\beta_0+2d-2)}}{(2\beta_0+d)(2\beta_0+2d)}.$$

10.3 Consistency of the wavelet periodogram

The wavelet periodogram provides an unbiased nonparametric estimator of the diagonal of the two-dimensional wavelet transform of the covariance function. In our setting, the wavelet periodogram at resolution level $j \in \mathbb{Z}$ and location $b \in L_j$ is defined as

$$S(j,b,b) = |X(\boldsymbol{\Psi}_{j,b})|^2 = \left| \int_{D\times D} \boldsymbol{\Psi}_{j,b}(x) a(\beta_0, x, y)\varepsilon(y)dxdy \right|^2 = \left| \int_D [A^{\beta_0}]^*(\boldsymbol{\Psi}_{j,b})(y)\varepsilon(y)dy \right|^2. \tag{10.3}$$

Proposition 1. *Let $\omega \in L^1(\mathbb{R}_+^d)$ be a weight function such as the integral in (10.4) is well-defined. Under conditions C1 and C2, the following limit holds in probability, as $j \longrightarrow \infty$:*

$$\mathscr{I}(j) = \int_{L_j} \omega(b)\left[S(j,b,b) - B_X^{\beta_0}(\boldsymbol{\Psi}_{j,b}, \boldsymbol{\Psi}_{j,b})\right] db \longrightarrow 0, \quad b \in L_j. \tag{10.4}$$

Proof of Proposition 1. Taking into account the unbiasedness property, in order to prove that (10.4) holds, it is sufficient to check that $\mathbb{E}(\mathscr{I}^2(j))$ tends to zero as $j \longrightarrow \infty$. Denote by X_W^N the normalized wavelet RF, which is given, at each resolution level $j \in \mathbb{Z}$, by

$$X_W^N(j,b) = \frac{X(\boldsymbol{\Psi}_{j,b})}{\left[B_X^{\beta_0}(\boldsymbol{\Psi}_{j,b}, \boldsymbol{\Psi}_{j,b})\right]^{1/2}}, \quad b \in L_j.$$

Consider also the function $F(z(b)) = [z(b)]^2$ applied to the values of the normalized wavelet RF. Function F admits an Hermite expansion with rank $r = 1$, which leads to the following expression for $\mathbb{E}(\mathscr{I}^2(j))$:

$$\mathbb{E}(\mathscr{I}^2(j)) = \sum_{k=1}^{\infty} \frac{C_k^2}{k!} \int_{L_j \times L_j} \omega(b)\omega(u) B_X^{\beta_0}(\boldsymbol{\Psi}_{j,b}, \boldsymbol{\Psi}_{j,b}) B_X^{\beta_0}(\boldsymbol{\Psi}_{j,u}, \boldsymbol{\Psi}_{j,u}) B_{X_W^N}^k(j,b,u)\, db\, du,$$

where C_k denotes the k-th Hermite coefficient of function F with respect to the k-th Hermite polynomial H_k, for $k \in \mathbb{N}$ and $B_{X_W^N}(j,b,u) = \mathbb{E}\left[X_W^N(j,b) X_W^N(j,u)\right]$. For j sufficiently large, from Corollary 1, the above equation can be approximated by:

$$\mathbb{E}(\mathscr{I}^2(j)) \simeq [2^{-j}]^{(4\beta_0 + 2d)} \sum_{k=1}^{\infty} \frac{C_k^2}{k!} I_k(j) \quad \text{with}$$

$$I_k(j) = \int_{L_j \times L_j} \omega(b)\omega(u) C_{\boldsymbol{\Psi}}^{B_X}(\beta_0, b, j) C_{\boldsymbol{\Psi}}^{B_X}(\beta_0, u, j) \frac{[C_{\boldsymbol{\Psi}}^{B_X}(\beta_0, b, u, j)]^k}{[C_{\boldsymbol{\Psi}}^{B_X}(\beta_0, b, j) C_{\boldsymbol{\Psi}}^{B_X}(\beta_0, u, j)]^{k/2}} db\, du. \tag{10.5}$$

For $k = 1$, as $j \longrightarrow \infty$, from Corollary 1, the integral $I_1(j)$ converges to:

$$I_1(\infty) = \widetilde{L}_d^{B_X}(\beta_0) \int_{\mathbb{R}^d \setminus \mathscr{B}_R^d(0)} \|x\|^{2\beta_0 - d} \omega * \omega(\infty, x) dx + \frac{\widetilde{L}_d^{B_X}(\beta_0)}{(2\beta_0 + d)(2\beta_0 + 2d)} \int_{\mathscr{B}_R^d(0)} \xi_\nu(x) \omega * \omega(\infty, x) dx,$$

which is finite for $\omega \in L^1(\mathbb{R}_+^d)$. Therefore, $\lim_{j \to \infty} (2^{-j})^{(4\beta_0 + 2d)} C_1^2 I_1(j) = 0$. For $k \geq 2$, the terms $I_k(j)$ are bounded by $G_k(j)$, applying Corollary 1 and Cauchy-Schwarz inequality. The function $G_k(j)$ converge to:

$$G_k(j) \stackrel{j\to\infty}{\longrightarrow} \widetilde{L}_d^{B_X}(\beta_0)\|\omega\|_{L^1(\mathbb{R}_+^d)} \left[\int_{\mathbb{R}^d\setminus\mathscr{B}_R^d(0)} \|x\|^{4\beta_0-2d}\omega*\omega(\infty,x)dx \right.$$
$$\left. + \frac{1}{[(2\beta_0+d)(2\beta_0+2d)]^2}\int_{\mathscr{B}_R^d(0)}[\xi_v^{B_X}(x)]^2\omega*\omega(\infty,x)dx\right]^{1/2},$$

which is finite for $\omega\in L^1\left(\mathbb{R}_+^d\right)$. Therefore, for any $k\geq 2$, $\lim_{j\to\infty}(2^{-j})^{(4\beta_0+2d)}\frac{C_k^2}{k!}I_k(j) = 0$. Hence, the integral in (10.5) goes to zero, as $j\longrightarrow\infty$, and weak consistency of the functional wavelet periodogram holds.

10.4 Minimum contrast estimator

For the class of linear RFs considered and in the wavelet scenario previously established, a minimum contrast estimator for β_0 is proposed. The methodology is similar to the one developed by Ruiz-Medina and Crujeiras (2011) for fractal RFs, and it is based on the wavelet periodogram introduced in the previous section. Define the contrast function:

$$K(\beta,\beta_0) = -\int_{\mathbb{R}^d}\left[\log\left(\frac{\widetilde{L}_d^{B_X}(\beta_0)}{\widetilde{L}_d^{B_X}(\beta)}\right) - \frac{\widetilde{L}_d^{B_X}(\beta_0)}{\widetilde{L}_d^{B_X}(\beta)}+1\right]\omega(b)db,$$

where ω is a suitable weight function. The sequence of random variables $\{U_j(\beta),\ j\in\mathbb{Z}\}$, given by:

$$U_j(\beta) = \int_{L_j}\left[\log\left(B_X^{\beta}(\boldsymbol{\Psi}_{j,b},\boldsymbol{\Psi}_{j,b})\right) + \frac{S(j,b,b)}{B_X^{\beta}(\boldsymbol{\Psi}_{j,b},\boldsymbol{\Psi}_{j,b})}\right]\omega(b)db, \quad \beta\in\Lambda,$$

defines a contrast process for the contrast function K, since the sequence $\{U_j(\beta) - U_j(\beta_0)\}$ converges in probability to $K(\beta,\beta_0)$, which is positive with a unique minimum at β_0. For each resolution level $j\in\mathbb{Z}$, the minimum contrast estimator $\widetilde{\beta}_j$ is then defined as the random variable satisfying

$$\widetilde{\beta}_j = \arg\min_{\beta\in\Lambda} U_j(\beta). \tag{10.6}$$

Under similar conditions to (A3-A4) of Ruiz-Medina and Crujeiras (2011) on the asymptotic, in scale, integrability order of ω, as well as on the existence of a suitable sequence of equicontinuous functions with respect to β, the following result is derived.

Proposition 2. *Under conditions C1-C2 and A3-A4 in Ruiz-Medina and Crujeiras (2011), as $j\longrightarrow\infty$, the minimum contrast estimator $\widetilde{\beta}_j\longrightarrow\beta_0$, in probability.*

10.5 Final comments

Asymptotic normality of the minimum contrast estimator can be obtained using central limit results for multiple stochastic integrals, from Nualart and Pecatti (2005). The previous result can be extended to the non-Gaussian case, in terms of Appell polynomials expansion, provided that the functionals involved admit such an expansion. Central limit results for integral non-linear functionals and quadratic forms involving Appell polynomials have been derived, for example, by Surgailis (2000) (considering linear, moving average sequeces with long-range dependence) and, recently, by Avram *et al.* (2010).

References

1. Avram, F., Leonenko, N., Sakhno, L.: On a Szegö type limit heorem, the Hölder-Young-Brascamp-Lieb inequality and the asymptotic theory of integrals and quadratic forms of stationary fields. ESAIM: Probability and Statistics **14**, 210–255 (2010)
2. Nualart, D., Peccati, G.: Central limit theorems for sequences of multiple stochastic integrals. Ann. Probab. **33**, 177–193 (2005)
3. Ruiz-Medina, M.D., Angulo, J.M., Anh, V.V.: Fractional generalized random fields on bounded domains. Stoch. Anal. Appl. **21**, 465–492 (2003)
4. Ruiz-Medina, M.D., Crujeiras, R.M.: Minimum contrast parameter estimation for fractal random fields based on the wavelet periodogram. Commun. Stat. Theor. M. To appear (2011)
5. Surgailis, D.: Long-range dependence and Appell rank. Ann. Probab. **28**, 478–497 (2000)

Chapter 11
Dimensionality Reduction for Samples of Bivariate Density Level Sets: an Application to Electoral Results

Pedro Delicado

Abstract A bivariate densities can be represented as a density level set containing a fixed amount of probability (0.75, for instance). Then a functional dataset where the observations are bivariate density functions can be analyzed as if the functional data are density level sets. We compute distances between sets and perform standard Multidimensional Scaling. This methodology is applied to analyze electoral results.

11.1 Introduction

The most important way of political participation for people in democratic countries is certainly to vote in electoral calls. Nevertheless the participation in elections is usually far for 100%: many people decide not going to vote for several reasons. A relevant question is if there exists some relationship between the political ideology of a given voter and its decision of going or not to vote in a particular election. In Spain it is given as a fact that potential left parties voters usually participate in elections less than right parties voters. In this work we analyze the relationship between position on the left-right political dimension and the willingness to vote. Given that individual data are not available we use aggregated data at level of electoral districts ("mesas electorales" in Spanish: lists of around 1000 people that vote at the same ballot box because they live in the same small area). Specifically we use electoral results from 2004 Spanish general elections.

For each electoral district the available information allows us to define these two variables: participation (proportion of potential voters that finally vote) and proportion of votes for right parties. Observe that this last variable is not exactly the same as the proportion of potential voters with right political ideology. Unfortunately we only know what is voting people that vote indeed. Nevertheless, if the size of the electoral district is small compared with the size of the city it is sensible to believe

Pedro Delicado
Universitat Politècnica de Catalunya, Spain, e-mail: pedro.delicado@upc.edu

that both quantities should be similar. We assume that, given the electoral district, the political orientation (left-right) is independent from the decision of voting or not.

We consider the 50 cities in Spain with the bigger numbers of electoral districts (157 districts or more). For each of these cities we have a list of observations of the bivariate random variable (*participation*, *proportion of votes for right parties*), an observation for each electoral district. We use then a kernel density estimator to obtain from this list an estimation of the joint distribution of these two variables in each of the 50 cities considered in our study. Therefore we have a functional dataset of length 50 consisting on bivariate densities.

A preliminary dimensionality reduction step is usually very useful to perform the exploratory analysis of functional datasets. Given that the dataset we are considering consists on bivariate densities, it is possible to adapt the dimensionality reduction techniques considered in Delicado (2011) for functional datasets formed by unidimensional densities. Nevertheless we propose here an alternative way.

A bivariate density $f(x,y)$ is frequently represented by some of its *density level sets*, defined as $L(c) = \{(x,y) \in \mathbf{R}^2 : f(x,y) \geq c\}$, for $c > 0$, or just their boundaries, in a contour plot. Bowman and Azzalini (1997) propose to display only the contour level plots that contain specific probabilities (they use 0.25, 0.50 or 0.75, reminiscing a boxplot) as a effective way to characterize the shape of a bivariate density. The roll of density level sets is also relevant in the area of set estimation (see Cuevas and Fraiman 2009).

Bowman and Azzalini (1997, Section 1.3) give a very nice illustration of the use of density level sets for exploratory data analysis. They study data on aircraft designs from periods 1914-1935, 1936-1955 and 1956-1984. They obtain the first two principal components and represent their joint density using a single level plot (that corresponding to probability 0.75) for each period. In a single graphic Bowman and Azzalini (1997, Figure 1.8) are able to summarize the way in which aircraft designs have changed over the last century.

We borrow this way to summarize a bivariate density (the density level plot corresponding to probability 0.75). Therefore our functional dataset is finally formed by 50 such density level sets. As an example Figure 11.1 shows the density level sets corresponding to the 5 largest municipalities in Spain, jointly with the density level set corresponding to the whole country as a reference. The standard correlation coefficient for each case has been annotated. It is clear that there is a considerable variability between different level sets. Moreover the relationship between participation and vote orientation is clearer when considering homogeneous sets of electoral districts (those corresponding to a specific city) than when considering the whole country (top left panel).

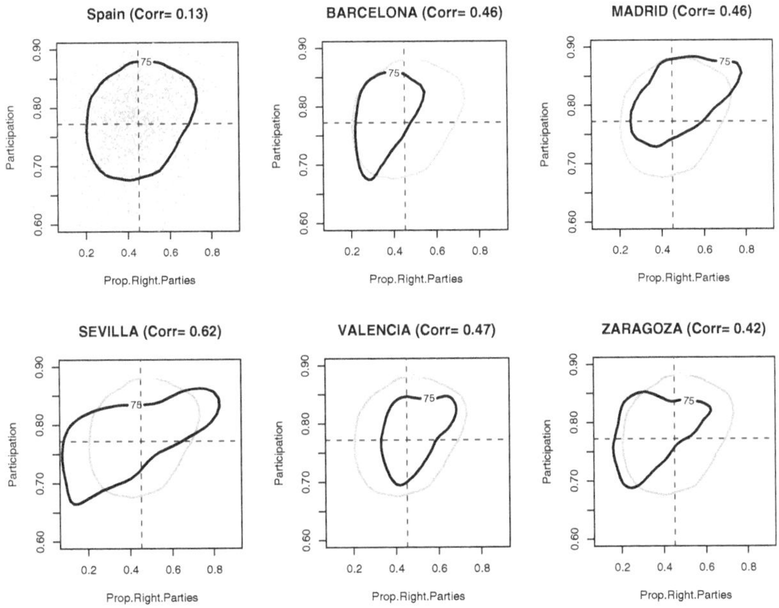

Fig. 11.1: Example of 6 density level sets.

11.2 Multidimensional Scaling for density level datasets

The functional data we are analyzing are sets (density level sets). When looking for a dimensionality reduction technique for this kind of data it is much more natural to turn to Multidimensional Scaling (MDS) than to some kind of Principal Component Analysis (PCA). The main reason is that there exist several well known definitions of distance between sets but there is not a clear Hilbert space structure on the set of sets allowing to define PCA for datasets of sets.

Two distances between sets used frequently (see Cuevas 2009, for instance) are the following:

Distance in measure: Given $U,V \subseteq \mathbf{R}^2$,

$$d_\mu(U,V) = \mu(U \,\Delta\, V),$$

where $U \,\Delta\, V = (U \cup V) - (U \cap V)$ is the symmetric difference of U and V, and μ is the Lebesgue measure in $\mathbf{R}^2$.

Hausdorff metric: Given $U,V \subseteq \mathbf{R}^2$,

$$d_H(U,V) = \inf\{\varepsilon > 0 : U \subseteq B(V,\varepsilon), V \subseteq B(U,\varepsilon)\},$$

where for $A \subseteq \mathbf{R}^2$, $B(A,\varepsilon) = \cup_{x \in A} B(x,\varepsilon)$, and $B(x,\varepsilon)$ is the closed ball with center x and radius ε in $\mathbf{R}^2$.

In this work we use distance in measure between density level sets. Once the distance matrix is calculated the MDS procedure follows in a standard way (see Borg and Groenen 2005, for instance).

11.3 Analyzing electoral behavior

Figure 11.2 represents the plane of the first two principle coordinates obtained from the MDS analysis of the distance in measure matrix between the 50 density level sets in our study. The labels used in this graphic indicate the province where the 50 big cities are placed (observe that some of them belong to the same province). The percentage of variability explained by these two principal coordinates is around 60%, so it could be interesting to explore additional dimensions. There is not any nonlinearity pattern neither clustering structure.

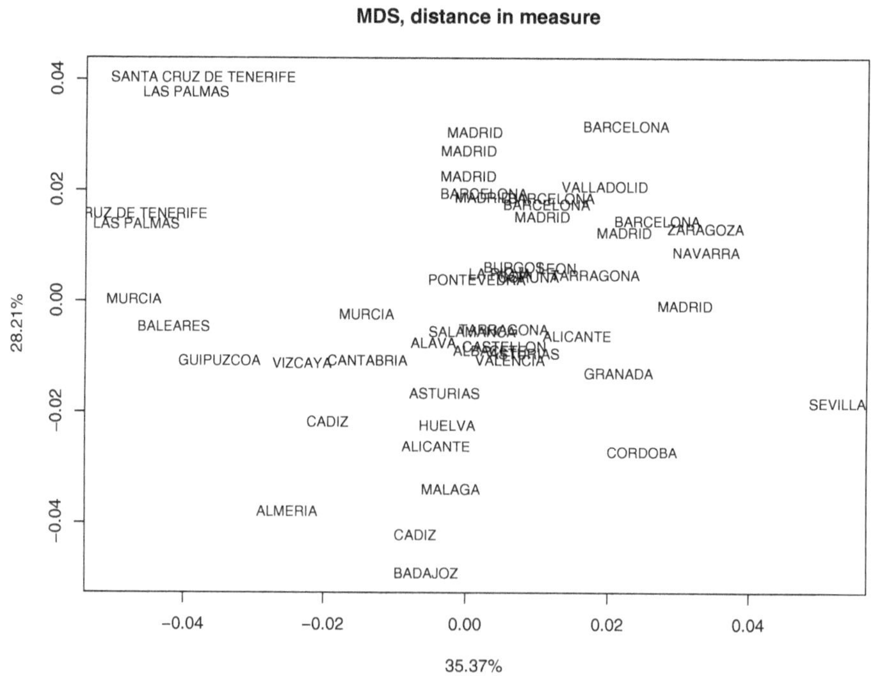

Fig. 11.2: Plane of the first two principle coordinates obtained from the MDS.

In order to have a better interpretation of these first two principle coordinates additional graphics are helpful. Jones and Rice (1992) propose the following way to represent functional principal coordinates (or principal components). They suggest picking just three functional data in the dataset: the data corresponding to the median principal coordinate score, and those corresponding to quantiles α and $(1-\alpha)$ of these score values (α close to zero guarantees that these functional data are representative of extreme values of principal component scores). Alternatively, functional data corresponding to the minimum and maximum scores could go with the median score functional data. This is exactly what we represent in Figure 11.3, using blue color for the minimum, black color for the median and red color for the maximum.

The first principal coordinate goes from negative relationship between participation and proportion of votes to right parties (a city in the province of Santa Cruz de Tenerife) to almost independence (a city in the province of Barcelona) to a positive relationship (Sevilla). The interpretation of the second principal coordinate is not so clear. We observe that the area of the density level sets decreases when moving from the minimum scores (Badajoz) to the maximum (a city in the province of Santa Cruz de Tenerife, different from that cited when talking about the first principal coordinate), but a deeper analysis should be done in order to establish a clearer interpretation.

References

1. Borg, I., Groenen, P.: Modern Multidimensional Scaling: Theory and Applications (Second Edition). Springer Verlag, New York (2005).
2. Bowman, A. W., Azzalini, A.: Applied Smoothing Techniques for Data Analysis. Oxford University Press, Oxford (1997).
3. Cuevas, A., Fraiman, R.: Set estimation. In: Kendall, W., Molchanov, I. (eds.) New Perspectives in Stochastic Geometry. Oxford University Press, Oxford (2009).
4. Cuevas, A.: Set estimation: Another bridge between statistics and geometry. Boletín de Estadítica e Investigación Operativa **25** (2), 71–85 (2009).
5. Delicado, P.: Dimensionality reduction when data are density functions. Comput. Stat. Data Anal. **55** (1), 401–420 (2011).
6. Jones, M.C., Rice, J.A.: Displaying the important features of large collections of similar curves. Amer. Statistician **46** (2), 140–145 (1992).

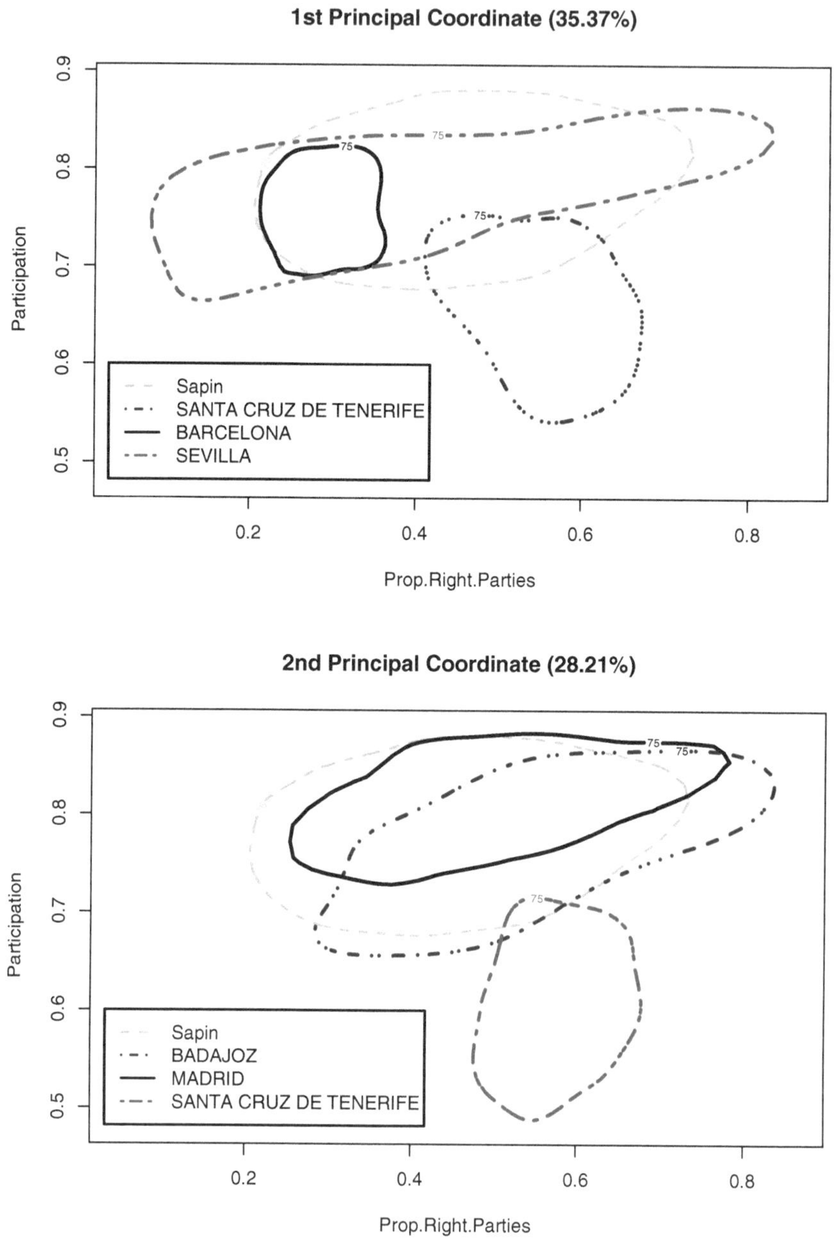

Fig. 11.3: Helping to the interpretation of the first two principle coordinates.

Chapter 12
Structural Tests in Regression on Functional Variable

Laurent Delsol, Frédéric Ferraty, Philippe Vieu

Abstract This work focuses on recent advances on the way general structural testing procedures can be constructed in regression on functional variable. Our test statistic is constructed from an estimator adapted to the specific model to be checked and uses recent advances concerning kernel smoothing methods for functional data. A general theoretical result states the asymptotic normality of our test statistic under the null hypothesis and its divergence under local alternatives. This result opens interesting prospects about tests for no-effect, for linearity, or for reduction dimension of the covariate. Bootstrap methods are then proposed to compute the threshold value of our test. Finally, we present some applications to spectrometric datasets and discuss interesting prospects for the future.

12.1 Introduction

A great variety of real world issues involve functional phenomena which may be represented as curves or more complex objects. They may for instance come from the observation of a phenomenon over time or more generally its evolution when the context of the study changes (e.g. growth curves, sound records, spectrometric curves, electrocardiograms, images). It is nowadays common to deal with a large amount of discretized observations of a given functional phenomenon that actually gives a relevant understanding of its dynamic and regularity. Classical multivariate statistical tools may be irrelevant in that context to take benefit from the underlying functional structure of these observations.

Laurent Delsol
Université d'Orléans, France, e-mail: laurent.delsol@univ-orleans.fr

Frédéric Ferraty
Institut de Mathématiques de Toulouse, France, e-mail: ferraty@math.univ-toulouse.fr

Philippe Vieu
Institut de Mathématiques de Toulouse, France, e-mail: vieu@math.univ-toulouse.fr

Recent advances in functional statistics offer a large panel of alternative methods to deal with functional variables (i.e. variables taking values in an infinite dimensional space) which become popular in real world studies. A general overview on functional statistics may be found in Ramsay and Silverman (1997, 2002, 2005), Bosq (2000), Ferraty and Vieu (2006), and more recently Ferraty and Romain (2010). This talk focuses on the study of regression models involving a functional covariate:

$$Y = r(\mathscr{X}) + \varepsilon,$$

where Y is a real valued random variable, $\mathscr{X}$ is a random variable taking values in a semi-metric space $(\mathscr{E}, d)$ and $\mathbb{E}[\varepsilon|\mathscr{X}] = 0$.
A lot of works have already been done on the estimation of the regression operator r through various versions of this model corresponding to structural assumptions on r. The most famous example is certainly the functional linear model introduced by Ramsay and Dalzell (1991):

$$Y = \alpha_0 + < \alpha, \mathscr{X} >_{\mathbb{L}^2([0;1])} + \varepsilon,\ (\alpha_0, \alpha) \in \mathbb{R} \times \mathbb{L}^2([0;1]).$$

This model has received a lot of attention and is still a topical issue. This is illustrated through the contributions of Cardot *et al.* (1999, 2000, 2007), Ramsay and Silverman (1997, 2005), Preda and Saporta (2005), Hall and Cai (2006), Crambes *et al.* (2009), or Ferraty and Romain (2010, Chapter 2) among others.
Several other examples of models based on a given structure of r have been considered. For instance Sood *et al.* (2009) studied a multivariate additive model based on the first coefficients of a functional P.C.A., Ait Saidi *et al.* (2008) focused on the functional single index model, Aneiros-Perez and Vieu (2009) investigated on the partial linear model.
On the other hand, nonparametric functional models in which only the regularity (Hölder) of r with respect to the semi-metric d is assumed, have been considered by Ferraty and Vieu (2000). Many references on recent contributions on this topic are given in Ferraty *et al.* (2002), Masry (2005), Ferraty and Vieu (2006), Delsol (2007, 2009) together with Ferraty and Romain (2011, Chapters 1, 4, and 5).

12.2 Structural tests

12.2.1 A general way to construct a test statistic

As discussed in the previous paragraph, a lot of work as been done on the estimation of the regression operator r. This work focuses on a different issue and proposes statistical tools for the construction of testing procedures allowing to check if r has a given structure (e.g. constant, linear, multivariate, ...). Such testing procedures are interesting by themselves to test the validity of an a priori assumption on the structure of the regression model. They are also complementary tools to estimation methods. They may be used as a preliminary step to check the validity of a structural assumption used to construct an estimator and may be relevant to test some structural assumption arising from the result of r estimation. To the best of our knowledge, the literature on this kind of problem is restricted to Cardot *et al.* (2003, 2004), Müller and Stadtmüller (2005) in the specific case of a linear model, Gadiaga and Ignaccolo (2005) on no effect tests based on projection methods, and Chiou and Müller (2007) on an heuristic goodness of fit test. Hence it seems no general theoretical background has been proposed to test the validity of the different modelizations discussed in the introduction part. In the remainder of this note $\mathscr{R}$ stand for a family of square integrable operators and w a weight function. Our aim is to present and discuss in this work a general methodology allowing to test the null hypothesis:

$$\mathscr{H}_0 : \{\exists r_0 \in \mathscr{R}, P(r(\mathscr{X}) = r_0(\mathscr{X})) = 1\}$$

under local alternatives of the form

$$\mathscr{H}_{1,n} : \{ \inf_{r_0 \in \mathscr{R}} \|r - r_0\|_{\mathbb{L}^2(wdP_{\mathscr{X}})} \geq \eta_n\}.$$

Extending the ideas of Härdle and Mammen (1993), we construct our test statistic from an estimator $\hat{r}$ adapted to the structural model (corresponding to the null hypothesis, i.e. induced by $\mathscr{R}$) we want to test and functional kernel smoothing tools (K denotes the kernel):

$$T_n = \int (\sum_{i=1}^{n} (Y_i - \hat{r}(\mathscr{X}_i))K\left(\frac{d(\mathscr{X}_i,x)}{h_n}\right))^2 w(x)dP_{\mathscr{X}}(x).$$

For technical reasons, we assume the estimator $\hat{r}$ is constructed on a sample D_1 independent from $D = (\mathscr{X}, Y_i)_{1\leq i \leq n}$. A theoretical result in Delsol *et al.* (2011) states under general assumptions the asymptotic normality of T_n under the null hypothesis and its divergence under the local alternatives. This result opens a large scope of potential applications of this kind of test statistic. Here are few examples:

- test of an a priori model: $\mathscr{R} = \{r_0\}$, $\hat{r} = r_0$.
- no effect test: $\mathscr{R} = \{r : \exists C \in \mathbb{R}, r \equiv C\}$, $\hat{r} = \overline{Y}_n$.
- test of a multivariate effect: $\mathscr{R} = \{r : r = g \circ V, V : \mathscr{E} \to \mathbb{R}^p \text{ known}, g : \mathbb{R}^p \to \mathbb{R}\}$, $\hat{r}$ multivariate kernel estimator constructed from $(Y_i, V(\mathscr{X}_i))_{1\leq i\leq n}$.
- linearity test: $\mathscr{R} = \{r : r = \alpha_0 + < \alpha, . >, (\alpha_0, \alpha) \in \mathbb{R} \times \mathbb{L}^2[0;1]\}$, $\hat{r}$ functional spline estimator (see Crambes *et al.* 2009).
- test of a functional single index model: $\mathscr{R} = \{r : r = g(< \alpha, . >), \alpha \in \mathscr{E}, g : \mathbb{R} \to \mathbb{R}\}$, $\hat{r}$ estimator proposed in Ait Saidi *et al.* (2008).

Other situations may also be considered whenever it is possible to provide an estimator $\hat{r}$ satisfying some conditions.

12.2.2 Bootstrap methods to get the threshold

The practical use of our test statistic requires the computation of the threshold value. One could propose to get it from the asymptotic distribution. However, the estimation of dominant bias and variance terms is not easy, that is why we prefer to use bootstrap procedures. The main idea is to generate, from the original sample, B samples for which the null hypothesis approximately holds. Then, compute on each of these samples the tests statistic and take as threshold the $1-\alpha$ empirical quantile of the values we have obtained.

We propose the following bootstrap procedure in which steps 2-4 are made separately on samples $D : (\mathscr{X}_i, Y_i)_{1\leq i\leq n}$ and $D_1 : (\mathscr{X}_i, Y_i)_{n+1\leq i\leq N}$. In the following lines $\hat{r}_K$ stands for the functional kernel estimator of the regression operator r computed from the whole dataset.

Bootstrap procedure:
Pre-treatment:

1. $\hat{\varepsilon}_i = Y_i - \hat{r}_K(X_i)$
2. $\tilde{\varepsilon}_i = \hat{\varepsilon}_i - \bar{\hat{\varepsilon}}$

Repeat B times steps 3-5:

3. Generate residuals (3 different methods NB, SNB or WB)

NB • $(\varepsilon_i^b)_{1\leq i\leq n}$ drawn with replacement from $(\tilde{\varepsilon}_i)_{1\leq i\leq n}$
SNB • $(\varepsilon_i^b)_{1\leq i\leq n}$ generated from a "smoothed" version $\tilde{F}_n$ of the empirical cumulative distribution function of $(\tilde{\varepsilon}_i)_{1\leq i\leq n}$ $(\varepsilon_i^b = \tilde{F}_n^{-1}(U_i), U_i \sim \mathscr{U}(0,1))$
WB • $(\varepsilon_i^b) = \tilde{\varepsilon}_i V_i$ where $V_i \sim P_W$ fulfills some moment assumptions: $E[V_i] = 0$, $E[V_i^2] = 1$ and $E[V_i^3] = 1$.

4. Generate bootstrap responses "corresponding" to $\mathscr{H}_0$

$$Y_i^b = \hat{r}(X_i) + \varepsilon_i^b$$

5. Compute the test statistic T_n^b from the bootstrap sample $(\mathscr{X}_i, Y_i^b)_{1\leq i\leq N}$

Compute the threshold value

6. For a test of level α, take as threshold the $1-\alpha$ quantile of the sample $(T_n^b)_{1\leq b\leq B}$.

Three examples of distributions P_W given in Mammen (1993) are considered. The different methods used to generate bootstrap residuals globally lead to similar results but some of them perform slightly better in terms of level or power. ¿From the results obtained in simulation studies, it seems relevant to use wild bootstrap methods (WB) which lead to more powerful tests and are by nature more robust to the heteroscedasticity of the residuals.

Finally the integral with respect to $P_{\mathscr{X}}$ which appears in T_n's definition may be approximated by Monte Carlo on a third subsample independent from D_1 and D_2.

12.3 Application in spectrometry

Spectrometric curves are an interesting example of functional data. They correspond to the measure of the absorption of a laserbeam emitted in direction of a product in function of its wavelength. Spectrometric curves have been used to give an estimation of the chemical content of a product without spending time and money in a chemical analysis (see for instance Borggaard and Thodberg, 1992). It is usual in chemometrics to make a pretreatment of the original curves (corresponding in some sense to considering derivatives). The approach described in this work may be used in this context to provide part of an answer to questions dealing with

- the validity of a model proposed by specialists.
- the existence of a link between one of the derivatives and the chemical content to predict.
- the nature of the link between the derivatives of the spectrometric curve and the chemical content of the product
- the validity of models in which the effect of the spectrometric curve is reduced to the the effect of some of its features (parts of the spectrum, few points).

The use of the proposed testing procedures to address such questions is briefly discussed through the study of real world data.

12.4 Discussion and prospects

Let us first discuss shortly the impact of the semi-metric d in our testing procedures. Assume d actually takes into account only some characteristics (e.g. derivatives, projections, ...) $\tilde{\mathscr{X}}$ of the explanatory curve $\mathscr{X}$. Because of its definition the test statistic T_n only depends on these characteristics. Hence the null and alternative hypothesis are actually made on the regression model

$$Y = r_d(\tilde{\mathscr{X}}) + \varepsilon_d,$$

with $\mathbb{E}[\varepsilon_d|\tilde{\mathscr{X}}] = 0$. Consequently, the use of a semi-metric based on first functional PCA scores will only be able to test assumptions on the regression model corresponding to these first scores and when a semi-metric based on derivatives is used structural assumptions concern the effect of the derivatives.

The general method described above is a first attempt in the construction of general structural testing procedures in regression on functional variable (see Delsol, 2008, and Delsol *et al.*, 2011 for a more detailed discussion). The use of these tests on spectrometric data provide relevant informations on the structure of the link between the spectrometric curve and the chemical content of a product. Such tools may be also useful in procedures that aim to extract informative features from the explanatory curve. However it seems relevant to try to improve our approach and propose other test statistics that does not require to split our sample into three subsamples what may cause troubles in practice. To this end, we are now considering the following test statistic:

$$T_{2,n} = \sum_{i \neq j} (Y_i - \hat{r}(\mathscr{X}_i))(Y_j - \hat{r}(\mathscr{X}_j)) K\left(\frac{d(\mathscr{X}_i, \mathscr{X}_j)}{h_n}\right) w(\mathscr{X}_i) w(\mathscr{X}_j)$$

The theoretical study of this new test statistic is in progress. However, in the case of no effect tests, it seems $T_{2,n}$ have the same kind of asymptotic properties than T_n. Moreover, the new statistic $T_{2,n}$ seems more powerful (from simulations made with the same value of n). To conclude, the structural procedures presented in this paper open a large potential scope of applications. They could be used in an interesting way as part of an algorithm allowing to extract informative features (parts, points, ...) of the explanatory curve. An other prospect concerns their use in the choice of the semi-metric d since they may used to test the regularity of r with respect to a semi-metric d_1 against its regularity with respect to d_2 if $d_1 \leq d_2$. We finally discuss potential improvements and conclude on potential prospects for the future.

References

1. Ait-Saïdi, A., Ferraty, F., Kassa, R., Vieu, P.: Cross-validated estimations in the single functional index model. Statistics **42**, 475–494 (2008)
2. Aneiros-Perez, G., Vieu, P.: Time series prediction: a semi-functional partial linear model. J. Multivariate Anal. **99**, 834–857 (2008)
3. Borggaard, C., Thodberg, H.H.: Optimal minimal neural interpretation of spectra. *Anal. Chem.*, 64, (5), 545–551 (1992)
4. Bosq, D.: Linear Processes in Function Spaces : Theory and Applications. Lecture Notes in Statistics, 149, Springer Verlag, New York (2000)
5. Cardot, H., Ferraty, F., Mas, A., Sarda, P.: Testing Hypotheses in the Functional Linear Model. Scand. J. Stat. **30**, 241–255 (2003)
6. Cardot, H., Ferraty, F., Sarda, P.: Functional Linear Model. Statist. Prob. Lett. **45**, 11–22 (1999)
7. Cardot, H., Ferraty, F., Sarda, P.: Etude asymptotique d'un estimateur spline hybride pour le modèle linéaire fonctionnel. (French) [Asymptotic study of a hybrid spline estimator for the functional linear model] C. R. Acad. Sci. Ser. I 330 (6), 501–504 (2000)
8. Cardot, H., Goia, A., Sarda, P.: Testing for no effect in functional linear regression models, some computational approaches. Commun. Stat. Simulat. C. **33** (1), 179–199 (2004)
9. Cardot, H., Crambes, C., Kneip, A., Sarda, P.: Smoothing splines estimators in functional linear regression with errors-in-variables. Comput. Stat. Data An. **51** (10), 4832–4848 (2007)
10. Chiou, J.M., Müller H.-G.: Diagnostics for functional regression via residual processes. Comput. Stat. Data An. **51** (10), 4849–4863 (2007)
11. Crambes, C., Kneip, A., Sarda, P.: Smoothing splines estimators for functional linear regression. Ann. Stat. **37**, 35–72 (2009)
12. Delsol, L. (2007) Régression non-paramétrique fonctionnelle : Expressions asymptotiques des moments. *Annales de l'I.S.U.P.*, LI, (3), 43–67.
13. Delsol, L. (2008) Régression sur variable fonctionnelle: Estimation, Tests de structure et Applications. *Thèse de doctorat de l'Université de Toulouse*.
14. Delsol, L.: Advances on asymptotic normality in nonparametric functional Time Series Analysis. Statistics **43** (1), 13–33 (2009)

15. Delsol, L., Ferraty, F., Vieu, P.: Structural test in regression on functional variables. J. Multivariate Anal. 102 (3), 422–447 (2011)
16. Ferraty F., Goia A., Vieu P.: Functional nonparametric model for time series : a fractal approach for dimension reduction. Test **11** (2), 317–344 (2002b)
17. Ferraty, F., Romain, Y.: The Oxford Handbook on Functional Data Analysis. Oxford University Press (2011)
18. Ferraty, F., Vieu, P.: Dimension fractale et estimation de la régression dans des espaces vectoriels semi-normés. C. R. Acad. Sci. Ser. I 330, 403–406 (2000)
19. Ferraty, F., Vieu, P.: Nonparametric Functional Data Analysis: Theory and Practice. Springer, New York (2006)
20. Gadiaga, D., Ignaccolo, R.: Test of no-effect hypothesis by nonparametric regression. Afr. Stat. **1** (1), 67–76 (2005)
21. Hall, P., Cai, T.T.: Prediction in functional linear regression. Ann. Stat. **34** (5), 2159–2179 (2006)
22. Härdle, W., Mammen, E.: Comparing Nonparametric Versus Parametric Regression Fits. Ann. Stat. **21** (4), 1926–1947 (1993)
23. Masry, E.: Nonparametric regression estimation for dependent functional data : asymptotic normality. Stoch. Process. Appl. 115 (1), 155–177 (2005)
24. Müller, H.-G., Stadtmüller, U.: Generalized functional linear models. Ann. Stat. **33** (2), 774–805 (2005)
25. Mammen, E.: Bootstrap and wild bootstrap for high-dimensional linear models. Ann. Stat. **21** (1), 255–285 (1993)
26. Preda, C., Saporta, G.: PLS regression on a stochastic process. Comput. Stat. Data An. **48** (1), 149–158 (2005)
27. Ramsay, J., Dalzell, C.: Some tools for functional data analysis. J. Roy. Stat. Soc. B **53**, 539–572 (1991)
28. Ramsay, J., Silverman, B.: Functional Data Analysis. Springer-Verlag, New York (1997)
29. Ramsay, J., Silverman, B.: Applied functional data analysis : Methods and case studies. Spinger Verlag, New York (2002)
30. Ramsay, J., Silverman, B.: Functional Data Analysis (Second Edition). Springer Verlag, New York (2005)
31. Sood, A., James, G., Tellis, G.: Functional Regression: A New Model for Predicting Market Penetration of New Products. Marketing Science **28**, 36–51 (2009)

Chapter 13
A Fast Functional Locally Modeled Conditional Density and Mode for Functional Time-Series

Jacques Demongeot, Ali Laksaci, Fethi Madani, Mustapha Rachdi

Abstract We study the asymptotic behavior of the nonparametric local linear estimation of the conditional density of a scalar response variable given a random variable taking values in a semi-metric space. Under some general conditions on the mixing property of the data, we establish the pointwise almost-complete convergence, with rates, of this estimator. Moreover, we give some particular cases of our results which can also be considered as novel in the finite dimensional setting: Nadaraya-Watson estimator, multivariate data and the independent and identically distributed data case. On the other hand, this approach is also applied in time-series analysis to the prediction problem via the conditional mode estimation.

13.1 Introduction

Let (X_i, Y_i) for $i = 1, \ldots, n$ be n pairs of random variables that we assume are drawn from the pair (X, Y) which is valued in $\mathscr{F} \times \mathbb{R}$, where $\mathscr{F}$ is a semi-metric space equipped with a semi-metric d.

Furthermore, we assume that there exists a regular version of the conditional probability of Y given X, which is absolutely continuous with respect to the Lebesgue measure on $\mathbb{R}$ and admits a bounded density, denoted by f^x. Local polynomial

---

Jacques Demongeot
Université J. Fourier, Grenoble, France, e-mail: Jacques.Demongeot@imag.fr

Ali Laksaci
Université Djillali Liabès, Sidi Bel Abbès, Algeria e-mail: alilak@yahoo.fr

Fethi Madani
Université P. Mendès France, Grenoble, France, e-mail: Fethi.Madani@imag.fr

Mustapha Rachdi
Université P. Mendès France, Grenoble, France, e-mail: Mustapha.Rachdi@upmf-grenoble.fr

smoothing is based on the assumption that the unknown functional parameter is smooth enough to be locally well approximated by a polynomial (cf. Fan and Gijbels, 1996).

In this paper, we consider the problem of the conditional density estimation by using a locally modeling approach when the explanatory variable X is of functional kind and when the observations $(X_i,Y_i)_{i\in\mathbb{N}}$ are strongly α-mixing (cf. for instance, Rio (2000), Ferraty and Vieu (2006), Ferraty et al. (2006) and the references therein). In functional statistics, there are several ways of extending the local linear ideas (cf. Barrientos-Marin et al. (2009), Baìllo and Grané (2009), Demongeot et al. (2010), El Methni and Rachdi (2011) and the references therein). Here we adopt the fast functional locally modeling, that is, we estimate the conditional density f^x by $\widehat{a}$ which is obtained by minimizing the following quantity:

$$\min_{(a,b)\in\mathbb{R}^2}\sum_{i=1}^{n}\left(h_H^{-1}H(h_H^{-1}(y-Y_i))-a-b\beta(X_i,x)\right)^2K(h_K^{-1}\delta(x,X_i)) \qquad (13.1)$$

where $\beta(.,.)$ is a known bi-functional operator from $\mathscr{F}^2$ into $\mathbb{R}$ such that, $\forall\xi\in\mathscr{F}$, $\beta(\xi,\xi)=0$, with K and H are kernels and $h_K=h_{K,n}$ (respectively $h_H=h_{H,n}$) is chosen as a sequence of positive real numbers and $\delta(.,.)$ is a function from $\mathscr{F}^2$ into $\mathbb{R}$ such that $|\delta(.,.)|=d(.,.)$. Clearly, by simple algebra, we get explicitly the following definition of $\widehat{f^x}$:

$$\widehat{f^x}(y)=\frac{\sum_{i,j=1}^{n}W_{ij}(x)H(h_H^{-1}(y-Y_i))}{h_H\sum_{i,j=1}^{n}W_{ij}(x)} \qquad (13.2)$$

where

$$W_{ij}(x)=\beta(X_i,x)\left(\beta(X_i,x)-\beta(X_j,x)\right)K(h_K^{-1}\delta(x,X_i))K(h_K^{-1}\delta(x,X_j))$$

with the convention $0/0=0$.

13.2 Main results

In what follows x denotes a fixed point in $\mathscr{F}$, N_x denotes a fixed neighborhood of x, S will be a fixed compact subset of $\mathbb{R}$, and $\phi_x(r_1,r_2)=\mathbb{P}(r_2\leq\delta(X,x)\leq r_1)$.

Our nonparametric model will be quite general in the sense that we will just need the following assumptions:

(H1) For any $r>0$, $\phi_x(r):=\phi_x(-r,r)>0$.

(H2) The conditional density f^x is such that $\exists b_1>0, b_2>0, \forall(y_1,y_2)\in S^2$ and $\forall(x_1,x_2)\in N_x\times N_x$:

$$|f^{x_1}(y_1) - f^{x_2}(y_2)| \leq C_x \left(|\delta^{b_1}(x_1,x_2)| + |y_1 - y_2|^{b_2} \right),$$

where C_x is a positive constant depending on x.

(H3) The function $\beta(.,.)$ is such that:

$$\forall y \in \mathscr{F},\ C_1 |\delta(x,y)| \leq |\beta(x,y)| \leq C_2 |\delta(x,y)|, \text{ where } C_1 > 0, C_2 > 0.$$

(H4) The sequence $(X_i, Y_i)_{i\in\mathbb{N}}$ satisfies:

$$\exists a > 0, \exists c > 0 \text{ such that } \forall n \in \mathbb{N},\ \alpha(n) \leq cn^{-a}$$

where α is the mixing coefficient, and

$$\max_{i\neq j} \mathbb{P}\left((X_i,X_j) \in B(x,h) \times B(x,h)\right) = \varphi_x(h) > 0$$

(H5) The conditional density of (Y_i, Y_j) given (X_i, X_j) exists and is bounded.

(H6) The kernel K is a positive, differentiable function and supported within $(-1,1)$.

(H7) The kernel H is a positive, bounded and Lipschitzian continuous function, satisfying that:

$$\int |t|^{b_2} H(t)dt < \infty \text{ and } \int H^2(t)dt < \infty.$$

(H8) The bandwidth h_K satisfies:

$$\exists n_0 \in \mathbb{N}, \text{ such that } \forall n > n_0,\ -\frac{1}{\phi_x(h_K)} \int_{-1}^{1} \phi_x(zh_K, h_K) \frac{d}{dz}\left(z^2 K(z)\right) dz > C_3 > 0$$

and

$$h_K \int_{B(x,h_K)} \beta(u,x) dP(u) = o\left(\int_{B(x,h_K)} \beta^2(u,x)\, dP(u) \right)$$

where $dP(x)$ is the probability measure of X.

(H9) The bandwidth h_H satisfies:

$$\lim_{n\to\infty} h_H = 0 \text{ and } \exists \beta_1 > 0 \text{ such that } \lim_{n\to\infty} n^{\beta_1} h_H = \infty$$

(H10)
$$\begin{cases} \lim_{n\to\infty} h_K = 0,\ \lim_{n\to\infty} \dfrac{\chi_x^{(1/2)}(h_K) \log n}{n\, h_H\, \phi_x^2(h_K)} = 0, \\[2ex] \text{and } \exists \eta_0 > \frac{3\beta_1+1}{a+1},\ Cn^{\frac{(3-a)}{(a+1)}+\eta_0} \leq h_H \chi_x^{1/2}(h_K) \\[2ex] \text{where } \chi_x(h) = \max(\phi_x^2(h), \varphi_x(h)) \end{cases}$$

Then, the following theorem gives the almost-complete convergence[1] (a.co.) of $\widehat{f}^x$.

---

[1] Let $(z_n)_{n\in\mathbb{N}}$ be a sequence of real random variables. We say that z_n converges almost-completely (a.co.) to 0 if, and only if, $\forall \varepsilon > 0$, $\sum_{n=1}^{\infty} \mathbb{P}(|z_n| > \varepsilon) < \infty$. Moreover, let $(u_n)_{n\in\mathbb{N}^*}$ be a sequence of

Theorem 13.1. *Under assumptions (H1)-(H10), we obtain that:*

$$\sup_{y\in S}|\widehat{f^x}(y)-f^x(y)| = O\left(h_K^{b_1}+h_H^{b_2}\right)+O_{a.co.}\left(\sqrt{\frac{\chi_x^{(1/2)}(h_K)\log n}{n\,h_H\,\phi_x^2(h_K)}}\right).$$

13.3 Interpretations and remarks

- *On the assumptions:* The hypotheses used in this work are not unduly restrictive and they are rather classical in the setting of nonparametric functional statistics. Indeed, the conditions (H1), (H3), (H6) and (H8) are the same as those used by Benhenni et al. (2007) and Rachdi and Vieu (2007). Specifically (H1) is needed to deal with the functional nonparametric characteristics of our model by controlling the concentration properties of the probability measure of the variable X. This latter is quantified, here, with respect the bi-functional operator δ which can be related to the topological structure of the functional space $\mathscr{F}$ by taking $d=|\delta|$. While (H3) is a mild regularity condition permitting to control the shape of the locating function β. Such condition is verified, for instance, if we take $\delta=\beta$. However, as pointed out in Barrientos-Marin et al. (2009), this consideration of $\delta=\beta$ is not very adequate in practice, because these bi-functional operators do not play the same role. We refer to Barrientos-Marin et al. (2009) for more discussions on these conditions and some examples of β and δ. As usually in nonparametric problems, the infinite dimension of the model is controlled by mean of the smoothness condition (H2). This condition is needed to evaluate the bias component of the rate of convergence. Notice that the first part of the assumption (H4) is a standard choice of the mixing coefficient in the time series analysis, while the second part of this condition measures the local dependence of the observations. Let us point out also that this last assumption has been exploited in the expression of the convergence rate. On the other hand, assumptions (H7), (H9) and (H10) are standard technical conditions in the nonparametric estimation literature. These assumptions are imposed for the sake of simplicity and brevity of the proofs.
- *Some particular cases:*
  - *The Nadaraya-Watson estimator:* In a first attempt we will look at what happens when $b=0$. It is clear that, in this particular case, the conditions (H3) and (H8) are not necessary to get our result and thus the Theorem 13.1 can be reformulated in the following way.

  Corollary 13.1. *Under assumptions (H1), (H2), (H4)-(H7), (H9) and (H10), we obtain that:*

positive real numbers; we say that $z_n=O_{a.co.}(u_n)$ if, and only if, $\exists\varepsilon>0$, $\sum_{n=1}^{\infty}I\!P(|z_n|>\varepsilon u_n)<\infty$. This kind of convergence implies both almost-sure convergence and convergence in probability.

$$\sup_{y\in S}|\widehat{f}^x_{NW}(y)-f^x(y)|=O\left(h_K^{b_1}+h_H^{b_2}\right)+O_{a.co.}\left(\sqrt{\frac{\chi_x^{(1/2)}(h_K)\log n}{n\,h_H\;\phi_x^2(h_K)}}\right),$$

where $\widehat{f}^x_{NW}(y)$ *is the popular Nadaraya-Watson estimator.*

– *The multivariate case:* In the vectorial case, when $\mathscr{F}=\mathrm{I\!R}^p$ for $p\geq 1$, if the probability density function of the random variable X (respectively, the joint density of (X_i,X_j)) is continuously differentiable, then $\phi_x(h)=O(h^p)$ and $\varphi_x(h)=O(h^{2p})$ which implies that $\chi_x(h)=O(h^{2p})$. Then our Theorem 13.1 leads straightforwardly to the following corollary.

Corollary 13.2. *Under assumptions (H2), (H3) and (H5)-(H10), we obtain that:*

$$\sup_{y\in S}|\widehat{f}^x(y)-f^x(y)|=O\left(h_K^{b_1}+h_H^{b_2}\right)+O_{a.co.}\left(\sqrt{\frac{\log n}{n\,h_H\;h_K^p}}\right).$$

We point out that, in the special case when $\mathscr{F}=\mathrm{I\!R}$ our estimator is identified to the estimator studied in Fan and Yim (2004) by taking $\beta(x,X)=|X-x|=\delta(x,X)$.

– *The i.i.d. and finite dimensional case:* the conditions (H4), (H5) and the last part of (H10) are automatically verified and: $\chi_x(h)=\varphi_x(h)=\phi_x^2(h)$. So, we obtain the following result.

Corollary 13.3. *Under assumptions (H1)-(H3) and (H6)-(H9), we have that:*

$$\sup_{y\in S}|\widehat{f}^x(y)-f^x(y)|=O\left(h_K^{b_1}+h_H^{b_2}\right)+O_{a.co.}\left(\sqrt{\frac{\log n}{n\,h_H\;\phi_x(h_K)}}\right).$$

- *Application to functional time-series prediction:* The most important application of our study, when the observations are dependent and of functional nature, is the prediction of future values of some continuous-time process by using the conditional mode $\theta(x)=\arg\sup_{y\in S}f^x(y)$ as a prediction tool. This latter is estimated by the random variable $\widehat{\theta}(x)$ which is such that:

$$\widehat{\theta}(x)=\arg\sup_{y\in S}\widehat{f}^x(y).$$

In practice, we proceed as follows: let $(Z_t)_{t\in[0,b[}$ be a continuous-time real valued random process. From Z_t we may construct N functional random variables $(X_i)_{i=1,\ldots,N}$ defined by:

$$\forall t\in[0,b[,\;X_i(t)=Z_{N^{-1}((i-1)b+t)}$$

and a real characteristic $Y_i=G(X_{i+1})$. So, we can predict the characteristic Y_N by the conditional mode estimator: $\widehat{Y}=\widehat{\theta}(X_N)$ given by using the $(N-1)$ pairs of $(X_i, Y_i)_{i=1,\ldots,N-1}$. Such a prediction is motivated by the following consistency result.

Corollary 13.4. *Under the hypotheses of Theorem 13.1, and if the function f^x is j-times continuously differentiable on the topological interior of S with respect to y, and that:*

$$\begin{cases} f^{x(l)}(\theta(x)) = 0, \ \textit{if}\ 1 \leq l < j \\ \textit{and}\ \ f^{x(j)}(\cdot)\ \textit{is uniformly continuous on}\ S \\ \textit{such that}\ |f^{x(j)}(\theta(x))| > C > 0, \end{cases} \tag{13.3}$$

then we get:

$$|\widehat{\theta}(x) - \theta(x)|^j = O(h_K^{b_1}) + O(h_H^{b_2}) + O_{a.co.}\left(\sqrt{\frac{\chi_x^{(1/2)}(h_K)\log n}{n\,h_H\,\phi_x^2(h_K)}}\right).$$

References

1. Barrientos-Marin, J., Ferraty, F., Vieu, P.: Locally Modelled Regression and Functional Data. J. Nonparametr. Stat. **22**, 617–632 (2010)
2. Benhenni, K., Griche-Hedli, S., Rachdi, M.: Estimation of the regression operator from functional fixed-design with correlated errors. J. Multivariate Anal. **101**, 476–490 (2010)
3. Benhenni, K., Ferraty, F., Rachdi, M., Vieu, P.: Local smoothing regression with functional data. Computation. Stat. **22**, 353–369 (2007)
4. Baìllo, A., Grané, A.: Local linear regression for functional predictor and scalar response. J. Multivariate Anal. **100**, 102–111 (2009)
5. Dabo-Niang, S., Laksaci, A.: Estimation non paramétrique du mode conditionnel pour variable explicative fonctionnelle. Pub. Inst. Stat. Univ. Paris **3**, Pages 27–42 (2007)
6. Demongeot, J., Laksaci, A., Madani, F., Rachdi, M.: Local linear estimation of the conditional density for functional data. C. R., Math., Acad. Sci. Paris **348**, 931-934 2010)
7. Fan, J., Yim, T.-H.: A cross-validation method for estimating conditional densities. Biometrika **91**, 819–834 (2004)
8. Fan, J., Gijbels, I.: Local Polynomial Modelling and its Applications. Chapman & Hall, London (1996)
9. Ferraty, F., Laksaci, A., Vieu, P.: Estimating some characteristics of the conditional distribution in nonparametric functional models. Stat. Infer. Stoch. Process. **9**, 47–76 (2006)
10. Ferraty, F., Vieu, P.: Nonparametric functional data analysis. Theory and Practice. Series in Statistics, Springer, New York (2006)
11. Laksaci, A.: Convergence en moyenne quadratique de l'estimateur à noyau de la densité conditionnelle avec variable explicative fonctionnelle. Pub. Inst. Stat. Univ. Paris **3**, 69–80 (2007)
12. Ouassou, I., Rachdi, M.: Stein type estimation of the regression operator for functional data. Advances and Applications in Statistical Sciences **1**, 233-250 (2010)
13. Rachdi, M., Vieu, P.: Nonparametric regression for functional data: automatic smoothing parameter selection. J. Statist. Plann. Inference **137**, 2784–2801 (2007)
14. Rio, E.: Théorie asymptotique des processus aléatoires faiblement dépendants. Collection Mathématiques et Applications, ESAIM, Springer (2000)

Chapter 14
Generalized Additive Models for Functional Data

Manuel Febrero-Bande, Wenceslao González-Manteiga

Abstract The aim of this paper is to extend the ideas of generalized additive models for multivariate data (with known or unknown link function) to functional data covariates. The proposed algorithm is a modified version of the local scoring and backfitting algorithms that allows for the non-parametric estimation of the link function. This algorithm would be applied to predict a binary response example.

14.1 Introduction

For multivariate covariates, a Generalized Linear Model (GLM) (McCullagh and Nelder, 1989) generalizes linear regression by allowing the linear model to be related with a response variable Y which is assumed to be generated from a particular distribution in the exponential family (normal, binomial, poisson,...). The response is connected with the linear combination of the covariates, $\mathbf{Z}$, through a link function. Generalized Additive Models (GAM) (Hastie and Tibshirani, 1990) are an extension of GLMs in which the linear predictor is not restricted to be linear in the covariates but is the sum of smoothing functions applied to the covariates. Some other alternatives are the Single Index Models (SIM) (Horowitz, 1998), and the GAM with an unknown link function (Horowitz, 2001), the latter nesting all the previous models. Our aim is to extend these ideas to the functional covariates. There are some previous works in this direction. The functional logit model is considered in Escabias et al. (2004, 2006) using principal components or functional PLS to represent the functional data. A similar idea is used in Müller and Yao (2008) to extend additive models to functional data. The aim of this paper is to extend the local scoring and backfitting algorithm to functional data in a non-parametric way.

Manuel Febrero-Bande
University of Santiago de Compostela, Spain, e-mail: manuel.febrero@usc.es

Wenceslao González-Manteiga
University of Santiago de Compostela, Spain, e-mail: wenceslao.gonzalez@usc.es

In Section 2 we describe some background in GLM and GAM focused in binary response regression models. If the link is supposed to be known, the procedure could be extended to other exponential distribution families. If not, some modifications should be done. Section 3 is devoted to describe a generalized version of the local scoring algorithm that allow us (a) to estimate non-parametrically the GAM (with unknown link function), and thus (b) to obtain the corresponding predictive equations. In the nonparametric estimation process, kernel smoothers are used, and the bandwidths are found automatically by generalized cross-validation. Finally, section 4 is devoted to applications.

14.2 Transformed Binary Response Regression Models

Let Y be a binary (0/1) response variable, and $\mathbf{Z} = \{\mathscr{X}^i\}_{i=1}^p$ a set of functional covariates with values in the product of the p infinite dimensional spaces $\mathbf{E} = \mathscr{E}^1 \times \ldots \times \mathscr{E}^p$. In this framework, denoting $p(\mathbf{Z}) = p(Y = 1|\mathbf{Z})$ and mimicking the generalized linear model (GLM) (McCullagh and Nelder, 1989), the model takes the form:

$$p(\mathbf{Z}) = \mathbf{H}(\eta_z) = \mathbf{H}(\beta_0 + \langle \mathbf{Z}, \beta \rangle) \tag{14.1}$$

where β is a functional parameter taking values in $\mathbf{E}$ and $\mathbf{H}$ is a fixed increasing monotone link function, describing the functional relationship between $p(\mathbf{Z})$ and the systematic component $\eta_z = \langle \mathbf{Z}, \beta \rangle$. Other possibility that does not asume linearity in the covariates is to adapt to functional context the GAM model. The GAM can be expressed as:

$$p(\mathbf{Z}) = \mathbf{H}(\eta_z) = \mathbf{H}\left(\beta_0 + \sum_{j=1}^{p} f_j(\mathscr{X}^j)\right) \tag{14.2}$$

where the partial function f_j's are assumed to be unknown but smooth. The above models make the hypothesis that the link function has a known form. This fixed form is, however, rarely justified. Respect to this, the semiparametric single index model (SIM)(Horowitz, 1998) generalizes the GLM (14.1) by allowing the link to be an arbitrary smooth function that has to be estimated from the data. The SIM can be expressed as:

$$p(\mathbf{Z}) = \mathbf{H}(\eta_z) = \mathbf{H}(\beta_0 + f(\langle \mathbf{Z}, \beta \rangle)). \tag{14.3}$$

The main goal of this paper is to propose an algorithm to solve this broader class of models to deal even in those practical situations in which there is not enough information either about the form of the link (as in the SIM) or about the shape of the partial functions (as in the GAM). Such a general formulation will be presented here as G-GAM (GAM with unknown link function) with the purpose of widening the assumptions regarding the link in generalized additive models.

14.3 GAM: Estimation and Prediction

A GAM (or a G-GAM) takes the form given in (14.2), where the link $\mathbf{H}$ is a known (an unknown) increasing monotone function. In this section we propose to adapt the techniques shown in Roca-Pardiñas et al (2004) in such a way that it will permit the non-parametric estimation of the partial functions f_j and, if needed the joint non-parametric estimation of the link $\mathbf{H}$, when the covariates are curves. But before estimating the partial functions and the link, some restrictions have to be imposed in order to ensure the GAM (G-GAM) identification. This is an usual topic in multivariate GAM and SIM models. In the GAM context, identification is guaranteed by introducing a constant β_0 into the model and requiring a zero mean for the partial functions ($E(f_j) = 0$). In the SIM and G-GAM, however, given that the link function is not fixed, it is necessary to establish futher conditions in order to avoid different combinations of $\mathbf{H}$ and f_js that could lead to the same model. In this paper, when estimating a GAM we impose the condition:

1. (General condition) $E\left[f_j\right] = 0, (j = 1,\ldots,p)$
2. (G-GAM only) $\beta_0 = 0$ and $E\left[\left(\Sigma_{j=1}^{p} f_j\right)^2\right] = 1.$

These are the same two conditions as in Roca-Pardiñas et al. (2004). Note that, from these conditions, the systematic component η_z becomes standardized.

The proposed algorithm is as follows:

For a given $(\mathbf{Z}, Y)$, the local scoring maximizes an estimation of the expected log-likelihood $E\left[l\{\eta_z;Y\}|\mathbf{Z}\right]$, being

$$l\{\eta_z;Y\} = Y\log\left[\mathbf{H}(\eta_z)\right] + (1-Y)\log\left[1-\mathbf{H}(\eta_z)\right] \tag{14.4}$$

by solving iteratively a reweighted least squares problem in the following way.

In each iteration, given the current guess $\hat{\eta}_Z^0$, the linearized response $\tilde{Y}$ and the weight $\tilde{W}$ are constructed as

$$\tilde{Y} = \hat{\eta}_Z^0 + \frac{Y-\mathbf{H}(\hat{\eta}_Z^0)}{\mathbf{H}'(\hat{\eta}_Z^0)}, \qquad \text{and} \qquad \hat{W} = Var(\tilde{Y}|\mathbf{Z})^{-1} = \frac{\mathbf{H}'(\hat{\eta}_Z^0)^2}{\mathbf{H}(\hat{\eta}_Z^0)\left(1-\mathbf{H}(\hat{\eta}_Z^0)\right)} \tag{14.5}$$

$\mathbf{H}'$ being the first derivative of $\mathbf{H}$. To estimate the f_js, we fit an additive regression model to $\tilde{Y}$, treating it as a response variable with associated weight $\hat{W}$. The resulting estimation of $\hat{\eta}_Z$ is $\hat{\eta}_Z^0$ of the next iteration. This procedure must be repeated until small changes in the systematic component. For the estimation of the f_js and $\mathbf{H}$ the following two alternating loops must be perfomed.

Loop 1. Let $\hat{\eta}_Z^0$, $\hat{p}^0(\mathbf{Z}) = \hat{\mathbf{H}}^0(\hat{\eta}_Z^0)$ and $\hat{\mathbf{H}}'^0(\hat{\eta}_Z^0)$ be the current estimates. Replacing functions $\mathbf{H}$ and $\mathbf{H}'$ by their current estimates, $\hat{\mathbf{H}}^0$ and $\hat{\mathbf{H}}'^0$, in formulas given in (14.5), $\hat{\eta}_Z = \beta_0 + \Sigma_{j=1}^{p} \hat{f}_j(\mathscr{X}^j)$ is then obtained by fitting an additive model of $\tilde{Y}$ on $\mathbf{Z}$ with weights $\hat{W}$. Here we use backfitting techniques based on

Nadaraya-Watson kernel estimators with bandwith automatically chosen by Generalized Cross-Validation.

Loop 2. (G-GAM only). Fixing $\hat{\eta}_Z$, the two estimates $\hat{p}^0(\mathbf{Z}) = \hat{\mathbf{H}}(\hat{\eta}_Z)$ and $\hat{\mathbf{H}}'(\hat{\eta}_Z)$ are then obtained by fitting a regression model of Y on $\mathbf{Z}$ weighted by $\left(\hat{p}^0(\mathbf{Z})\left(1-\hat{p}^0(\mathbf{Z})\right)\right)^{-1}$. Here we use linear local kernel estimators in order to have estimations of the first derivative.

These two loops are repeated until the relative change in deviance is negligible.

At each iteration of the estimation algorithm, the partial functions are estimated by applying Nadaraya-Watson weighted kernel smoothers to the data $\{\mathscr{X}^j, R^j\}$ with weights $\hat{W}$, R^j being the residuals associated to $\mathscr{X}^j$ obtained by removing the effect of the other covariates. In this paper, for each $\hat{f}_j$ the corresponding bandwidth h_j is selected automatically by minimizing, in each of the cycles of the algorithm, the weighted GCV error criterion whereas the bandwidth for estimating the link function (if needed) is found minimizing the cross-loglikelihood error criterion (analogous to (14.4)).

14.4 Application

In this section, we present an application of GAM model (14.2) to the Tecator dataset. This data set was widely used in examples with functional data (see Ferraty & Vieu, 2006) to predict the content of fat content on samples of finely chopped meat. For each food sample, the spectrum of the absorbances recorded on a Tecator Infratec Food and Feed Analyzer working in the wavelength range 850-1050 mm by the near-infrared transmission (NIT) principle is provided also with the fat, protein and moisture contents, measured in percent and determined by analytic chemistry. We had $n = 215$ independent observations usually divided into two data sets: the training sample with the first 165 observations and the testing sample with the others. In this study, we are trying to predict $Y = I\{\text{Fat} > 65|\mathbf{Z}\}$ where $\mathbf{Z} = \left(\mathscr{A}, \mathscr{A}''\right)$ being $\mathscr{A}$ the absorbances and $\mathscr{A}''$ its second derivative. The use of the second derivative is justified by previous works (see for example Aneiros-Pérez & Vieu, 2006, among others) where those models with information about the second derivative have better prediction results.

So, in this case the model can be expressed:

$$E(Y=1|\mathbf{Z}) = p(\mathbf{Z}) = p(\mathscr{A}, \mathscr{A}'') = \mathbf{H}(\eta_z) = \mathbf{H}\left(\beta_0 + f_1(\mathscr{A}) + f_2(\mathscr{A}'')\right) \quad (14.6)$$

where $\mathbf{H}$ is the logit link.

The curves and the second derivative are shown in figure 14.1. Here, the red group (fat over 65%) is clearly quite well separated when considering the second derivative and quite mixing when considering the spectrum itself. This suggests that the relevant information about high percentage of fat is mainly related with

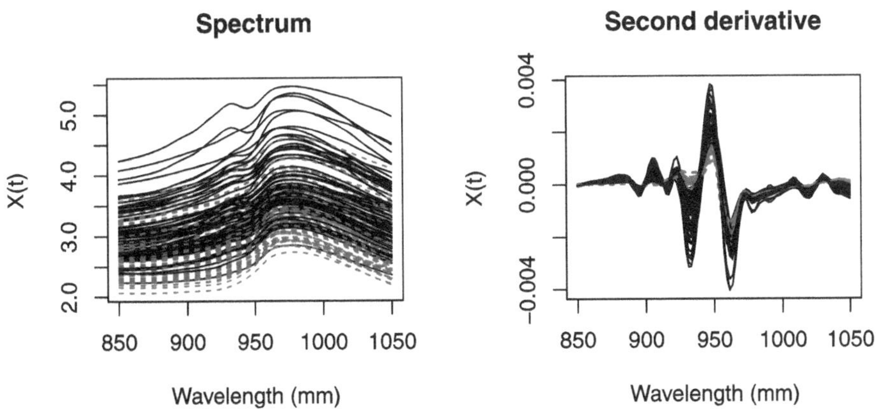

Fig. 14.1: Spectrum and second derivative of training sample coloured by binary response

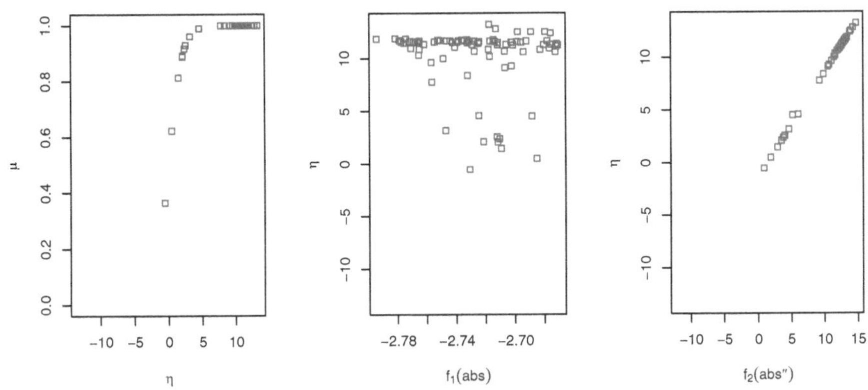

Fig. 14.2: Final results with the effects of every functional covariate

the second derivative. This impression could be confirmed in figure 14.2 where the contribution of every functional covariate to η are shown in central and right plots. The spectrum curves show a chaotic behaviour respect to η whereas the second derivative of these curves shows a clearly increasing pattern. Indeed, the trace of the smoothing matrices S_1, S_2 associated with f_1, f_2 are respectively 1.64 and 67.12 which indicates a higher contribution of the second covariate. Classifying every observation according to the estimated probability, the percentage of good classification in the training sample is 96.36% which raises to 98% in the testing sample.

We have also repeated this analysis 200 times changing at random which data are included in the training sample and keeping the size of the training sample in 165 observations. The results are summarized in table 14.1 and are quite promising. As a conclusion, we have proposed an algorithm to estimate a wide class of regression models for functional data with response belonging to the exponential family. Nev-

Sample	Min.	1st. Qu.	Median	Mean	3rd. Qu.	Max.
Training	91.5%	96.4%	97.0%	97.2%	98.2%	100%
Testing	86.0%	94.0%	96.0%	95.6%	98.0%	100%

Table 14.1: Percentage of good classification

ertheless, some two questions arise in the application to real data: (i) the algorithm is quite high consuming specially when the link function have to be estimated and the convergence is slow and, (ii) the error criteria for automatic choice of bandwidths must be revised in order to work properly. It seems that GCV and CV criteria give small bandwidths.

Acknowledgements This research has been funded by project MTM2008-03010 from Ministerio de Ciencia e Innovación, Spain.

References

1. Aneiros-Pérez, G., Vieu, P.: Semi-functional partial linear regression. Stat. Probabil. Lett. **76**, 1102–1110 (2006)
2. Cardot, H., Sarda, P.: Estimation in generalized linear models for functional data via penalized likelihood. J. Multivariate Anal. **92** (1), 24–41 (2005)
3. Escabias, M., Aguilera, A.M., Valderrama, M.J.: Principal component estimation of functional logistic regression: discussion of two different approaches. J. Nonparametr. Stat. **16** (3-4), 365–384 (2004)
4. Escabias, M., Aguilera, A.M., Valderrama, M.J.: Functional PLS logit regression model. Comput. Stat. Data An. **51**, 4891–4902 (2006)
5. Ferray, F., Vieu, P.: Nonparametric functional data analysis. Springer, New York (2006)
6. Horowitz, J.L.: Semiparametric Methods in Econometrics. Lecture Notes in Statistics, 131, Springer Verlag (1998)
7. Horowitz, J.L.: Nonparametric estimation of a generalized additive model with an unknown link function. Econometrica **69**, 499–514 (2001)
8. McCullagh, P., Nelder, J.A.: Generalized Linear Models. Chapman & Hall (1989)
9. Müller, H.G., Yao, F.: Functional Additive Model. J. Am. Stat. Assoc. **103** (484), 1534–1544 (2008)
10. Roca-Pardiñas, J., González-Manteiga, W., Febrero-Bande, M., Prada-Sánchez, J.M., Cadarso-Suárez, C.: Predicting binary time series of SO_2 using generalized additive models with unknown link function. Environmetrics **15**, 729–742 (2004)

Chapter 15
Recent Advances on Functional Additive Regression

Frédéric Ferraty, Aldo Goia, Enersto Salinelli, Philippe Vieu

Abstract We introduce a flexible approach to approximate the regression function in the case of a functional predictor and a scalar response. Following the Projection Pursuit Regression principle, we derive an additive decomposition which exploits the most interesting projections of the prediction variable to explain the response. The goodness of our procedure is illustrated from theoretical and pratical points of view.

15.1 The additive decomposition

Let (X,Y) be a centered r.v. with values in $H \times \mathbb{R}$ where $H = \left\{h : \int_I h^2(t)\,dt < +\infty\right\}$, I interval of $\mathbb{R}$, is a separable Hilbert space equipped with the inner product $\langle g, f\rangle = \int_I g(t) f(t)\,dt$ and induced norm $\|g\|^2 = \langle g, g\rangle$. The regression problem is stated in a standard way as

$$Y = r[X] + \mathscr{E}$$

with $r[X] = \mathbb{E}[Y|X]$. As usual, we assume $\mathbb{E}[\mathscr{E}|X] = 0$ and $\mathbb{E}\left[\mathscr{E}^2|X\right] < \infty$.
We approximate the unknown regression functional r by a finite sum of terms

$$r[X] \approx \sum_{j=1}^{m} g_j^*\left(\langle\theta_j^*, X\rangle\right) \tag{15.1}$$

Frédéric Ferraty
Institut de Mathématiques de Toulouse, France, e-mail: ferraty@math.univ-toulouse.fr

Aldo Goia
Università del Piemonte Orientale, Novara, e-mail: aldo.goia@eco.unipmn.it

Enersto Salinelli
Università del Piemonte Orientale, Novara, e-mail: ernesto.salinelli@eco.unipmn.it

Philippe Vieu
Institut de Mathématiques de Toulouse, France, e-mail: vieu@math.univ-toulouse.fr

where $\theta_j^* \in H$ with $\left\|\theta_j^*\right\|^2 = 1$, g_j^*, for $j = 1,\dots,m$, are real univariate functions and m is a positive integer to determine. The aim is to project X onto the predictive directions θ_1^*, $\theta_2^*,\dots$ that are the most interesting for explaining Y and, at the same time, describing the relation with Y by using a sum of functions g_j^*. We make this by looking at the pairs $\left(\theta_j^*, g_j^*\right)$ iteratively. At first step, we determine θ_1^* by solving

$$\min_{\|\theta_1\|^2=1} \mathbb{E}\left[(Y - \mathbb{E}[Y|\langle\theta_1, X\rangle])^2\right].$$

Once θ_1^* is obtained, we have $g_1^*(u) = \mathbb{E}[Y|\langle\theta_1^*, X\rangle = u]$. If we set $\mathscr{E}_{1,\theta_1^*} = Y - g_1^*(\langle\theta_1^*, X\rangle)$, then $\mathscr{E}_{1,\theta_1^*}$ and $\langle\theta_1^*, X\rangle$ are uncorrelated. So, in an iterative way, we can define

$$\mathscr{E}_{j,\theta_j^*} = Y - \sum_{s=1}^{j} g_s^*(\langle\theta_s^*, X\rangle) \qquad j = 1,\dots,m$$

with at each stage $\mathbb{E}\left[\mathscr{E}_{j,\theta_j^*}|\left\langle\theta_j^*, X\right\rangle\right] = 0$. Then, one can obtain for $j > 1$ the j-th direction θ_j^* by solving the minimum problem

$$\min_{\|\theta_j\|^2=1} \mathbb{E}\left[\left(\mathscr{E}_{j-1,\theta_{j-1}^*} - \mathbb{E}\left[\mathscr{E}_{j-1,\theta_{j-1}^*}|\langle\theta_j, X\rangle\right]\right)^2\right]$$

and then define the j-th component as $g_j^*(u) = \mathbb{E}\left[\mathscr{E}_{j-1,\theta_{j-1}^*}|\langle\theta_j^*, X\rangle = u\right]$.
By this way, the directions θ_j^* entering in (41.8) are explicitely constructed and so, after the m-th step, one has the additive decomposition with $\mathbb{E}\left[\mathscr{E}_{m,\theta_m^*}|\langle\theta_m^*, X\rangle\right] = 0$:

$$Y = \sum_{j=1}^{m} g_j^*\left(\langle\theta_j^*, X\rangle\right) + \mathscr{E}_{m,\theta_m^*}.$$

15.2 Construction of the estimates

We illustrate how to estimate the functions g_j^* and θ_j^* from a sample (X_i, Y_i), $i = 1,\dots,n$, drawn from (X, Y). We base the procedure on an alternating optimization strategy combining a spline approximation of directions and the Nadaraya-Watson kernel regression estimate.
Denote by $\mathscr{S}_{d,N}$ the $(d+N)$-dimensional space of spline functions defined on I with degree d and with $N-1$ interior equispaced knots (with $d > 2$ and $N > 1$, integers). Let $\{B_{d,N,s}\}$ be the normalized B-splines. For $j = 1,\dots,m$, the spline approximation of θ_j is represented as $\gamma_j^T \mathbf{B}_{d_j,N_j}(t)$, where $\mathbf{B}_{d_j,N_j}(t)$ is the vector of all the B-splines and γ_j is the vector of coefficients satisfying the normalization condition

$$\gamma_j^T \int_I \mathbf{B}_{d_j,N_j}(t)\,\mathbf{B}_{d_j,N_j}(t)^T\,dt\,\gamma_j = 1. \tag{15.2}$$

The estimation procedure is based on the following steps:

- **Step 1 -** Initialize the algorithm by setting $m = 1$ and current residuals $\widehat{\mathscr{E}}_{m-1,\widehat{\gamma}_{m-1},i} = Y_i$, $i = 1,\ldots,n$.
- **Step 2 -** Choose the dimension $N_m + d_m$ of $\mathscr{S}_{d_m,N_m}$ and fix the initial direction setting the vector of inital coefficients $\gamma_m^{(0)}$ satisfying (15.2).
Find an estimate $\widehat{g}^{-i}_{m,\gamma_m^{(0)}}$ of g_m using the Nadaraya-Watson kernel regression approach excluding the i-th observation X_i:

$$\widehat{g}^{-i}_{m,\gamma_m^{(0)}}(z) = \sum_{l\neq i} \frac{K_m\left(\frac{z-\left(\gamma_m^{0}\right)^T \mathbf{b}_{m,l}}{h_m}\right)}{\sum_{l\neq i} K_m\left(\frac{z-\left(\gamma_m^{0}\right)^T \mathbf{b}_{m,l}}{h_m}\right)} \widehat{\mathscr{E}}_{m-1,\widehat{\gamma}_{m-1},l}$$

where $\mathbf{b}_{m,l} = \left\langle \mathbf{B}_{d_m,N_m}, X_l \right\rangle$.
Then, compute an estimate $\widehat{\gamma}_m$ by minimizing

$$CV_m(\gamma_m) = \frac{1}{n}\sum_{i=1}^{n}\left[\left(\widehat{\mathscr{E}}_{m-1,\widehat{\gamma}_{m-1},i} - \widehat{g}^{-i}_{m,\gamma_m^{(0)}}\left(\gamma_m^T \mathbf{b}_{m,i}\right)\right)^2\right]$$

over the set of vectors $\gamma_m \in \mathbb{R}^{N_m+d_m}$ satisfying (15.2). Update $\gamma_m^{(0)} = \widehat{\gamma}_m$, and repeat the cycle until the convergence: the alghoritm terminates when the variation of CV_m passing from the previous to the current iteration (normalized by the variance of current residuals) is positive and less than a prespecified threshold.
- **Step 3 -** Let u_n be a positive sequence tending to zero as n grows to infinity. If the penalized criterion of fit

$$GCV(m) = \frac{1}{n}\sum_{i=1}^{n}\left[\left(\widehat{\mathscr{E}}_{m-1,\widehat{\gamma}_{m-1},i} - \sum_{j=1}^{m} \widehat{g}^{-i}_{m,\widehat{\gamma}_m}\left(\widehat{\gamma}_m^T \mathbf{b}_{m,i}\right)\right)^2\right](1+u_n)$$

does not decrease, then stop the algorithm. Otherwise, construct the next set of residuals

$$\widehat{\mathscr{E}}_{m,\widehat{\gamma}_m,i} = \widehat{\mathscr{E}}_{m-1,\widehat{\gamma}_{m-1},i} - \widehat{g}_{m,\widehat{\gamma}_m}\left(\widehat{\gamma}_m^T \mathbf{b}_{m,i}\right),$$

update the term counter $m = m+1$, and go to Step 2.

Once the m^* most predictive directions θ_j^* and functions g_j^* which approximate the link between the functional regressor and the scalar response are estimated, it could be possible to improve the prediction performances, by using a boosting procedure with a final full nonparametric step: we compute the residuals

$$Y_i - \sum_{j=1}^{m^*} \widehat{g}_{j,\widehat{\theta}_j}\left(\left\langle \widehat{\theta}_j, X_i \right\rangle\right)$$

and we estimate the regression function between these residuals and the whole functional regressors X_i by using the Nadaraya-Watson type estimator.

15.3 Theoretical results

We resume the most important theoretical results. At the first step, supposing that the directional parameters $\theta_1, \ldots, \theta_m$ are fixed/known, we state that one can estimate each component involved in the additive decomposition without being affected by the dimensionality of the problem. In fact the rates of convergences obtained are the optimal ones for univariate regression problems. More precisely, assuming that *i)* the functions g_{j,θ_j} satisfy a Hölder condition of order β and has $q_j > 0$ continuous derivatives, *ii)* each kernel K_j has support $(-1,1)$, is of order k_j (with $k_j \geq q_j$ and $k_j < k_{j-1}$) and each bandwith h_j satisfies $h_j \sim \left(\frac{1}{n}\right)^{\frac{1}{2k_j+1}}$, then for $n \longrightarrow \infty$ one has:

$$\sup_{u\in\mathscr{C}} \left| \widehat{g}_{j,\theta_j}(u) - g_{j,\theta_j}(u) \right| = O\left(\left(\frac{\log n}{n} \right)^{\frac{\beta}{2\beta+1}} \right), \quad a.s.$$

and

$$\mathbb{E}\left[\int_{\mathscr{C}} \left(\widehat{g}_{j,\theta_j}(u) - g_{j,\theta_j}(u) \right)^2 du \right] \sim \left(\frac{1}{n} \right)^{\frac{2k_j}{2k_j+1}}$$

where $\mathscr{C}$ is a compact subset of $\mathbb{R}$.
This first result can be used for deriving the optimality of the estimate $\widehat{\theta}_1, \ldots, \widehat{\theta}_m$ for any fixed value m, as n grows to infinity. In particular we prove that the estimated directions $\widehat{\theta}_j$, $j = 1, \ldots, m$, are L_2-asymptotically optimal in the sense that they minimize, as n grows to infinity, the following L_2 theoretical measure of accuracy:

$$MISE_j(\theta_j) = \mathbb{E}\left[\int_{\mathscr{C}} \left(g_{j,\theta_j^*}(u) - \widehat{g}_{j,\theta_j}(u) \right)^2 du \right].$$

In fact, under suitable hypothesis on the approximation space in which we work and on the distribution of the functional variable X, one has for any $j = 1, \ldots, m$:

$$\frac{MISE_j(\widehat{\theta}_j)}{MISE_j(\widetilde{\theta}_j)} \longrightarrow 1, \quad a.s., \quad \text{as } n \longrightarrow \infty$$

where $\widetilde{\theta}_j$ is the theoretical L_2-optimal value of θ_j defined as $\widetilde{\theta}_j = \arg\min_{\theta_j \in \Theta} MISE_j(\theta_j)$.

15.4 Application to real and simulated data

The methodology developed (named FPPR in the sequel) is applied to real and simulated data in order to asses its performances. For each case considered, we compute the estimates on a training set and the goodness of prediction is evaluated on a testing sample by using the Mean Square Error of Prediction (MSEP):

$$MSEP = \frac{1}{n^{out}} \sum_{i=1}^{n^{out}} (y_i - \widehat{y}_i)^2$$

where, y_i and $\widehat{y}_i$ are the true value and the corresponding prediction, and n^{out} is the size of the testing sample. The results are compared with those obtained by the functional linear model (FLM) and the nonparametric method (NPM) based on the Nadaraya-Watson approach.

About the simulation study, we present here only a significant example among many, and we consider the model

$$Y_i = \int_{-1}^{1} X_i(t) \log |X_i(t)| \, dt + \sigma \mathscr{E}_i , \quad i = 1, \ldots, 300$$

where curves X_i are generated according to

$$X_i(t) = a_i + b_i t^2 + c_i \exp(t) + \sin(d_i t), \qquad t \in [-1, 1]$$

with a_i (respectively b_i, c_i and d_i) uniformly distributed on $(0,1)$ (respectively on $(0,1)$, $(-1,1)$ and $(-2\pi, 2\pi)$). We work with both dense and sparse design of measurement locations, corresponding to 100 and respectively 6 equispaced points. The r.v. $\mathscr{E}_i$ are i.i.d. with zero mean and unitary variance, and σ is equal to ρ times ($\rho = 0.1, 0.3$) the standard deviation of the regression functional. We consider two distributions of error: the standard normal $\mathscr{N}(0,1)$ and the standardized gamma $\gamma(4,1)$, which is right-skewed. We base our study on samples of 300 couples (X_i, Y_i): we use the first 200 as training-set and the remaining 100 as testing-set. The Table 15.1 provides both the MSEP and MSEP divided by the empirical variance of Y's (in brackets). We can note that it is sufficient a one step FPPR (eventually followed by a full nonparametric on residuals) to achieve superior results with respect to the NPM approach.

For the application to real data, we refer to the Tecator data-set, a benchmark for testing regression models. The data-set consists of 215 Near Infrared (NIR) absorbance spectra of finely chopped pure meat samples, recorded on a Tecator Infratec Food Analyzer in the wavelength range 850-1050 nm. Each functional observation is discretized over 100 channel spectrum of absorbance; to every curve corresponds a content in percentage of water, fat and protein determined by analytic chemistry. Our goal is to predict the fat content on the basis of its NIR absorbance spectrum. The data set has been split in a training-set including the first 160 elements and a testing-set with the remaining 55 ones. Since spectrometric curves suffer from a cal-

Method	$\rho=0.1,\ \mathcal{N}(0,1)$ Dense	Sparse	$\rho=0.1,\ \gamma(1,4)$ Dense	Sparse	$\rho=0.3,\ N(0,1)$ Dense	Sparse	$\rho=0.3,\ \gamma(1,4)$ Dense	Sparse
FLM	0.0849	0.0817	0.1059	0.1020	0.1378	0.1337	0.1838	0.1786
	(0.2629)	(0.2530)	(0.2741)	(0.2638)	(0.3626)	(0.3519)	(0.3679)	(0.3576)
NPM	0.0423	0.0422	0.0627	0.0652	0.0953	0.0957	0.1426	0.1466
	(0.1310)	(0.1306)	(0.1622)	(0.1688)	(0.2509)	(0.2520)	(0.2856)	(0.2936)
FPPR ($m=1$)	0.0389	0.0400	0.0502	0.0507	0.0846	0.0854	0.1252	0.1107
	(0.1205)	(0.1238)	(0.1298)	(0.1313)	(0.2228)	(0.2248)	(0.2507)	(0.2217)
FPPR & NPM	0.0304	0.0320	0.0370	0.0380	0.0803	0.0817	0.1086	0.0979
	(0.0942)	(0.0990)	(0.0956)	(0.0983)	(0.2114)	(0.2151)	(0.2174)	(0.1959)

Table 15.1: MSEP and Relative MSEP for the simulated data.

ibration problem intrinsic to the NIR spectrometer analyzer, the second derivative spectra is used.

We have run our procedure and we have stopped the algorithm at $\widehat{m}=2$. A pure non-parametric step on the residuals after these two steps has been performed. The out-of-sample performances, collected in Table 15.2, show that our method is equivalent to the nonparametric estimation and, using the boosting procedure, we have the best results.

Method	FLM	NPM	FPPR (Step 1)	FPPR (Steps 1 & 2)	FPPR & NPM
MSEP	7.174	1.915	3.289	2.037	1.647

Table 15.2: MSEP for the Tecator data.

To conclude, this additive functional regression is a good predictive tool (comparable with the nonparametric approach in some situations) while providing interesting outputs for describing the relationship: the predictive directions and the additive components.

References

1. Ferraty, F., Goia, A., Salinelli, E., Vieu, P.: Additive Functional Regression based on Predictive Directions. WP 13/10, Dipartimento di Scienze Economiche e Metodi Quantitativi, Università del Piemonte Orientale A. Avogadro (2010)
2. Ferraty, F., Vieu, P.: Nonparametric functional data analysis. Springer, New York (2006)
3. Friedman, J.H., Stuetzle, W.: Projection Pursuit Regression. J. Am. Stat. Assoc. **76**, 817–823 (1981)
4. Hall, P. (1989). On projection Pursuit Regression. Ann. Stat. **17** (2), 573–588 (1989)
5. James, G.M., Silverman, B.W.: Functional Adaptive Model Estimation. J. Am. Stat. Assoc. **100** (470), 565–576 (2005)
6. Müller, H.G., Yao, F.: Functional Additive Model. J. Am. Stat. Assoc. **103** (484), 1534–1544 (2008)
7. Ramsay, J.O., Silverman, B.W.:Functional Data Analysis (Second Edition). Springer Verlag, New York (2005)

Chapter 16
Thresholding in Nonparametric Functional Regression with Scalar Response

Frédéric Ferraty, Adela Martínez-Calvo, Philippe Vieu

Abstract In this work, we have focused on the nonparametric regression model with scalar response and functional covariate, and we have analyzed the existence of underlying complex structures in data by means of a thresholding procedure. Several thresholding functions are proposed, and a cross-validation criterion is used in order to estimate the threshold value. Furthermore, a simulation study shows the effectiveness of our method.

16.1 Introduction

Many recent contributions have studied the functional regression model with scalar response from both *parametric* viewpoint (see Ramsay and Silverman (2005)) and *nonparametric* one (see Ferraty and Vieu (2006)). In this work, we have considered the more general nonparametric framework, and we have studied the regression model given by

$$Y = r(X) + \varepsilon, \tag{16.1}$$

where Y is a real random variable, X is a random variable valued in a separable Hilbert space $(\mathscr{H}, \langle \cdot, \cdot \rangle)$, $r : \mathscr{H} \to \mathbb{R}$ is the regression operator, and ε is a real centered random variable such that $\mathbb{E}(\varepsilon^2) = \sigma_\varepsilon^2$.

Sometimes we are confronted with complex regression structures which are unlikely detectable using standard graphical or descriptive techniques (for instance, the existence of several subsamples of curves or different regression models in the

Frédéric Ferraty
Institut de Mathématiques de Toulouse, France, e-mail: ferraty@math.univ-toulouse.fr

Adela Martínez-Calvo
Universidade de Santiago de Compostela, Spain, e-mail: adela.martinez@usc.es

Philippe Vieu
Institut de Mathématiques de Toulouse, France, e-mail: vieu@math.univ-toulouse.fr

sample). The objective of this work is to present an exploratory method that allows us to discover certain kind of hidden structures. Our approach analyzes the existence of threshold in the covariable X and/or the real response Y and, when the threshold exists, estimate its value by means of a cross-validation procedure. Moreover, the cross-validation criterion can be plotted and used as a graphical support to decide if there is any type of threshold in the data. We have tested our method with a simulation study and several real data applications. However, space restrictions forced us to reduce the simulation results and remove real data applications from this paper.

16.2 Threshold estimator

The key of our procedure is to rewrite the regression operator $r(x) = \mathbb{E}(Y|X = x)$ as the sum of two components as follows. First of all, let us fix a function Ψ : $\mathscr{H} \times \mathbb{R} \to \mathscr{E}$, where $\mathscr{E}$ is a beforehand fixed space, and an associated set of pairs $\{(E_1^\tau, E_2^\tau)\}_{\tau \in T_n}$ such that $E_s^\tau \subset \mathscr{E}$ for $s \in S = \{1,2\}$, and

$$P(\Psi(X,Y) \in \bigcap_{s \in S} E_s^\tau) = 0, \quad P(\Psi(X,Y) \in \bigcup_{s \in S} E_s^\tau) = 1, \quad \forall \tau \in T_n.$$

¿From now on, let $\{(X_i,Y_i)\}_{i=1}^n$ be a sample of independent and identically distributed pairs as (X,Y). For each observation (X_i,Y_i), let us define $\delta_{i,s}^\tau = 1_{\{\Psi(X_i,Y_i) \in E_s^\tau\}}$ and $Y_{i,s}^\tau = Y_i \delta_{i,s}^\tau$, for $i \in \{1,\dots,n\}$ and $s \in S$.

Consequently, the regression model (16.1) can be expressed as

$$Y_i = \sum_{s \in S} Y_{i,s}^\tau = \sum_{s \in S} r_s^\tau(X_i) + \varepsilon_i = r(X_i) + \varepsilon_i, \quad \text{for } i = 1,\dots,n,$$

where $r_s^\tau(x) = \mathbb{E}(Y 1_{\{\Psi(X,Y) \in E_s^\tau\}} | X = x)$ for $s \in S$. Once we have written the regression operator as $r(x) = \sum_{s \in S} r_s^\tau(x)$, a new family of estimates can be built considering separately each component, that is,

$$\hat{r}^\tau(x) := \sum_{s \in S} \hat{r}_s^\tau(x), \quad \forall \tau \in T_n, \tag{16.2}$$

where each $\hat{r}_s^\tau$ is constructed from $\{(X_i, Y_{i,s}^\tau)\}_{i=1}^n$. In particular, for each $s \in S$, we have used the following kernel-type estimator

$$\hat{r}_s^\tau(x) = \frac{\sum_{i=1}^n Y_{i,s}^\tau K(h_s^{-1}||X_i - x||)}{\sum_{i=1}^n K(h_s^{-1}||X_i - x||)},$$

where $||\cdot|| = \langle \cdot,\cdot \rangle^{1/2}$ is the induced norm of $\mathscr{H}$, K is a kernel function, and h_s is a sequence of bandwidths such that $h_s \in H_n \subset \mathbb{R}^+$.

Let us remark that, when the same bandwidth is selected for the two components (i.e., $h = h_1 = h_2$), the proposed estimator (16.2) is just the standard kernel-type estimator given by

$$\hat{r}(x) = \frac{\sum_{i=1}^{n} Y_i K(h^{-1}||X_i - x||)}{\sum_{i=1}^{n} K(h^{-1}||X_i - x||)}, \tag{16.3}$$

which was studied in the recent literature (see Ferraty and Vieu (2006) or Ferraty et al. (2007)).

Threshold function: some examples

Our method needs the user to select a threshold function in advance, and this choice should be done as far as possible in accordance with the pattern that user wants to find in the data. Some interesting threshold functions can be taken into consideration when $\mathscr{E} = \mathbb{R}$, and $E_1^\tau = (-\infty, \tau]$ and $E_2^\tau = (\tau, +\infty)$ for each $\tau \in T_n$.

When we suspect there is a threshold connected to the response, we can consider functions which only depend on Y, $\Psi(x,y) = f(y)$. According to the kind of structure we want to detect, we will select the most adequate function (for instance, $f(y) = |y|$, $f(y) = \log(y)$, $f(y) = \exp(y)$, $f(y) = \cos(y), \ldots$).

If we look for a threshold related to the covariable, then $\Psi(x,y) = g(x)$ and we can use any norm or semi-norm on $\mathscr{H}$. For example, one can consider the following family of threshold functions

$$g_d(x) = ||x^{(d)}||,$$

where $x^{(d)}$ is the d-derivative of the curve x (if $d = 0$, g is the norm of $\mathscr{H}$). On the other hand, if we select an orthonormal basis of $\mathscr{H}$, $\{e_j\}_{j=1}^{+\infty}$, and project the data onto the first J elements, we can define

$$g_J(x) = ||x_J||_J,$$

where $x_J = (\langle x, e_1 \rangle, \ldots, \langle x, e_J \rangle)^t$, and $|| \cdot ||_J$ is a norm in $\mathbb{R}^J$ (e.g., $g_J(x) = \sqrt{x_J^t M x_J}$ with M a fixed $J \times J$-matrix). Furthermore, other type of datasets can lead us to choose $g(x) = \max_t |x(t)|$, $g(x) = \int x(t)dt$, $g(x) = ||x - x_0||$ for a fixed $x_0 \in \mathscr{H}, \ldots$

Obviously, we can select more complicated Ψ such as threshold functions which depend simultaneously on X and Y, or related with the projection on several directions. However, we must bear in mind that these options probably imply an increment of the computational cost of estimating process.

16.3 Cross-validation criterion: a graphical tool

In our estimator (16.2), there are clearly three parameters which need to be estimated: the threshold τ, and the two bandwidths (h_1, h_2). From now on, we are going to simplify notation using $\omega \equiv (\tau, h_1, h_2)$, and $\Omega \equiv T_n \times H_n \times H_n$. To obtain adequate values for ω, we propose to use one of the most extended techniques in the literature: a cross-validation method. In our case, the aim is to find $\omega \in \Omega$ such that minimizes the next cross-validation criterion

$$CV(\omega) = \frac{1}{n}\sum_{j=1}^{n}(Y_j - \hat{r}^{\tau,(-j)}(X_j))^2,$$

being

$$\hat{r}^{\tau,(-j)}(x) = \sum_{s\in S}\hat{r}_s^{\tau,(-j)}(x) = \sum_{s\in S}\frac{\hat{r}_{s,N}^{\tau,(-j)}(x)}{\hat{r}_{s,D}^{\tau,(-j)}(x)} = \sum_{s\in S}\frac{\frac{1}{n}\sum_{i\neq j}Y_{i,s}^{\tau}\Delta_{i,s}(x)/\mathbb{E}(\Delta_{0,s}(x))}{\frac{1}{n}\sum_{i\neq j}\Delta_{i,s}(x)/\mathbb{E}(\Delta_{0,s}(x))}.$$

Hence, we estimate ω by $\omega_{CV} = \arg\min_{\omega\in\Omega} CV(\omega)$.

Moreover, selecting a grid of possible τ values and plotting $CV(\omega_{CV}(\tau))$, where $\omega_{CV}(\tau) = \arg\min_{h_1,h_2\in H_n} CV(\omega)$, we are going to obtain a constant graphic if there is no threshold in the data, or a convex curve with minimum in τ_0 when the theshold exists for $\tau = \tau_0$. As a result, depicting CV criterion as function of τ becomes in a graphical tool in order to analyse the existence of threshold in data.

The optimality of our cross-validation procedure with respect to the mean integrated squared error given by

$$MISE(\omega) = \mathbb{E}((r(X_0) - \hat{r}^{\tau}(X_0))^2),$$

is shown in the following theorem which ensures that ω_{CV} approximates the optimal choice in terms of $MISE$ criterion (see Ait-Saïdi et al. (2008) for a similar result in the single-functional index model context).

Theorem 16.1. *Under certain hypotheses,*

$$\frac{MISE(\omega^*)}{MISE(\omega_{CV})} \to 1 \quad a.s.$$

where $\omega^* = \arg\min_{\omega\in\Omega} MISE(\omega)$ *and* $\omega_{CV} = \arg\min_{\omega\in\Omega} CV(\omega)$.

Furthermore, the next result shows that CV and $MISE$ criteria have similar *shape* when they are taken as functions of τ. Thanks to this fact, we can deduce the behaviour of the $MISE$ criterion, which can not be obtained from a practical point of view, by means of the analysis of the CV criterion that can be derived from the data.

Theorem 16.2. *Under hypotheses of Theorem 16.1,*

$$\sup_{\tau\in T_n}\left|\frac{CV(\omega_{CV}(\tau)) - MISE(\omega^*(\tau)) - \hat{\sigma}_{\varepsilon}^2}{MISE(\omega^*(\tau))}\right| \to 0 \quad a.s.$$

where $\omega_{CV}(\tau) = \arg\min_{h_1,h_2\in H_n} CV(\omega)$, $\omega^*(\tau) = \arg\min_{h_1,h_2\in H_n} MISE(\omega)$ *for each* $\tau \in T_n$, *and* $\hat{\sigma}_{\varepsilon}^2$ *is defined as* $\hat{\sigma}_{\varepsilon}^2 = \frac{1}{n}\sum_{i=1}^{n}\varepsilon_j^2$.

16.4 Simulation study

In this study, we have consider $\mathscr{H} = L^2[0,\pi]$, and $||\cdot||_{L^2}$ the standard L^2-norm. We have drawn $ns = 200$ samples of size $n = 200$ from

$$\begin{cases} Y_i = r_1(X_i) + \varepsilon_i = \max_{t\in[0,\pi]} |X_i(t)| + \varepsilon_i, \; i = 1,\ldots,n_1, \\ Y_i = r_2(X_i) + \varepsilon_i = ||X_i||_{L^2} + \varepsilon_i, \qquad i = n_1+1,\ldots,n = n_1+n_2, \end{cases}$$

being $n_1 = n_2 = 100$, and $\varepsilon_i \sim \mathscr{N}(0,\sigma_\varepsilon)$ with $\sigma_\varepsilon = 0.01$. The covariables X_i were simulated as

$$X_i(t) = a_i\sqrt{2/\pi}\cos(2t), \quad t\in[0,\pi], \quad i = 1,\ldots,n,$$

where

$$a_i \sim \mathscr{U}(0,1) \quad \text{for} \quad i = 1,\ldots,n_1, \qquad a_i \sim \mathscr{U}(m,m+1) \quad \text{for} \quad i = n_1+1,\ldots,n,$$

with $m \in \{0,1/2,1\}$. The impossibility of dealing with continuous curves in practice leads us to discretize $\{X_i\}_{i=1}^n$ in a equidistant grid composed of $p = 100$ values in $[0,\pi]$.

We have calculated the standard nonparametric estimator $\hat{r}$ given by (16.3), and the estimator based on our procedure for the following threshold function

$$\Psi(X_i) = ||X_i||_{L^2} = \left[\int_0^\pi X_i^2(t)dt\right]^{1/2} = |a_i|,$$

and for $E_1^\tau = (-\infty,\tau]$ and $E_2^\tau = (\tau,+\infty)$. Furthermore, we have also built another estimator for the regression operator as follows. We consider the threshold value $\hat{\tau}$ estimated during the calculation of $\hat{r}^\tau$, and define $\hat{I}_s = \{i\in\{1,\ldots,n\}|\Psi(X_i)\in E_s^{\hat{\tau}}\}$ for $s\in S$. For a new observation X_{n+1}, we obtain $\hat{s}\in S$ such that $\Psi(X_{n+1})\in E_{\hat{s}}^{\hat{\tau}}$, and we predict the response value Y_{n+1} by

$$\hat{Y}_{n+1} = \hat{r}^{\hat{\tau}}(X_{n+1}) = \sum_{i\in\hat{I}_{\hat{s}}} Y_i K(\tilde{h}_{\hat{s}}^{-1}||X_i - X_{n+1}||) / \sum_{i\in\hat{I}_{\hat{s}}} K(\tilde{h}_{\hat{s}}^{-1}||X_i - X_{n+1}||).$$

Let us observe that $\Psi(X_i) = a_i \in [0,1]$ for $i = 1,\ldots,n_1$, whereas

- if $m = 0$, $\Psi(X_i) = a_i \in [0,1]$ for $i = n_1+1,\ldots,n$,
- if $m = 1/2$, $\Psi(X_i) = a_i \in [1/2,3/2]$ for $i = n_1+1,\ldots,n$, and
- if $m = 1$, $\Psi(X_i) = a_i \in [1,2]$ for $i = n_1+1,\ldots,n$.

Hence, we get that there is no threshold when $m = 0$. The case $m = 1/2$ is an intermediate situation, since the images of Ψ for each subsample are overlapped, so perhaps values in the interval $[1/2,1]$ could be detected as threshold. Finally, $\tau = 1$ is the threshold value for $m = 1$.

The cross-validation criteria for the 200 simulated samples are plotted in Figure 16.1, where each column correspond to the different values for m. As we expected, when $m = 0$ the curves are almost constant and no threshold is detected.

If $m = 1/2$, the *CV* criteria seems to detect something in $[1/2, 1]$ for some curves. Finally, for $m = 1$, the threshold is correctly estimated.

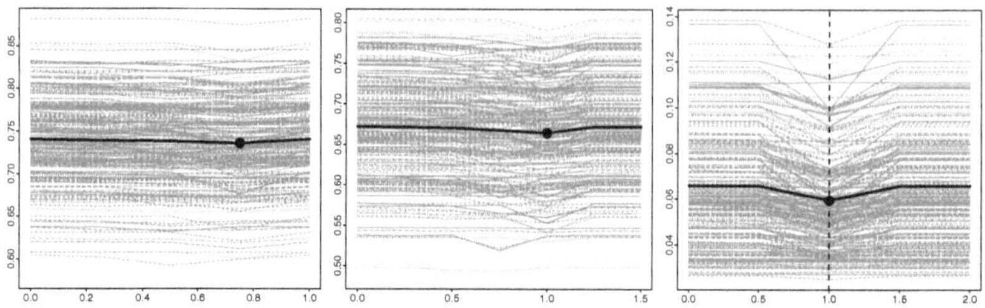

Fig. 16.1: *CV* criteria for $ns = 200$ samples (grey curves) for the three different cases. Black solid line is the mean of the cross-validation curves, and the black point its minimum. Vertical dashed line indicates the threshold value (when it exists).

To assess the performance of all the computed estimates in terms of prediction error, for each learning sample $\{(X_i, Y_i)\}_{i=1}^{n}$, we have also generated a testing sample $\{(X_i, Y_i)\}_{i=n+1}^{2n}$. We have constructed by means of the learning sample the different estimators $\tilde{r} \in \{\hat{r}, \hat{r}^{\tau}, \hat{r}^{\hat{\tau}}\}$, and we have used the testing one to calculate the mean squared error of prediction

$$MSEP = \sum_{i=n+1}^{2n} (Y_i - \tilde{r}(X_i))^2.$$

This quantity has been obtained for each replication, and we show the mean of these values in Table 16.1. We can conclude that the errors of prediction are similar for the three estimators when there is no threshold ($m = 0$), whereas $\hat{r}^{\hat{\tau}}$ produces smaller MSEP values than the standard kernel estimator when the threshold exists ($m = 1$).

m	$\hat{r}$	$\hat{r}^{\tau}$	$\hat{r}^{\hat{\tau}}$
0	0.00372	0.00378	0.00378
1/2	0.00339	0.00346	0.00338
1	0.00035	0.00034	0.00019

Table 16.1: Mean of MSEP.

Acknowledgements First and third authors wish to thanks all the participants of the working group STAPH on Functional and Operatorial Statistics in Toulouse for their continuous supports and comments. The work of the second author was supported by Ministerio de Ciencia e Innovación (grant MTM2008-03010), and by Consellería de Innovación e Industria (regional grant PGIDIT07PXIB207031PR), and Consellería de Economía e Industria (regional grant 10MDS207015PR), Xunta de Galicia.

References

1. Ait-Saïdi, A., Ferraty, F., Kassa, R., Vieu, P.: Cross-validated estimations in the single-functional index model. Statistics **42** (6), 475–494 (2008)
2. Ferraty, F., Mas, A., Vieu, P.: Nonparametric regression on functional data: inference and practical aspects. Aust. N.Z. J. Stat. **49** (3), 267–286 (2007)
3. Ferraty, F., Vieu, P.:Nonparametric functional data analysis: theory and practice. Series in Statistics, Springer, New York (2006)
4. Ramsay, J. O., Silverman, B. W.: Functional Data Analysis (Second Edition). Springer Verlag, New York (2005)

Chapter 17
Estimation of a Functional Single Index Model

Frédéric Ferraty, Juhyun Park, Philippe Vieu

Abstract Single index models have been mostly studied as an alternative dimension reduction technique for nonparametric regression with multivariate covariates. The index parameter appearing in the model summarizes the effect of the covariates in a finite dimensional vector. We consider an extension to a functional single index parameter which is of infinite dimensional, as a summary of the effect of a functional explanatory variablc on a scalar response variable and propose a new estimator based on the idea of functional derivative estimation.

17.1 Introduction

We are concerned with the functional regression model where the response variable Y is a scalar variable and the explanatory variable X is functional in the class of $L_2(\mathscr{I})$. Denote the mean response of Y given X by

$$m(X) = \mathrm{E}[Y|X],$$

and consider the regression model

$$Y = m(X) + \varepsilon,$$

where m is a smooth functional from $L_2(\mathscr{I})$ to the real line. The linear regression model assumes that

Frédéric Ferraty
Institut de Mathématiques de Toulouse, France, e-mail: ferraty@math.univ-toulouse.fr

Juhyun Park
Lancaster University, Lancaster, U.K. e-mail: juhyun.park@lancaster.ac.uk

Philippe Vieu
Institut de Mathématiques de Toulouse, France, e-mail: vieu@math.univ-toulouse.fr

$$m(X) = \beta_0 + <X, \beta> = \beta_0 + \int_{\mathscr{I}} X(t)\beta(t)\,dt$$

and the coefficient function $\beta(\cdot)$ is used as a summary of the effect of X. When the functional form m is completely unspecified, the regression problem becomes nonparametric functional regression. In this article we focus on studying a functional single index model, a semiparametric approach where the effect of the regressor X is captured through a linear predictor under an unknown link function.

In classical multivariate regression with a scalar response variable Y and a d-dimensional covariate X, the single index model assumes

$$m(X) = r_\theta(X) = r(\theta^T X),$$

where r is a unknown link function and θ is a d-dimensional vector. This is a flexible semiparametric approach generalizing the multiple linear regression and providing an effective dimension reduction technique compared to the fully nonparametric approach which is subject to the curse of dimensionality.

Extending this relationship to the case of a functional covariate X defines a functional single index model the same way as

$$m(X) = r_\theta(X) = r(<X, \theta>).$$

Here r_θ is a smooth functional from $L_2(\mathscr{I})$ to the real line, whereas r is a smooth function on the real line. Similarly to the multivariate regression, this model is a generalization of the functional linear regression with an identity or a known link function r and provides a useful alternative to the fully nonparametric functional regression approach.

We have used the term *functional* to emphasize the functional nature of the index parameter θ. There are other extensions of the single index model to the functional regression in the literature. A different version of a functional single index model appears in Jiang and Wang (2011) where the term is used when both X and Y are functional but the index parameter θ there refers to a vector valued parameter. Li et al. (2010) uses a type of single index model with functional covariates and a scalar response variable however the index parameter of their interest is also a vector valued parameter. Although both models are developed with more complex scenarios with functional variables in mind, they do not bear resemblance to the model referred here.

The main contribution of our work is to investigate the problem of estimating the functional parameter θ based on the idea of functional derivative estimation studied by Hall et al. (2010). It turns out that we can naturally extend the definition of the *average* derivative for the single index model proposed in Härdle and Stoker (1989) to the functional case, presented in Section 2. However, the underlying estimating equation that was the basis of the construction of the estimator for the multivariate case does not work for the functional case and we need the new approach tailored to a functional variable. The directional derivative estimation is reviewed in Section 3 and the new estimator for the functional single index model is proposed in Section

4. A detailed theoretical analysis as well as numerical examples will be skipped here but will be presented in the main talk. We will also discuss extensions of our approach in functional regression with several functional variables. We believe that this view sheds new lights on the usage of the functional single index model in broader applications.

17.2 Index parameter as an *average* derivative

In multiple regression problem, θ is known to be related to the average derivative m' (Härdle and Stoker, 1989), that is,

$$\theta = \mathrm{E}[m'(X)], \tag{17.1}$$

where m' is the vector of partial derivatives and the expectation is taken with respect to the marginal distribution of X. Based on this relationship and some further manipulation, they constructed a two-stage estimator where $\hat{\theta}$ is an empirical average of the derivative estimator and $\hat{r}$ is a one dimensional nonparametric smoother.

At first glance, it seems natural to apply this relationship to the functional case but care needs to be taken, as this requires a generalization of the finite dimensional relationship to an infinite dimensional one. In functional regression, the derivative of m should be understood as a derivative in functions space. Recently Hall et al. (2010) studied the problem of derivative estimation in functions space, where the directional derivative of m at x is defined to be the linear operator m_x satisfying

$$m(x+\delta u) = m(x) + \delta m_x(u) + o(\delta).$$

Applying the same idea, we can extend the relationship (17.1) for the single index model to the functional case in the following sense:

$$\begin{aligned} m_x(u) &= \lim_{\delta \to 0} \frac{1}{\delta}\{m(x+\delta u) - m(x)\} \qquad (17.2)\\ &= \lim_{\delta \to 0} \frac{1}{\delta}\{r(<x+\delta u, \theta>) - r(<x, \theta>)\} \\ &= <u, \theta r'(<x, \theta>)>. \end{aligned}$$

By Riesz representation theorem, there exists an element $m_x^* \in L_2(\mathscr{I})$ that satisfies the relation

$$m_x(u) = <u, m_x^*>,$$

where $m_x^* = \theta r'(<x, \theta>)$. Therefore, we obtain the following equality:

$$\mathrm{E}[m_X^*] = \theta \cdot \mathrm{E}[r'(<X, \theta>)] = const \cdot \theta,$$

in the same spirit as in the result (17.1) but now for the case of functional predictor. As shown in Härdle and Stoker (1989), the constant is related to parametrization and may be set to be 1 by reparametrizing the functions r and θ accordingly.

Although the idea of the average derivative is naturally linked to the derivative of the marginal density for the multivariate regression, which hence leads to construction of an estimator (Härdle and Stoker, 1989), its extension to the functional case is not straightforward. In fact the notion of marginal density is not well understood for functional variables and is very difficult to define precisely (Delaigle and Hall, 2010). Instead we rely on the development of the directional derivatives for the operator to construct a direct estimator for the functional single index model.

17.3 Estimation of the directional derivatives

Let $\{\psi_j, j = 1,2,\cdots\}$ be an orthonormal basis function in $L_2(\mathscr{I})$. For any $\beta \in L_2(\mathscr{I})$, we may write $\beta = \sum_{j=1}^{\infty} < \beta, \psi_j > \psi_j$. Then we have

$$m_x(\beta) = \sum_{j=1}^{\infty} m_x(< \beta, \psi_j > \psi_j) = \sum_{j=1}^{\infty} < \beta, \psi_j > m_x(\psi_j) =< \beta, \sum_{j=1}^{\infty} m_x(\psi_j)\psi_j >,$$

with the second last equality following from the linearity of the operator m_x. Thus, we may write

$$m_x^* = \sum_{j=1}^{\infty} m_x(\psi_j)\psi_j = \sum_{j=1}^{\infty} \gamma_{x,j}\psi_j,$$

where $\gamma_{x,j} = m_x(\psi_j)$.

Suppose that $\{(X_i, Y_i), i = 1,2,\cdots,n\}$ be an i.i.d. sample from (X,Y). In practice, the ψ_j's denote the eigenfunctions derived from the functional principal component analysis of the process X. A standard estimator $\hat{\psi}_j$ of ψ_j is obtained by achieving a spectral analysis of the empirical covariance operator of X. Hence, a consistent estimator of $\gamma_{x,j}$ for a fixed direction is proposed in Hall et al. (2010), which is defined to be

$$\hat{\gamma}_{x,j} = \frac{\sum\sum_{i_1,i_2}^{(j)} (Y_{i_1} - Y_{i_2}) K_j(i_1,i_2|x)}{\sum\sum_{i_1,i_2}^{(j)} \hat{\xi}_j(i_1,i_2) K_j(i_1,i_2|x)},$$

with

$$K_j(i_1,i_2|x) = K_1\left(h_1^{-1}\|x - X_{i1}\|\right) K_1\left(h_1^{-1}\|x - X_{i2}\|\right) K_2\left(1 - \frac{\hat{\xi}_j(i_1,i_2)^2}{\|X_{i1} - X_{i2}\|^2}\right),$$

$\hat{\xi}_j(i_1,i_2) = \int (X_{i_1} - X_{i_2})\hat{\psi}_j$ measures the difference in the projection of the pair (X_{i1}, X_{i2}) of functional trajectories onto the direction of $\hat{\psi}_j$, $K_1(.)$ and $K_2(.)$ are kernel functions, and where $\sum\sum_{i_1,i_2}^{(j)}$ denotes summation over pairs $(i1,i2)$ such that

$\hat{\xi}_j(i_1,i_2) > 0$. The mechanism of this estimator is the following. For given x and the direction $\hat{\psi}$, define the δ-neighborhood of x in the direction of $\hat{\psi}$ and select the sub-sample falling in the δ-neighborhood. The numerator is the mean difference between two responses, which can be approximated by the weighted average difference in responses (Y_{i1},Y_{i2}) of each pair of the sample $\{(X_{i_1},Y_{i_1}),(X_{i_2},Y_{i_2})\}$. The denominator can be approximated by the weighted average distance between (X_{i1},X_{i2}). The weights are determined according to the closeness to the direction of $\hat{\psi}$ as well as that to x.

17.4 Estimation for functional single index model

Viewing θ as the average of the directional derivatives, its estimator $\hat{\theta}$ can be constructed from the empirical counterpart. Specifically we consider the two-stage estimator where at the first step, the estimator $\hat{\theta}$ is obtained by

$$\hat{\theta} = n^{-1}\sum_{i=1}^{n}\hat{m}^*_{X_i},$$

where $\hat{m}^*_X = \sum_{j=1}^{k_n}\hat{\gamma}_{X,j}\hat{\psi}_j$ with $k_n \to \infty$ as $n \to \infty$. Given $\hat{\theta}$ and x in $L_2(\mathscr{I})$, define a new random variable $Z_i =< X_i,\hat{\theta} >$ and a real number $z =< x,\hat{\theta} >$. In the second step, the estimator of the link function r is obtained from a one-dimensional nonparametric kernel regression with $\{(Z_i,Y_i) : i = 1,\ldots,n\}$ as

$$\hat{r}(x) = \frac{\sum_{i=1}^{n} Y_i W_h\big(Z_i - z\big)}{\sum_{i=1}^{n} W_h\big(Z_i - z\big)},$$

where $W_h(\cdot) = W(\cdot/h)$ is a symmetric kernel weight function defined on a real line.

Properties of the estimators: Given a consistent estimator of θ, it can be easily shown that $\hat{r}$ is a consistent estimator. However, notice that the consistency of $\hat{\gamma}_{x,j}$ is not sufficient to guarantee the consistency of $\hat{\theta}$. We can prove, under some regularity conditions, that $\hat{\theta}$ is indeed a consistent estimator.

References

1. Delaigle, A., Hall, P.: Defining probability density for a distribution of random functions. Ann. Stat. **38** (2), 1171–1193 (2010)
2. Hall, P., Müller, H. G., Yao, F.: Estimation of functional derivatives. Ann. Stat. **37**, 3307-3329 (2009)
3. Härdle, W., Stoker, T. M.: Investing smooth multiple regression by the method of average derivatives. J. Am. Stat. Assoc. **84** (408), 986–995 (1989)
4. Jiang, C.-R., Wang, J.-L.: Functional single index models for longitudinal data. Ann. Stat. **39** (1), 362–388 (2011)

5. Li, Y., Wang, N., Carroll, R. J.: Generalized functional linear models with semiparametric single-index interactions. J. Am. Stat. Assoc. **105** (490), 621–633 (2010)

Chapter 18
Density Estimation for Spatial-Temporal Data

Liliana Forzani, Ricardo Fraiman, Pamela Llop

Abstract In this paper we define a nonparametric density estimator for spatial-temporal data and under mild conditions we prove its consistency and obtain strong orders of convergence.

18.1 Introduction

Spatial-temporal models arise when data are collected across time as well as space. More precisely, a spatial-temporal model is a random hypersurface which evolves at regular intervals of time (for instance, every day or every week). The data analysis therefore exhibits spatial dependence, but also observations at each point in time are typically not independent but form a time series of random surfaces. For this kind of models, over the last decade there has been very rapid growth in research mainly looking at parametric ones (see for instance the recent book by Tang et al., 2008).

The goal of this paper is to define a nonparametric estimator for the marginal density function of a random field in this setup and give its order of convergence. Our approach falls in the functional data framework which, as it is well known, is a very important topic in modern statistics and a great effort is being done to provide statistical tools for its analysis (see for instance, Ramsay and Silverman, 2002, Ramsay and Silverman, 2005, Ferraty and Vieu, 2006, González Manteiga and Vieu, 2007,

Liliana Forzani
Instituto de Matemática Aplicada del Litoral - CONICET, Argentina, e-mail: liliana.forzani@gmail.com

Ricardo Fraiman
Universidad de San Andrés, Argentina, and Universida de la República, Uruguay, e-mail: rfraiman@udesa.edu.ar

Pamela Llop
Instituto de Matemática Aplicada del Litoral - CONICET, Argentina, e-mail: lloppamela@gmail.com

and the handbook by Ferraty and Romain, 2011). In this context, the problem of estimating the marginal density function when a single sample path is observed continuously over $[0,T]$ has been studied starting in Rosenblatt (1970), Nguyen (1979), and mainly by Castellana and Leadbetter (1986), where it is shown that for continuous time processes a parametric speed of convergence is attained by kernel type estimates. More recently, it has also been considered by Blanke and Bosq (1997), Blanke (2004), Kutoyants (2004) among others. In particular, Labrador (2008) propose a k-nearest neighbor type estimate using local time ideas and, based in this estimator, Llop et al. (2011) showed that a parametric $\sqrt{n}$ speed of convergence is attained when independent samples are available.

For random fields defined in the d-dimensional integer lattice $\mathbb{Z}^d$ with values in $\mathbb{R}^N$, Tran and Yakowitz (1993) showed the asymptotic normality of the k-nearest neighbour estimator under random field stationary and mixing assumptions. For kernel estimators, Tran (1990) proved the asymptotic normality under dependence assumptions, the L_1 convergence of this type of estimators have been studied by Carbon et al. (1996) and Hallin et al. (2004). Hallin et al. (2001), without assuming dependence conditions by assuming linearity, showed the multivariate asymptotic normality of the kernel density estimator at any k-tuple of sites and also computed their limiting covariance matrix. The uniform consistency of this estimator was studied by Carbon et al. (1997).

In this paper we computed a marginal density estimator for a random field $\mathscr{X}(\mathbf{s})$ verifying the model

$$\mathscr{X}(\mathbf{s}) = \mu(\mathbf{s}) + e(\mathbf{s}), \quad \mathbf{s} \in \mathbf{S} \subset \mathbb{R}^p \tag{18.1}$$

where $\mu(\mathbf{s})$ stands for the mean function, and $e(\mathbf{s})$ is a zero mean, first-order stationary random field with density unknown function f_e. Throughout this work, we will assume that $e(\mathbf{s})$ admits a local time (see Geman and Horowitz, 1981). Using the ideas given in Llop et al. (2011) we will introduce a k–nearest neighbor type estimate based on the occupation measure and prove its consistency both, for the stationary and the nonstationary cases.

18.2 Density estimator

In this section we define and give the order of convergence for a marginal density estimator for a random field verifying (18.1) when $\{\mathscr{X}_1(\mathbf{s}),\dots,\mathscr{X}_T(\mathbf{s})\}$ random fields with the same distribution as $\mathscr{X}(\mathbf{s})$ are given. For $\mathbf{s}$ fixed, the errors $\{e_1(\mathbf{s}),\dots,e_T(\mathbf{s})\}$ have the geometrically α-mixing dependence property, i.e., there exists a non-increasing sequence of positive numbers $\{\alpha(r), r \in \mathbb{N}\}$ with $\alpha(r) \to 0$ when $r \to \infty$ such that $\alpha(r) \le a\rho^r$, with $0 < \rho < 1$, $a > 0$ and

$$|P(A \cap B) - P(A)P(B)| \le \alpha(r),$$

for $A \in \mathscr{M}_{t_1}^{t_u}$ and $B \in \mathscr{M}_{l_1}^{l_v}$ where $\mathscr{M}_{t_a}^{t_b} = \sigma\{e_t(\mathbf{s}), t_a \le t \le t_b\}$ is the σ-algebra generated by $\{e_t(\mathbf{s})\}_{t=1}^{T}$ and $1 \le t_1 \le \ldots \le t_u < t_u + r = l_1 \le \ldots \le l_v \le T$.

18.2.1 Stationary case: $\mu(s) = \mu$ constant

First let us observe that if $\mu(\mathbf{s})$ is constant the sequence $\{\mathscr{X}_1(\mathbf{s}),\ldots,\mathscr{X}_T(\mathbf{s})\}$ inherits all the properties of the sequence $\{e_1(\mathbf{s}),\ldots,e_T(\mathbf{s})\}$. This means that $\mathscr{X}(\mathbf{s})$ is a first order stationary random field which admits a local time, and the sequence $\{\mathscr{X}_1(\mathbf{s}),\ldots,\mathscr{X}_T(\mathbf{s})\}$ is a geometrically mixing sequence of random fields. It is clear however, that this is not the case if $\mu(\mathbf{s})$ is not constant, we will consider this problem in the next section.

As $\mathscr{X}(\mathbf{s})$ has the same properties as $e(\mathbf{s})$ its density estimator $\hat{f}_{\mathscr{X}}$, will be computed in the same way as $\hat{f}_e$ and the results given for $\hat{f}_e$ will hold for $\hat{f}_{\mathscr{X}}$. The estimator for the density function of $e(\mathbf{s})$ is defined as follows.

For $\{k_T\}$, $k_T/T < |\mathbf{S}|$, a real number sequence that converges to infinity, we define the random variable $h_T^e \doteq h_T^e(x)$ such that

$$k_T = \sum_{t=1}^{T} \int_{\mathbf{S}} \mathbb{I}_{I_{(x,h_T^e(x))}}(e_t(\mathbf{s}))\,\mathbf{ds}, \tag{18.2}$$

where $I_{(x,r)} = [x-r, x+r]$ and the marginal density estimator for f_e is defined as

$$\widehat{f_e}(x) \doteq \frac{k_T}{2T|\mathbf{S}|h_T^e(x)}. \tag{18.3}$$

If the process $e(\mathbf{s})$ admits a local time, then $\widehat{f_e}$ is well defined since h_T^e exists and it is unique (see for instance Llop et al., 2011).

18.2.2 Non-stationary case: $\mu(s)$ any function

Now let us observe that if $\mu(\mathbf{s})$ is non constant, the sequence $\{\mathscr{X}_1(\mathbf{s}),\ldots,\mathscr{X}_T(\mathbf{s})\}$ is not a first order stationary random field any more. That means that its density function is diferent for any point of the space. It will be denoted by $f_{\mathscr{X}_\mathbf{s}}$. We define the density estimator for $f_{\mathscr{X}_\mathbf{s}}$ as

$$\widehat{f_{\mathscr{X}_\mathbf{s}}}(x) = \widehat{f_u}(x - \bar{\mathscr{X}}_T(\mathbf{s})),$$

where

$$\widehat{f_u}(x) \doteq \frac{k_T}{2n|\mathbf{S}|h_T^u(x)}, \tag{18.4}$$

with $u = \{\mathscr{U}_{T1},\ldots,\mathscr{U}_{TT}\}$ given by

$$\mathscr{U}_{Tt}(\mathbf{s}) = \mathscr{X}_t(\mathbf{s}) - \bar{\mathscr{X}}_T(\mathbf{s}) = e_t(\mathbf{s}) - \bar{e}_T(\mathbf{s}).$$

Here $\{e_1(\mathbf{s}),\ldots,e_T(\mathbf{s})\}$ is a sequence with the same distribution as the stationary random field $e(\mathbf{s})$ and h_T^u is defined as (18.2) replacing $\{e_1(\mathbf{s}),\ldots,e_T(\mathbf{s})\}$ by u.

18.2.3 Hypothesis

In order to obtain the rate of convergence of the estimators defined on (18.3) and (18.4), we will assume the following set of assumptions.

H1. $\{e_t(\mathbf{s}), 1 \leq t \leq T, \mathbf{s} \in \mathbf{S}\}$ is a random field sequence with the same distribution than $e(\mathbf{s})$ that admits a local time.

H2. $e(\mathbf{s})$ is a stationary process with unknown density function f_e strictly positive.

H3. The density f_e is Lipschitz with constant K.

H4. For each $\mathbf{s}$ fixed, the sequence $\{e_t(\mathbf{s}), 1 \leq t \leq T\}$ of random variables is geometrically α-mixing.

H5. $\{k_T\}$ and $\{v_T\}$ are sequences of positive numbers such that $v_T\left(\frac{k_T}{T}\right) = o(1)$ and $\sum_{T=1}^{\infty} \exp\left\{-a\frac{1}{T^{1/4}}\left(\frac{k_T}{v_T}\right)^{1/2}\right\} < \infty$ for each $a > 0$.

18.2.4 Asymptotic results

Theorem 18.1. ***Rates of convergence:*** *Suppose H1-H4 holds. Then,*

I. Stationary case: Suppose furthermore that H5 holds. Then, for each $x \in \mathbb{R}$,

$$\lim_{T \to \infty} v_T(\widehat{f_e}(x) - f_e(x)) = 0 \quad a.co.$$

II. Non-stationary case: Let us choose two sequences of real positive numbers $\{k_T\}$ *and* $\{v_T\}$ *both of them going to infinity such that* $v_T(T/k_T)|\bar{e}_T(\mathbf{s})| \to 0$ *a.co. For those sequences* $\{k_T\}$ *and* $\{v_T\}$ *let us assume H5. Then, for each* $x \in \mathbb{R}$,

$$\lim_{n \to \infty} v_T(\widehat{f_{\mathscr{X}_s}}(x) - f_{\mathscr{X}_s}(x)) = 0 \qquad a.co.$$

Remarks:

(a) In the stationary case, we can choose k_T such that $v_T = T^{\gamma}$ for any $\gamma < \frac{1}{4}$. More precisely, let $k_T = T^{\beta}$ and $v_T = T^{\gamma}$. For conditions $(k_T/T)v_T = o(1)$ and $\frac{1}{T^{1/4}}\left(\frac{k_T}{v_T}\right)^{1/2} \to \infty$ holds it is enough that $\beta < 1 - \gamma$ and $\beta > \gamma + \frac{1}{2}$. Then, given $\gamma < \frac{1}{4}$, we can choose β such that these conditions are true.

(b) In the non-stationary case, we can choose k_T such that $v_T = T^{\gamma}$ for any $\gamma < \frac{1}{4}$. More precisely, if $k_T = T^{\beta}$ and $v_T = T^{\gamma}$, for conditions $(k_T/T)v_T = o(1)$ and $\frac{1}{T^{1/4}}\left(\frac{k_T}{v_T}\right)^{1/2} \to \infty$ hols it is enough that $\beta < 1 - \gamma$ and $\beta > \gamma + \frac{1}{2}$. In addition, for $v_T(T/k_T)|\bar{e}_T(\mathbf{s})| \to 0$ a.co. holds, if there exists $M > 0$ such that $|e(\mathbf{s})| < M$ with probability one, since $\bar{e}_T(\mathbf{s}) = o(T^{-\alpha})$ with $\alpha < 1/2$ using Bernstein's inequality we need that $\beta > \gamma + \frac{1}{2}$. Therefore, given $\gamma < \frac{1}{4}$ we can choose β such that these conditions are true.

The proof of part I is a consequence of the Bernstein inequality for α-mixing process. Since for each $\mathbf{s}$ fixed, the random variables $\{\mathscr{U}_{T1}(\mathbf{s}),\ldots,\mathscr{U}_{TT}(\mathbf{s})\}$ are identically distributed but not necessarily α-mixing dependent, the proof of part II will be not a direct consequence of part I however it will be a consequence of part I and a result that prove the Lispschitz continuity of h_T^e and h_T^u.

References

1. Blanke, D.: Adaptive sampling schemes for density estimation. J. Stat. Plan. Infer. **136** (9), 2898–2917 (2004)
2. Blanke, D., Bosq, D.: Accurate rates of density estimators for continuous-time processes. Statist. Probab. Lett. **33** (2), 185–191 (1997)
3. Carbon, M., Hallin, M., Tran, L.: Kernel density estimation for random fields: the L_1 theory. J. Nonparametr. Stat. **6** (2-3), 157–170 (1996)
4. Carbon, M., Hallin, M., Tran, L.: Kernel density estimation for random fields (density estimation for random fields). Statist. Probab. Lett. **36** (2), 115–125 (1997)
5. Castellana, J. V., Leadbetter, M. R.: On smoothed probability density estimation for stationary processes. Stoch. Process. Appl. **21** (2), 179–193 (1986)
6. Ferraty, F., Romain, Y.: The Oxford Handbook of Functional Data Analysis. Oxford University Press (2011)
7. Ferraty, F., Vieu, P.: Nonparametric Functional Data Analysis: Theory and Practice. Springer, New York (2006)
8. Geman, D., Horowitz, J.: Smooth perturbations of a function with a smooth local time. T. Am. Math. Soc. **267** (2), 517–530 (1981)
9. González Manteiga, W., Vieu, P.: Statistics for functional data. Comput. Stat. Data Anal. **51**, 4788–4792 (2007)
10. Hallin, M., Lu, Z., Tran, L.: Density estimation for spatial linear processes. Bernoulli **7** (4), 657–668 (2001)
11. Hallin, M., Lu, Z., Tran, L.: Kernel density estimation for spatial processes: the L_1 theory. Ann. Stat. **88**, 61–75 (2004)
12. Kutoyants, Y.: On invariant density estimation for ergodic diffusion processes. SORT. **28** (2), 111–124 (2004)
13. Labrador, B.: Strong pointwise consistency of the k_T-occupation time density estimator. Statist. Probab. Lett. **78** (9), 1128–1137 (2008)
14. Llop, P., Forzani, L., Fraiman, R.: On local times, density estimation and supervised classification from functional data. J. Multivariate Anal. **102** (1), 73–86 (2011)
15. Nguyen, H.: Density estimation in a continuous-time stationary markov process. Ann. Stat. **7** (2), 341–348 (1979)
16. Ramsay, J., Silverman, B.: Applied Functional Data Analysis. Method and case studies. Series in Statistics, Springer, New York (2002)
17. Ramsay, J., Silverman, B. (2005). Functional Data Analysis (Second Edition). Series in Statistics, Springer, New York (2005)
18. Rosenblatt, M.: Density estimates and markov sequences. Nonparametric Techniques in Statistical Inference. Cambridge Univ. Press. Mathematical Reviews, 199–210 (1970)
19. Tang, X., Liu, Y., Zhang, J., Kainz, W.: Advances in Spatio-Temporal Analysis. ISPRS Book Series, Vol. 5 (2008)
20. ran, L., Yakowitz, S.: Nearest neighbor estimators for random fields. J. Multivariate Anal. **44** (1), 23–46 (1993)
21. ran, L. T.: Kernel density estimation on random fields. J. Multivariate Anal. **34** (1), 37–53 (1990)

Chapter 19
Functional Quantiles

Ricardo Fraiman, Beatriz Pateiro-López

Abstract A new projection-based definition of quantiles in a multivariate setting is proposed. This approach extends in a natural way to infinite-dimensional Hilbert and Banach spaces. Sample quantiles estimating the corresponding population quantiles are defined and consistency results are obtained. Principal quantile directions are defined and asymptotic properties of the empirical version of principal quantile directions are obtained.

19.1 Introduction

The fundamental one-dimension concept of quantile function of a probability distribution is a well known device going back to the foundations of probability theory. The quantile function is essentially defined as the inverse of a cumulative distribution function. More precisely, given a real valued random variable X with distribution P_X, the α-quantile ($0 < \alpha < 1$) is defined as

$$Q_X(\alpha) =: Q(P_X, \alpha) = \inf\{x \in \mathbb{R} : F(x) \geq \alpha\}, \tag{19.1}$$

where F denotes the cumulative distribution function of X.

In spite of the fact that the generalization of the concept of quantile function to a multivariate setting is not straightforward, since the lack of a natural order in the d-dimensional space makes the definition of multivariate quantiles difficult, a huge literature has been devoted to this problem in the last years. Different methodological approaches have been proposed, from those based on the concept of data depth, to

Ricardo Fraiman
Universidad de San Andrés, Argentina, and Universida de la República, Uruguay, e-mail: rfraiman@udesa.edu.ar

Beatriz Pateiro-López
Universidad de Santiago de Compostela, Spain, e-mail: beatriz.pateiro@usc.es

those based on the geometric configuration of multivariate data clouds, see Chaudhuri (1996). We refer to the survey by Serfling (2002) for a complete overview and a exhaustive comparison of the different methodologies. Our proposal in this work is based on a directional definition of quantiles, indexed by an order $\alpha \in (0,1)$ and a direction u in the unit sphere. An important contribution in this sense has been made recently by Kong and Mizera (2008). For a given α, they define directional quantiles by projecting the probability distribution onto the straight line defined by each vector on the unit sphere, extending in a very simple way the one-dimensional concept of quantile. A shortcoming of their approach is the lack of any reasonable form of equivariance of the resulting quantile contours, even with respect to translation, since their definition of quantiles heavily depends on the choice of an origin. Anyway, as we shall see, it is easy to modify the simple and more intuitive definition of directional quantiles given by Kong and Mizera (2008) in order to attain the main equivariance properties that are adequate for a quantile function.

On the other hand, beyond the lack of a widely accepted definition of multivariate quantiles there is also an increasing need for quantile functions valid for infinite-dimensional data (a problem recently posed by Jim Ramsay) in connection with the increasing demand of statistical tools for functional data analysis (FDA) where the available data are functions $x = x(t)$ defined on some real interval (say $[0,1]$). See e.g., Ferraty and Romain (2011), Ferraty (2010), Ramsay and Silverman (2005) or Ferraty and Vieu (2006) for general accounts on FDA.

Therefore, the goal of this work is to provide an intuitive definition of directional quantiles that allows us to describe the behaviour of a probability distribution in finite and infinite-dimensional spaces.

19.2 Quantiles in Hilbert spaces.

In the remainder of this paper, $\mathscr{X}$ will denote a functional random variable valued in some infinite-dimensional space $\mathscr{E}$. We do not bother to distinguish in our notation between functions, scalar quantities and non-random elements of $\mathscr{E}$ and we use standard letters for all cases. Since we will still need to introduce multivariate variables in some definitions and examples, we adopt the convention of writing vectors as boldface lower case letters and matrices in boldface upper case.

Let $\mathscr{H}$ be a separable Hilbert space where $\langle \cdot, \cdot \rangle$ denotes the inner product and $\|\cdot\|$ denotes the induced norm in $\mathscr{H}$. Let $\mathscr{X}$ be a random element in $\mathscr{H}$ with distribution $P_{\mathscr{X}}$ and such that $\mathbb{E}(\|\mathscr{X}\|) < \infty$. Our extension of the concept of quantiles to multidimensional and infinite-dimensional spaces is based on a directional definition of quantiles. Thus, we denote $\mathbb{B} = \{u \in \mathscr{H} : \|u\| = 1\}$ the unit sphere in $\mathscr{H}$ and define, for $0 < \alpha < 1$, the α-quantile in the direction of $u \in \mathbb{B}$, $Q_{\mathscr{X}}(\alpha, u) \in \mathscr{H}$, as

$$Q_{\mathscr{X}}(\alpha, u) = Q_{\langle \mathscr{X} - \mathbb{E}(\mathscr{X}), u \rangle}(\alpha) u + \mathbb{E}(\mathscr{X}). \tag{19.2}$$

In some sense, this definition reminds us the quantiles definition (in a finite-dimensional setting) given by Kong and Mizera (2008). They define directional quantiles as the quantiles of the projections of the probability distribution into the directions of the unit sphere. However note that, in (19.2), the α-quantile in the direction of $u \in \mathbb{B}$ is defined from the α-quantile of the corresponding projection of $\mathscr{Z} = \mathscr{X} - \mathbb{E}(\mathscr{X})$. Centering the random element before projecting is essential in order to obtain quantile functions fulfilling desirable equivariance properties. Now, let $P_{\mathscr{Z}}(u)$ denote the probability distribution of the random variable $\langle \mathscr{Z}, u \rangle$. Following the notation introduced in (19.1) for the univariate case, the α-quantile in (19.2) can also be written as

$$Q_{\mathscr{X}}(\alpha, u) = Q(P_{\mathscr{Z}}(u), \alpha)u + \mathbb{E}(\mathscr{X}). \tag{19.3}$$

For convenience, we will use both the notations (19.2) and (19.3) throughout this paper. For fixed α, the quantile function $Q_{\mathscr{X}}(\alpha, \cdot)$ indexed by u in the unit sphere naturally yields quantile contours $\{Q_{\mathscr{X}}(\alpha, u),\ u \in \mathbb{B}\}$.

Equivariance properties. The quantiles defined by (19.2) fulfill the following equivariance properties: *location equivariance, equivariance under unitary operators* and *equivariance under homogeneous scale transformations.*

Quantile contours in the multivariate setting. The preceding definition of quantiles in a separable Hilbert space applies directly to the Euclidean space $\mathbb{R}^d$. As in the general case, these directional quantiles yield quantile contours $\{Q_{\mathbf{X}}(\alpha, \boldsymbol{u}),\ \boldsymbol{u} \in \mathbb{B}\}$. Figure 19.1 illustrates our definition of quantile contours in the finite dimensional case.

19.2.1 Sample quantiles

In order to define the sample version of the quantiles, let us first consider the univariate case. Given the observations $X_1, \ldots, X_n$, denote by P_n the empirical measure, that is, the random measure that puts equal mass at each of the n observations. For $0 < \alpha < 1$, the sample α-quantile, $Q(P_n, \alpha)$, is defined as

$$Q(P_n, \alpha) = \inf\{x \in \mathbb{R} : F_n(x) \geq \alpha\}, \tag{19.4}$$

where F_n denotes the sample cumulative distribution function, $F_n(x) = \frac{1}{n}\sum_{i=1}^{n} \mathbb{I}_{\{X_i \leq x\}}$. Clearly, if $X_1, X_2, \cdots, X_n$, are independent and identically distributed observations from a random variable X with distribution P_X, then $Q(P_n, \alpha)$ will act as an estimate of $Q_X(\alpha)$ based on those observations.

For the general setting, let $\mathscr{X}$ be a random element in $\mathscr{H}$ with probability distribution $P_{\mathscr{X}}$ such that $\mathbb{E}(\|\mathscr{X}\|) < \infty$. Then, let $\mathscr{Z} = \mathscr{X} - \mathbb{E}(\mathscr{X})$ with distribution $P_{\mathscr{Z}}$. Given $\mathscr{X}_1, \ldots, \mathscr{X}_n$ a random sample of elements identically distributed as $\mathscr{X}$, denote $\mathscr{Z}_{ni} = \mathscr{X}_i - \bar{\mathscr{X}},\ i = 1, \ldots, n$. Now, for $u \in \mathbb{B}$, let $P_n(u)$ denote the empirical

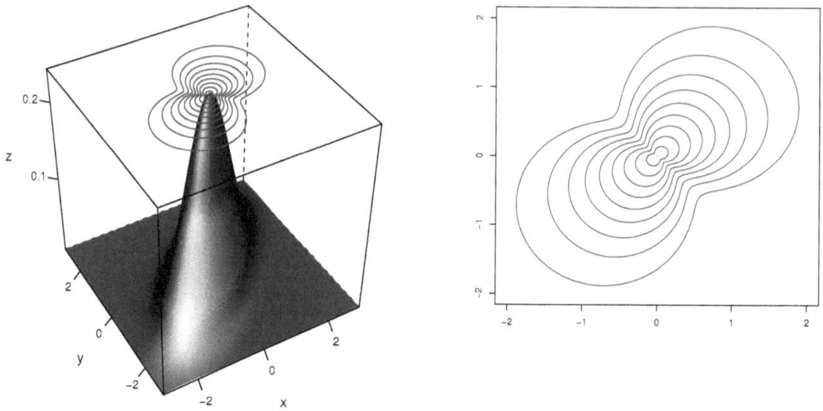

Fig. 19.1: (a) A three-dimensional view of a Normal distribution in $\mathbb{R}^2$ with zero mean and covariance matrix $\boldsymbol{\Sigma} = (\sigma_{ij})$, $\sigma_{ii} = 1$, $\sigma_{ij} = 0.75$, $i \neq j$, with two-dimensional quantile contours for $\alpha = 0.5, 0.55, \ldots, 0.95$ projected onto the top. (b) Two-dimensional view of the quantile contours.

measure of the observations $\langle \mathscr{X}_{n1}, u \rangle, \ldots, \langle \mathscr{X}_{nn}, u \rangle$. We define the empirical version of the quantiles in (19.2) by replacing the univariate α-quantile, $Q_{\langle \mathscr{X} - \mathbb{E}(\mathscr{X}), u \rangle}(\alpha)$, with the sample α-quantile $Q(P_n(u), \alpha)$ as given in (19.4). That is, we define

$$\hat{Q}_{\mathscr{X}}(\alpha, u) = Q(P_n(u), \alpha)u + \bar{\mathscr{X}} \tag{19.5}$$

where now $Q(P_n(u), \alpha) = \inf\{x \in \mathbb{R} : F_n^u(x) \geq \alpha\}$ and $F_n^u(x) = \frac{1}{n} \sum_{i=1}^n \mathbb{I}_{\{\langle \mathscr{X}_{ni}, u \rangle \leq x\}}$.

19.2.2 Asymptotic behaviour

Before we tackle the asymptotic behaviour of the sample quantiles $\hat{Q}_{\mathscr{X}}(\alpha, u)$ in (19.5), we will need some auxiliary results on the convergence of the empirical measure $P_n(u)$. Classical results on the consistency of the univariate sample quantiles are obtained as a consequence of the consistency of the empirical distribution function. However, the consistency of the empirical distribution function relies on the assumption of independent and identically distributed random variables, which is not the case in our setting. Note that in the definition of $Q(P_n(u), \alpha)$, the empirical distribution function is computed from the observations $\langle \mathscr{X}_{n1}, u \rangle, \ldots, \langle \mathscr{X}_{nn}, u \rangle$, which are clearly not independent. For each $h \in \mathscr{H}$ denote by $F^h(t)$ the probability distribution function of the random variable $\langle \mathscr{X}, h\rangle$. We obtain the following sharp result.

Proposition 1 *Let $\mathscr{H}$ be a separable Hilbert space. Then,*

$$\lim_{n\to\infty} \sup_{\|h\|=1, t\in\mathbb{R}} |F_n^h(t) - F^h(t)| = 0 \ a.s.$$

if and only if

$$\lim_{\varepsilon\to 0} \sup_{\|h\|=1, t\in\mathbb{R}} P(\{x \in \mathscr{H} : |\langle h, x\rangle - t| < \varepsilon\}) = 0. \tag{19.6}$$

It can be proved that, for the Euclidean space $\mathbb{R}^d$, Condition (19.6) is straightforwardly satisfied.

Based on the previous results we show the pointwise consistency of $Q(P_n(u), \alpha)$ to $Q(P_{\mathscr{X}}(u), \alpha)$ (for each fixed direction u), the uniform convergence of $Q(P_n(u), \alpha)$ and the uniform convergence of the sample quantiles to the population version under mild conditions.

19.3 Principal quantile directions

One of the goals of the multivariate data analysis is the reduction of dimensionality. The use of principal components is often suggested for such dimensionality reduction. More recently, the PCA methods were extended to functional data and used for many different statistical purposes, see Ramsay and Silverman (2005).

A way to summarize the information in the quantile functions is to consider principal quantile directions for a given level α, defined as follows. The first principal quantile direction is the one that maximizes the norm of the centered quantile function $Q_{\mathscr{X}}(\alpha, u) - \mathbb{E}(\mathscr{X})$, i.e. the direction $u_1 \in \mathbb{B}$ satisfying

$$u_1 = \arg\max_{u\in\mathbb{B}} \left| Q_{\langle \mathscr{X} - \mathbb{E}(\mathscr{X}), u\rangle}(\alpha) \right|. \tag{19.7}$$

The k–principal quantile direction is defined as the direction $u_k \in \mathbb{B}$ satisfying

$$u_k = \arg\max_{u\in\mathbb{B}, u\perp H_{k-1}} \left| Q_{\langle \mathscr{X} - \mathbb{E}(\mathscr{X}), u\rangle}(\alpha) \right|, \tag{19.8}$$

where H_{k-1} is the linear subspace generated by $u_1, \dots, u_{k-1}$.

Proposition 2 *Let $\boldsymbol{X}$ be a random vector with finite expectation and elliptically symmetric distribution. Then, the principal quantile directions defined by (19.7) and (19.8) coincide with the principal components.*

Proposition 3 *Let $\mathscr{X} = \{X(t),\ t \in [0,1]\}$ be a Gaussian process in $L^2[0,1]$ with covariance function*

$$\gamma(s,t) = \mathrm{Cov}(X(t), X(s)),$$

which we assume to be square integrable. Then, the principal quantile directions defined by (19.7) and (19.8) coincide with the principal components. Moreover,

$$\max_{u\in\mathbb{B},u\perp H_{k-1}} \left|Q_{\langle \mathscr{X}-\mathbb{E}(\mathscr{X}),u\rangle}(\alpha)\right| = \Phi^{-1}(\alpha)\sqrt{\lambda_k},$$

where Φ stands for the cumulative distribution function of a standard Normal random variable and $\lambda_1 \geq \lambda_2, \ldots$ is the sequence of eigenvalues of the covariance operator.

19.3.1 Sample principal quantile directions

The first sample principal quantile direction is defined as the one that maximizes the norm of the centered empirical quantile function $Q(P_n(u),\alpha)$, i.e. the direction $\hat{u}_1 \in \mathbb{B}$ satisfying

$$\hat{u}_1 = \arg\max_{u\in\mathbb{B}}|Q(P_n(u),\alpha)|.$$

The sample k–principal quantile direction is defined as the direction $\hat{u}_k \in \mathbb{B}$ satisfying

$$\hat{u}_k = \arg\max_{u\in\mathbb{B},u\perp H_{k-1}}|Q(P_n(u),\alpha)|,$$

where H_{k-1} is the linear subspace generated by $\hat{u}_1,\ldots,\hat{u}_{k-1}$.

19.3.2 Consistency of principal quantile directions

Let us denote $\mathbb{F}_1 = \{u \in \mathbb{B} : u = \arg\max_{u\in\mathbb{B}}|Q(P_{\mathscr{X}}(u),\alpha)|\}$, $\mathbb{F}_{1n} = \{u \in \mathbb{B} : u = \arg\max_{u\in\mathbb{B}}|Q(P_n(u),\alpha)|\}$, and consider the following additional assumption.

Assumption C1. Given $\varepsilon > 0$ and $u_1 \in \mathbb{F}_1$, there exists $\delta > 0$ such that $|Q(P_{\mathscr{X}}(u),\alpha)| < |Q(P_{\mathscr{X}}(u_1),\alpha)| - \delta \;\; \forall u \notin B(\mathbb{F}_1,\varepsilon)$, where $B(\mathbb{F}_1,\varepsilon) = \cup_{u\in\mathbb{F}_1} B(u,\varepsilon)$, being $B(u,\varepsilon)$ the ball with centre u and radius ε. In the finite-dimensional case, Assumption C1 will hold if for instance $Q(P_{\mathscr{X}}(u),\alpha)$ is a continuous function of u.

Proposition 4 *Under the additional Assumption C1 we have that*
i) Given $\varepsilon > 0$, $u_n \in \mathbb{F}_{1n}$ implies that $u_n \in B(\mathbb{F}_1,\varepsilon)$ if $n \geq n_0$ a.s.
ii) If the principal population quantile directions are unique then,

$$\lim_{n\to\infty} \|\hat{u}_k - u_k\| = 0 \;\; a.s. \;\; \forall k \geq 1.$$

References

1. Chaudhuri, P.: On a geometric notion of quantiles for multivariate data. J. Am. Stat. Assoc. **91**, 862–872 (1996)
2. Ferraty, F.: Statistical Methods and Problems in Infinite-dimensional Spaces. Special Issue of J. Multivariate Anal. **101**, 305–490 (2010)
3. Ferraty, F., Romain, Y.: Oxford Handbook of Functional Data Analysis. Oxford University Press (2011)
4. Ferraty, F., Vieu, P.: Nonparametric Functional Data Analysis: Theory and Practice. Springer, New York (2006)
5. Fraiman, R., Pateiro–López, B.: Quantiles for functional data. *Submitted* (2010)
6. Kong, L., Mizera, I.: Quantile tomography: Using quantiles with multivariate data. arXiv:0805.0056v1 (2008)
7. Ramsay, J. O., Silverman, B. W.: Functional Data Analysis. Springer Verlag (2005)
8. Serfling, R.: Quantile functions for multivariate analysis: approaches and applications. Statist. Neerlandica **56**, 214–232 (2002)

Chapter 20
Extremality for Functional Data

Alba M. Franco-Pereira, Rosa E. Lillo, Juan Romo

Abstract The statistical analysis of functional data is a growing need in many research areas. In particular, a robust methodology is important to study curves, which are the output of experiments in applied statistics. In this paper we introduce some new definitions which reflect the "extremality" of a curve. Given a collection of functions, these definitions establish the "extremality" of an observation and provide a natural ordering for sample curves.

20.1 Introduction

The analysis of functional data is receiving a steadily increasing attention in recent years (see, e.g., Ramsay and Silverman (2005)). In particular, a robust methodology is important to study curves, which are the output of experiments in applied statistics. A natural tool to analyze functional data aspects is the idea of statistical depth. It has been introduced to measure the 'centrality' or the 'outlyingness' of an observation with respect to a given dataset or a population distribution.

The notion of depth was first considered for multivariate data to generalize order statistics, ranks, and medians to higher dimensions. Several depth definitions for multivariate data have been proposed and analyzed by Mahalanobis (1936), Tukey (1975), Oja (1983), Liu (1990), Singh (1991), Fraiman and Meloche (1999), Vardi and Zhang (2000), Koshevoy and Mosler (1997), and Zuo (2003), among others. Direct generalization of current multivariate depths to functional data often leads

Alba M. Franco-Pereira
Universidad de Vigo, Spain, e-mail: alba.franco@uvigo.es

Rosa E. Lillo
Universidad Carlos III de Madrid, Spain, e-mail: rosaelvira.lillo@uc3m.es

Juan Romo
Universidad Carlos III de Madrid, Spain e-mail: juan.romo@uc3m.es

to either depths that are computationally intractable or depths that do not take into account some natural properties of the functions, such as shape. For that reason, several specific definitions of depth for functional data have been introduced; see, for example, Vardi and Zhang (2000), Fraiman and Muniz (2001), Cuevas, Febrero and Fraiman (2007), Cuesta-Albertos and Nieto-Reyes (2008), Cuevas and Fraiman (2009) and López-Pintado and Romo (2009, 2011). The definition of depth for curves provides criteria for ordering the sample curves from the center-outward (from the deepest to the most extreme). Laniado, Lillo and Romo (2010) introduced a new concept of "extremality" to measure the "farness" of a multivariate point with respect to a data cloud or to a distribution. In this paper, we extend this idea to define 'extremality' of a curve within a set of functions.

The half-graph depth for functional data introduced by López-Pintado and Romo (2011) is based on the notion of 'half graph' of a curve. The half-graph depth gives a natural criterion to measure the centrality of a function within a sample of curves. Here we introduce two definitions which are based on a similar idea to measure the extremality of a curve.

20.2 Two measures of extremality for functional data

We recall the definitions of hypograph and hypergraph given in López-Pintado and Romo (2011). Let $C(I)$ be the space of continuous functions defined on a compact interval I. Consider a stochastic process X with sample paths in $C(I)$ and distribution P. Let $x_1(t),\dots,x_n(t)$ be a sample of curves from P. The graph of a function x is the subset of the plane $G(x)=\{(t,x(t)):t\in I\}$. The hypograph (hg) and the hypergraph (Hg) of a function x in $C(I)$ are given by

$$\begin{aligned} hg(x) &= \{(t,y)\in I\times\mathbb{R}: y\leq x(t)\},\\ Hg(x) &= \{(t,y)\in I\times\mathbb{R}: y\geq x(t)\}. \end{aligned}$$

Next, we introduce the two following concepts that measure the extremality of a curve within a set of curves.

Definition 20.1. The hyperextremality of x with respect to a set of functions $x_1(t),\dots,x_n(t)$ is

$$HEM_n(x) = 1-\frac{\sum_{i=1}^n I_{\{G(x_i)\subset hg(x)\}}}{n} = 1-\frac{\sum_{i=1}^n I_{\{x_i(t)\leq x(t),t\in I\}}}{n}. \tag{20.1}$$

Hence, the hyperextremality of x is one minus the proportion of functions in the sample whose graph is in the hypograph of x; that is, one minus the proportion of curves in the sample below x. The population version of $HEM_n(x)$ is

$$HEM(x) = 1-P(G(X)\subset hg(x)) = 1-P(X(t)\leq x(t),t\in I). \tag{20.2}$$

Definition 20.2. The hypoextremality of x with respect to a set of functions $x_1(t),\dots,x_n(t)$ is

$$hEM_n(x) = 1 - \frac{\sum_{i=1}^{n} I_{\{G(x_i) \subset Hg(x)\}}}{n} = 1 - \frac{\sum_{i=1}^{n} I_{\{x_i(t) \geq x(t), t \in I\}}}{n}. \tag{20.3}$$

Hence, the hypoextremality of x is one minus the proportion of functions in the sample whose graph is in the hypergraph of x; that is, one minus the proportion of curves in the sample above x.

The population version of $hEM_n(x)$ is

$$hEM(x) = 1 - P(G(X) \subset Hg(x)) = 1 - P(X(t) \geq x(t), t \in I). \tag{20.4}$$

It is straightforward to check that, given a curve x, the larger the hyperextremality or the hypoextremality of x is, the more extreme is the curve x. Therefore, both concepts measure the extremality of the curves, but from a different perspective.

20.3 Finite-dimensional versions

The concepts of hypograph and hypergraph introduced in the previous section can be adapted to finite-dimensional data. Consider each point in $\mathbb{R}^d$ as a real function defined on the set of indexes $\{1,...,d\}$, the hypograph and hypergraph of a point $x = (x(1), x(2), ..., x(d))$ can be expressed, respectively, as

$$\begin{aligned} hg(x) &= \{(k,y) \in \{1,...,d\} \times \mathbb{R} : y \leq x(k)\}, \\ Hg(x) &= \{(k,y) \in \{1,...,d\} \times \mathbb{R} : y \geq x(k)\}. \end{aligned}$$

Let X be a d-dimensional random vector with distribution function F_X. Let $X \leq x$ and $X \geq x$ be the abbreviations for $\{X(k) \leq x(k), k = 1,...,d\}$ and $\{X(k) \geq x(k), k = 1,...,d\}$, respectively. If we particularize our extreme measures to the finite-dimensional case, we obtain

$$HEM(x,F) = 1 - P(X \leq x) = 1 - F_X(x),$$

and

$$hEM(x) = 1 - P(X \geq x) = F_X(x);$$

that is, the hyperextremality (hypoextremality) of a d-dimensional point x indicates the probability that a point is componentwise greater (smaller) that x. Let $x_1, \ldots, x_n$ be a random sample from X, the sample version of our extreme measures are

$$HEM_n(x) = 1 - \frac{\sum_{i=1}^{n} I_{\{x_i \leq x\}}}{n},$$

and

$$hEM_n(x) = 1 - \frac{\sum_{i=1}^{n} I_{\{x_i \geq x\}}}{n}. \tag{20.5}$$

Let C_x^u be a convex cone with vertex x obtained by moving the nonnegative orthant and translating the origin to x. Then, the finite dimensional version of the hyperextremality can be also seen as the probability that the vector x belongs to C_x^u where $u = (1,1)$ and the hypoextremality can be also seen as the probability that the vector x belongs to C_x^u where $u = (-1,-1)$. Therefore, the hyperextremality and the hypoextremality coincide with the extreme measure for multivariate data introduced by Laniado, Lillo and Romo (2010), which is computationally feasible and useful for studying high dimensional observations.

References

1. Cuesta-Albertos, J., Nieto-Reyes, A.: The random Tukey depth. Comput. Stat. Data Anal. **52**, 4979–4988 (2008)
2. Cuevas, A., Febrero, M., Fraiman, R.: Robust estimation and classification for functional data via projection-based depth notions. Computation. Stat. **22**, 481–496 (2007)
3. Cuevas, A., Fraiman, R.: On depth measures and dual statistics. A methodology for dealing with general data. J. Multivariate Anal. **100**, 753–766 (2009)
4. Fraiman, R., Meloche, J.: Multivariate L-estimation. TEST **8**, 255–317 (1999)
5. Fraiman, R., Muniz, G.: Trimmed means for functional data. TEST **10**, 419–440 (2001)
6. Koshevoy, G., Mosler, K.: Zonoid trimming for multivariate distributions. Ann. Stat. **25**, 1998–2017 (1997)
7. Laniado, H., Lillo, R. E., Romo, J.: Multivariate extremality measure. Working paper **10-19**, Statistics and Econometrics Series 08, Universidad Carlos III de Madrid (2010)
8. Liu, R.: On a notion of data depth based on random simplices. Ann. Stat. **18**, 405–414 (1990)
9. López-Pintado, S., Romo, J.: A half-graph depth for functional data. Comput. Stat. Data Anal., to appear (2011)
10. López-Pintado, S., Romo, J.: On the concept of depth for functional data. J. Am. Stat. Assoc. **104**, 718–734 (2009)
11. Mahalanobis, P. C.: On the generalized distance in statistics. *Proceedings of National Academy of Science of India* **12**, 49–55 (1936)
12. Oja, H.: Descriptive statistics for multivariate distributions. Stat. Probab. Lett. **1**, 327–332 (1983)
13. Ramsay, J. O., Silverman, B. W.: Functional data analysis (Second Edition). Springer Verlag (2005)
14. Singh, K.: A notion of majority depth. Unpublished document (1991)
15. Tukey, J.: Mathematics and picturing data. Proceedings of the 1975 International Congress of Mathematics **2**, 523–531 (1975)
16. Vardi, Y., Zhang, C. H.: The multivariate L_1-median and associated data depth. Proceedings of the National Academy of Science USA **97**, 1423–1426 (2000)
17. Zuo, Y.: Projection based depth functions and associated medians. Ann. Stat. **31**, 1460–1490 (2003)

Chapter 21
Functional Kernel Estimators of Conditional Extreme Quantiles

Laurent Gardes, Stéphane Girard

Abstract We address the estimation of "extreme" conditional quantiles *i.e.* when their order converges to one as the sample size increases. Conditions on the rate of convergence of their order to one are provided to obtain asymptotically Gaussian distributed kernel estimators. A Weissman-type estimator and kernel estimators of the conditional tail-index are derived, permitting to estimate extreme conditional quantiles of arbitrary order.

21.1 Introduction

Let (X_i, Y_i), $i = 1, \dots, n$ be independent copies of a random pair (X, Y) in $E \times \mathbb{R}$ where E is a metric space associated to a distance d. We address the problem of estimating $q(\alpha_n|x) \in \mathbb{R}$ verifying $\mathbb{P}(Y > q(\alpha_n|x)|X = x) = \alpha_n$ where $\alpha_n \to 0$ as $n \to \infty$ and $x \in E$. In such a case, $q(\alpha_n|x)$ is referred to as an extreme conditional quantile in contrast to classical conditional quantiles (known as regression quantiles) for which $\alpha_n = \alpha$ is fixed in $(0,1)$. While the nonparametric estimation of ordinary regression quantiles has been extensively studied, see for instance the seminal papers (Roussas, 1969), (Stone, 1977) or (Ferraty and Vieu, 2006, Chapter 5) less attention has been paid to extreme conditional quantiles despite their potential interest. Here, we focus on the setting where the conditional distribution of Y given $X = x$ has an infinite endpoint and is heavy-tailed, an analytical characterization of this property being given in the next section. We show, under mild conditions, that extreme conditional quantiles $q(\alpha_n|x)$ can still be estimated through a functional kernel estimator of $\mathbb{P}(Y > .|x)$. We provide sufficient conditions on the rate of convergence of α_n to 0 so that our estimator is asymptotically Gaussian distributed. Making use of

Laurent Gardes
INRIA Rhône-Alpes and LJK, Saint-Imier, France, e-mail: Laurent.Gardes@inrialpes.fr

Stéphane Girard
INRIA Rhône-Alpes and LJK, Saint-Imier, France, e-mail: Stephane.Girard@inrialpes.fr

this, some functional kernel estimators of the conditional tail-index are introduced and a Weissman type estimator (Weissman, 1978) is derived, permitting to estimate extreme conditional quantiles $q(\beta_n|x)$ where $\beta_n \to 0$ arbitrarily fast.

21.2 Notations and assumptions

The conditional survival function (csf) of Y given $X = x$ is denoted by $\bar{F}(y|x) = \mathbb{P}(Y > y|X = x)$. The kernel estimator of $\bar{F}(y|x)$ is defined for all $(x,y) \in E \times \mathbb{R}$ by

$$\hat{\bar{F}}_n(y|x) = \sum_{i=1}^{n} K(d(x,X_i)/h)Q((Y_i - y)/\lambda) \Bigg/ \sum_{i=1}^{n} K(d(x,X_i)/h), \qquad (21.1)$$

with $Q(t) = \int_{-\infty}^{t} q(s)ds$ where $K : \mathbb{R}^+ \to \mathbb{R}^+$ and $q : \mathbb{R} \to \mathbb{R}^+$ are two kernel functions, and $h = h_n$ and $\lambda = \lambda_n$ are two nonrandom sequences such that $h \to 0$ as $n \to \infty$. In this context, h and λ are called window-width. This estimator was considered for instance in (Ferraty and Vieu, 2006, page 56). In Theorem 1, the asymptotic distribution of (21.1) is established when estimating small tail probabilities, *i.e* when $y = y_n$ goes to infinity with the sample size n. Similarly, the kernel estimators of conditional quantiles $q(\alpha|x)$ are defined via the generalized inverse of $\hat{\bar{F}}_n(.|x)$:

$$\hat{q}_n(\alpha|x) = \inf\{t,\ \hat{\bar{F}}_n(t|x) \le \alpha\}, \qquad (21.2)$$

for all $\alpha \in (0,1)$. Many authors are interested in this type of estimator for fixed $\alpha \in (0,1)$: weak and strong consistency are proved respectively in (Stone, 1977) and (Gannoun, 1990), asymptotic normality being established when E is finite dimensional by (Stute, 1986), (Samanta, 1989), (Berlinet *et al.*, 2001) and by (Ferraty *et al.*, 2005) when E is a general metric space. In Theorem 2, the asymptotic distribution of (21.2) is investigated when estimating extreme quantiles, *i.e* when $\alpha = \alpha_n$ goes to 0 as the sample size n goes to infinity. The asymptotic behavior of such estimators depends on the nature of the conditional distribution tail. In this paper, we focus on heavy tails. More specifically, we assume that the csf satisfies

$$\text{(A1): } \bar{F}(y|x) = c(x)\exp\left\{-\int_1^y \left(\frac{1}{\gamma(x)} - \varepsilon(u|x)\right)\frac{du}{u}\right\},$$

where $\gamma(.)$ is a positive function of the covariate x, $c(.)$ is a positive function and $|\varepsilon(.|x)|$ is continuous and ultimately decreasing to 0. (A1) implies that the conditional distribution of Y given $X = x$ is in the Fréchet maximum domain of attraction. In this context, $\gamma(x)$ is referred to as the conditional tail-index since it tunes the tail heaviness of the conditional distribution of Y given $X = x$. Assumption (A1) also yields that $\bar{F}(.|x)$ is regularly varying at infinity with index $-1/\gamma(x)$. *i.e* for all $\zeta > 0$,

$$\lim_{y\to\infty} \frac{\bar{F}(\zeta y|x)}{\bar{F}(y|x)} = \zeta^{-1/\gamma(x)}. \qquad (21.3)$$

The function $\varepsilon(.|x)$ plays an important role in extreme-value theory since it drives the speed of convergence in (21.3) and more generally the bias of extreme-value estimators. Therefore, it may be of interest to specify how it converges to 0. In (Gomes *et al.*, 2000), the auxiliary function is supposed to be regularly varying and the estimation of the corresponding regular variation index is addressed. Some Lipschitz conditions are also required:

(A2): There exist κ_ε, κ_c, $\kappa_\gamma > 0$ and $u_0 > 1$ such that for all $(x,x') \in E^2$ and $u > u_0$,

$$\begin{aligned}\left|\log c(x) - \log c(x')\right| &\leq \kappa_c d(x,x'),\\ \left|\varepsilon(u|x) - \varepsilon(u|x')\right| &\leq \kappa_\varepsilon d(x,x'),\\ \left|1/\gamma(x) - 1/\gamma(x')\right| &\leq \kappa_\gamma d(x,x').\end{aligned}$$

The last assumptions are standard in the kernel estimation framework.

(A3): K is a function with support $[0,1]$ and there exist $0 < C_1 < C_2 < \infty$ such that $C_1 \leq K(t) \leq C_2$ for all $t \in [0,1]$.

(A4): q is a probability density function (pdf) with support $[-1,1]$.

One may also assume without loss of generality that K integrates to one. In this case, K is called a type I kernel, see (Ferraty and Vieu, 2006, Definition 4.1). Finally, let $B(x,h)$ be the ball of center x and radius h. The small ball probability of X is defined by $\varphi_x(h) = \mathbb{P}(X \in B(x,h))$. Under (A3), for all $\tau > 0$, the τth moment is defined by $\mu_x^{(\tau)}(h) = \mathbb{E}\{K^\tau(d(x,X)/h)\}$.

21.3 Main results

Let us first focus on the estimation of small tail probabilities $\bar{F}(y_n|x)$ when $y_n \to \infty$ as $n \to \infty$. The following result provides sufficient conditions for the asymptotic normality of $\hat{\bar{F}}_n(y_n|x)$.

Theorem 21.1. *Suppose (A1) – (A4) hold. Let $x \in E$ such that $\varphi_x(h) > 0$ and introduce $y_{n,j} = a_j y_n$ for $j = 1,\dots,J$ with $0 < a_1 < a_2 < \cdots < a_J$ and where J is a positive integer. If $y_n \to \infty$ such that $n\varphi_x(h)\bar{F}(y_n|x) \to \infty$, $n\varphi_x(h)\bar{F}(y_n|x)(\lambda/y_n)^2 \to 0$ and $n\varphi_x(h)\bar{F}(y_n|x)(h\log y_n)^2 \to 0$ as $n \to \infty$, then*

$$\left\{\sqrt{n\mu_x^{(1)}(h)\bar{F}(y_n|x)}\left(\frac{\hat{\bar{F}}_n(y_{n,j}|x)}{\bar{F}(y_{n,j}|x)} - 1\right)\right\}_{j=1,\dots,J}$$

is asymptotically Gaussian, centered, with covariance matrix $C(x)$ where $C_{j,j'}(x) = a_{j\wedge j'}^{1/\gamma(x)}$ for $(j,j') \in \{1,\dots,J\}^2$.

Note that $n\varphi_x(h)\bar{F}(y_n|x) \to \infty$ is a necessary and sufficient condition for the almost sure presence of at least one sample point in the region $B(x,h) \times (y_n,\infty)$ of $E \times \mathbb{R}$. Thus, this natural condition states that one cannot estimate small tail probabilities out of the sample using $\hat{\hat{F}}_n$. This result may be compared to (Einmahl, 1990) which establishes the asymptotic behavior of the empirical survival function in the unconditional case but without assumption on the distribution. Letting $\sigma_n(x) = (n\mu_x^{(1)}(h)\alpha_n)^{-1/2}$, the asymptotic normality of $\hat{q}_n(\alpha_n|x)$ when $\alpha_n \to 0$ as $n \to \infty$ can be established under similar conditions.

Theorem 21.2. *Suppose (A1) – (A4) hold. Let $x \in E$ such that $\varphi_x(h) > 0$ and introduce $\alpha_{n,j} = \tau_j\alpha_n$ for $j = 1,\dots,J$ with $\tau_1 > \tau_2 > \cdots > \tau_J > 0$ and where J is a positive integer. If $\alpha_n \to 0$ such that $\sigma_n(x) \to 0$, $\sigma_n^{-1}(x)\lambda/q(\alpha_n|x) \to 0$ and $\sigma_n^{-1}(x)h\log\alpha_n \to 0$ as $n \to \infty$, then*

$$\left\{\sigma_n^{-1}(x)\left(\frac{\hat{q}_n(\alpha_{n,j}|x)}{q(\alpha_{n,j}|x)} - 1\right)\right\}_{j=1,\dots,J}$$

is asymptotically Gaussian, centered, with covariance matrix $\gamma^2(x)\Sigma$ where $\Sigma_{j,j'} = 1/\tau_{j\wedge j'}$ for $(j,j') \in \{1,\dots,J\}^2$.

The functional kernel estimator of extreme quantiles $\hat{q}_n(\alpha_n|x)$ requires a stringent condition on the order α_n of the quantile, since by construction it cannot extrapolate beyond the maximum observation in the ball $B(x,h)$. To overcome this limitation, a Weissman type estimator (Weissman, 1978) can be derived:

$$\hat{q}_n^{\mathrm{W}}(\beta_n|x) = \hat{q}_n(\alpha_n|x)(\alpha_n/\beta_n)^{\hat{\gamma}_n(x)}.$$

Here, $\hat{q}_n(\alpha_n|x)$ is the functional kernel estimator of the extreme quantile considered so far and $\hat{\gamma}_n(x)$ is a functional estimator of the conditional tail-index $\gamma(x)$. As illustrated in the next theorem, the extrapolation factor $(\alpha_n/\beta_n)^{\hat{\gamma}_n(x)}$ allows to estimate extreme quantiles of order β_n arbitrary small.

Theorem 21.3. *Suppose (A1)–(A4) hold. Let us introduce*

- *$\alpha_n \to 0$ such that $\sigma_n(x) \to 0$, $\sigma_n^{-1}(x)\lambda/y_n \to 0$ and $\sigma_n^{-1}(x)h\log\alpha_n \to 0$ as $n \to \infty$,*
- *(β_n) such that $\beta_n/\alpha_n \to 0$ as $n \to \infty$,*
- *$\hat{\gamma}_n(x)$ such that $\sigma_n^{-1}(x)(\hat{\gamma}_n(x) - \gamma(x)) \xrightarrow{d} \mathcal{N}(0,v^2(x))$ where $v^2(x) > 0$.*

Then, for all $x \in E$,

$$\frac{\sigma_n^{-1}(x)}{\log(\alpha_n/\beta_n)}\left(\frac{\hat{q}_n^{\mathrm{W}}(\beta_n|x)}{q(\beta_n|x)} - 1\right) \xrightarrow{d} \mathcal{N}(0,v^2(x)).$$

Note that, when K is the pdf of the uniform distribution, this result is consistent with (Gardes *et al.*, 2010, Theorem 3), obtained in a fixed-design setting.
Let us now give some examples of functional estimators of the conditional tail-index. Let $\alpha_n \to 0$ and $\tau_1 > \tau_2 > \cdots > \tau_J > 0$ where J is a positive integer. Two

additionary notations are introduced for the sake of simplicity: $u = (1,\ldots,1)^t \in \mathbb{R}^J$ and $v = (\log(1/\tau_1),\ldots,\log(1/\tau_J))^t \in \mathbb{R}^J$. The following family of estimators is proposed

$$\hat{\gamma}_n(x) = \frac{\varphi(\log \hat{q}_n(\tau_1 \alpha_n|x),\ldots,\log \hat{q}_n(\tau_J \alpha_n|x))}{\varphi(\log(1/\tau_1),\ldots,\log(1/\tau_J))},$$

where $\varphi : \mathbb{R}^J \to \mathbb{R}$ denotes a twice differentiable function verifying the shift and location invariance conditions $\varphi(\theta v) = \theta\varphi(v)$ for all $\theta > 0$ and $\varphi(\eta u + x) = \varphi(x)$ for all $\eta \in \mathbb{R}$ and $x \in \mathbb{R}^J$. For instance, introducing the auxiliary function $m_p(x_1,\ldots,x_J) = \sum_{j=1}^J (x_j - x_1)^p$ for all $p > 0$ and considering $\varphi_{\text{H}}(x) = m_1(x)$ gives rise to a kernel version of the Hill estimator (Hill, 1975):

$$\hat{\gamma}_n^{\text{H}}(x) = \sum_{j=1}^J [\log \hat{q}_n(\tau_j \alpha_n|x) - \log \hat{q}_n(\alpha_n|x)] \Big/ \sum_{j=1}^J \log(1/\tau_j).$$

Generalizations of the kernel Hill estimator can be obtained with $\varphi(x) = m_p(x)/m_1^{p-1}(x)$, see (Gomes and Martins, 2001, equation (2.2)), $\varphi(x) = m_p^{1/p}(x)$, see e.g. (Segers, 2001, example (a)) or $\varphi(x) = m_{p\theta}^{1/\theta}(x)/m_{p-1}(x)$, $p \geq 1$, $\theta > 0$, see (Caeiro and Gomes, 2002). In the case where $J = 3$, $\tau_1 = 4$, $\tau_2 = 2$ and $\tau_3 = 1$, the function

$$\varphi_{\text{P}}(x_1,x_2,x_3) = \log\left(\frac{\exp x_2 - \exp x_1}{\exp x_3 - \exp x_2}\right)$$

leads us to a kernel version of Pickands estimator (Pickands, 1975)

$$\hat{\gamma}_n^{\text{P}}(x) = \frac{1}{\log 2} \log\left(\frac{\hat{q}_n(\alpha_n|x) - \hat{q}_n(2\alpha_n|x)}{\hat{q}_n(2\alpha_n|x) - \hat{q}_n(4\alpha_n|x)}\right).$$

We refer to (Gijbels and Peng, 2000) for a different variant of Pickands estimator in the context where the distribution of Y given $X = x$ has a finite endpoint. The asymptotic normality of $\hat{\gamma}_n(x)$ is a consequence of Theorem 2.

Theorem 21.4. *Under assumptions of Theorem 2 and if $\sigma_n^{-1}(x)\varepsilon(q(\tau_1\alpha_n|x)|x) \to 0$ as $n \to \infty$, then, $\sigma_n^{-1}(x)(\hat{\gamma}_n(x) - \gamma(x))$ converges to a centered Gaussian random variable with variance*

$$V(x) = \frac{\gamma^2(x)}{\varphi^2(v)} (\nabla\varphi(\gamma(x)v))^t \Sigma (\nabla\varphi(\gamma(x)v)).$$

As an illustration, in the case of the kernel Hill and Pickands estimators, we obtain

$$V_{\text{H}}(x) = \gamma^2(x)\left(\sum_{j=1}^J \frac{2(J-j)+1}{\tau_j} - J^2\right) \Big/ \left(\sum_{j=1}^J \log(1/\tau_j)\right)^2.$$

$$V_{\text{P}}(x) = \frac{\gamma^2(x)(2^{2\gamma(x)+1}+1)}{4(\log 2)^2(2^{\gamma(x)}-1)^2}.$$

Clearly, $V_{\mathrm{P}}(x)$ is the variance of the classical Pickands estimator, see for instance (de Haan and Ferreira, 2006, Theorem 3.3.5). Focusing on the kernel Hill estimator and choosing $\tau_j = 1/j$ for each $j = 1,\dots,J$ yields $V_{\mathrm{H}}(x) = \gamma^2(x)J(J-1)(2J-1)/(6\log^2(J!))$. In this case, $V_{\mathrm{H}}(x)$ is a convex function of J and is minimum for $J = 9$ leading to $V_{\mathrm{H}}(x) \simeq 1.25\gamma^2(x)$.

References

1. Berlinet, A., Gannoun, A., Matzner-Løber, E.: Asymptotic normality of convergent estimates of conditional quantiles. Statistics **35**, 139–169 (2001)
2. Caeiro, F., Gomes, M.I.: Bias reduction in the estimation of parameters of rare events. Theor. Stoch. Process. **8**, 67–76 (2002)
3. Einmahl, J.H.J.: The empirical distribution function as a tail estimator. Stat. Neerl. **44**, 79–82 (1990)
4. Ferraty, F., Vieu, P.: Nonparametric functional data analysis. Springer (2006)
5. Ferraty, F., Rabhi, A., Vieu, P.: Conditional quantiles for dependent functional data with application to the climatic *El Nino* Phenomenon. Sankhyā **67** (2), 378–398 (2005)
6. Gannoun, A.: Estimation non paramétrique de la médiane conditionnelle, médianogramme et méthode du noyau. Publications de l'Institut de Statistique de l'Université de Paris XXXXVI, 11–22 (1990)
7. Gardes, L., Girard, S., Lekina, A.: Functional nonparametric estimation of conditional extreme quantiles. J. Multivariate Anal. **101**, 419–433 (2010)
8. Gijbels, I., Peng, L.: Estimation of a support curve via order statistics. Extremes **3**, 251–277 (2000)
9. Gomes, M.I., Martins, M.J., Neves, M.: Semi-parametric estimation of the second order parameter, asymptotic and finite sample behaviour. Extremes **3**, 207–229 (2000)
10. Gomes, M.I., Martins, M.J.: Generalizations of the Hill estimator - asymptotic versus finite sample behaviour. J. Stat. Plan. Infer. **93**, 161–180 (2001)
11. de Haan, L., Ferreira, A.: Extreme Value Theory: An Introduction. Springer Series in Operations Research and Financial Engineering, Springer, 2006.
12. Hill, B.M.: A simple general approach to inference about the tail of a distribution. Ann. Stat. **3**, 1163–1174 (1975)
13. Pickands, J.: Statistical inference using extreme order statistics. Ann. Stat. **3**, 119–131 (1975)
14. Roussas, G.G.: Nonparametric estimation of the transition distribution function of a Markov process. Ann. Math. Stat. **40**, 1386–1400 (1969)
15. Samanta, T.: Non-parametric estimation of conditional quantiles. Stat. Probab. Lett. **7**, 407–412 (1989)
16. Segers, J.: Residual estimators. J. Stat. Plan. Infer. **98**, 15–27 (2001)
17. Stone, C.J.: Consistent nonparametric regression (with discussion). Ann. Stat. **5**, 595–645 (1977)
18. Stute, W.: Conditional empirical processes. Ann. Stat. **14**, 638–647 (1986)
19. Weissman, I.: Estimation of parameters and large quantiles based on the k largest observations. J. Am. Stat. Assoc. **73**, 812–815 (1978)

Chapter 22
A Nonparametric Functional Method for Signature Recognition

Gery Geenens

Abstract We propose to use nonparametric functional data analysis techniques within the framework of a signature recognition system. Regarding the signature as a random function from $\mathbb{R}$ (time domain) to $\mathbb{R}^2$ (position (x,y) of the pen), we tackle the problem as a genuine nonparametric functional classification problem, in contrast to currently used biometrical approaches. A simulation study on a real data set shows good results.

22.1 Introduction

The problem of automatic signature recognition has attracted attention for a long time, since signatures are well established in our everyday lives as the most common means of personal identification, with applications in commerce, banking transactions or any other official use. There is therefore a clear need for accurate and reliable signature recognition systems, and it is no surprise that many digital procedures aiming at discriminating forgeries from genuine signatures have been proposed in biometrics, pattern recognition and engineering literature. Impedovo and Pirlo (2008) comprehensively summarize the most valuable results up to 2008, and Impedovo et al (2010) complete that study with the most recent works.

However, it turns out that even those methods which claim to be functional or dynamic are actually based on a finite number of parameters describing the temporal evolution of some considered characteristics, like pen pressure or azimuth for example. Never, to our knowledge, has the problem been addressed from a purely functional point-of-view, that is, keeping the whole "signature-function" as the object of central interest. Ramsay and Silverman (1997) and Ramsay (2000) present handwriting analysis as an important application of functional data analysis, but do

Gery Geenens
University of New South Wales, sydney, Australia, e-mail: ggeenens@unsw.edu.au

not really focus on the signature recognition problem. In contrast, this work mainly aims at using modern functional data analysis tools, like nonparametric functional regression ideas, to think up, develop and implement an efficient signature recognition system, and to check whether this exclusively statistical method is able to match the currently used pattern recognition and biometrical methods in terms of simplicity, ease of implementation and, of course, efficiency at exposing fakes.

22.2 Signatures as random objects

The method that we propose is based on the natural idea of modelling a signature as a random function

$$\mathscr{S} : \mathscr{T} \subset \mathbb{R}^+ \to \mathscr{P} \subset \mathbb{R}^2 : t \to \mathscr{S}(t) = (\mathscr{X}(t), \mathscr{Y}(t))$$

where $\mathscr{S}(t) = (\mathscr{X}(t), \mathscr{Y}(t))$ represents the position of the pen in $\mathscr{P}$, a given portion of the two-dimensional plane, at time $t \in \mathscr{T}$, the considered time domain. We therefore assume that the signature $\mathscr{S}$ lies in an appropriate infinite-dimensional functional space, say Σ. The random nature of the so-defined object obviously accounts for the natural variability between successive signatures from one writer. The benefit of working directly with a whole function is evident : it automatically gives access to some features 'hidden' in $\mathscr{S}$. In particular, the first and second derivative vectors of $\mathscr{S}(t)$ provide information about the temporal evolution of the speed and acceleration of the pen during the signing process. Precisely, we propose to analyze that acceleration. It is commonly admitted that the acceleration of the pen is mainly dictated by the movement of the wrist of the person signing. Besides, it is quite clear that the "genuine" wrist movement is very hard, if not impossible, to detect and reproduce even for a skilled faker. Unlike the drawing itself, or any other global characteristic, this movement and the acceleration it induces are consequently unique to every single person and should be very efficient discriminatory elements. Of course, analyzing second derivatives of random functions requires functional data analysis (FDA) methods : linking FDA to the signature recognition problem is what this work attempts to do.

Suppose we observe a realization ς of the random object $\mathscr{S}$, and we have to make a decision as to whether this observed signature is a fake or not. This is obviously nothing else but a classification problem. The decision will be based on an estimation of the probability of ς being a fake, that is

$$\pi(\varsigma) = P(Z = 1 | \mathscr{S} = \varsigma),$$

where Z is a binary random variable, taking the value 1 if $\mathscr{S}$ is a forgery and 0 if it is a genuine signature. Note that, due to the binary nature of Z, this conditional probability can also be written

$$\pi(\varsigma) = E(Z|\mathscr{S} = \varsigma),$$

so that $\pi(\varsigma)$ can be estimated by functional regression methods. Here, "functional regression" refers to situations where the predictor itself is a whole function, as opposed to the more classical situation of "vectorial regression", when we wish to predict a response from a (finite) set of univariate predictors. In this functional case, it appears that fitting any parametric problem would be very hazardous. Indeed, none of the classical parametric models for binary regression, e.g. logit, probit, etc., possess any physical interpretation in our application. Besides, there is no visual guide available as any graphical representation is inconceivable in an infinite-dimensional space like Σ. As graphical representations like scatter-plots or residual plots are usually the primary tools to define and validate a suitable parametric regression model, it turns out that the risk of model misspecification is even higher in this setting than in classical parametric regression. Consequently, we turn to nonparametric regression methods. The theoretical foundation of Nonparametric Functional Analysis has been quite recently initiated by Ferraty and Vieu (2006). Since then, a wide literature in the field has rapidly come up, see Ferraty and Romain (2011) for a comprehensive and up-to-date reference.

22.3 A semi-normed functional space for signatures

It is the case that any nonparametric regression method is essentially local, see classical texts like Wand and Jones (1995) or Fan and Gijbels (1996). Consequently, this means that only information 'close' to the observed signature ς will be used to estimate $\pi(\varsigma)$. Therefore, a notion of closeness (or similarity) between two signatures in the considered functional space has to be properly defined. Ferraty and Vieu (2006) suggest to work in a semi-normed space as an elegant way to account for the proximity between functions. Unlike a distance, a semi-distance, say δ, is such that $\delta(\varsigma_1, \varsigma_2) = 0$ does not imply that $\varsigma_1 = \varsigma_2$, for two functional objects ς_1 and ς_2. Being less stringent than a genuine distance, a semi-distance dictates that two functional objects which might be different but which share some common characteristics are close. An appropriate choice of semi-distance therefore allows one to focus on features of the functional objects that we know to be particularly relevant in the considered context, whilst avoiding (or at least reducing) an extreme impact of the so-called 'curse of dimensionality' on the quality of the estimators, see the above references for a detailed discussion about this infamous phenomenon.

For the reasons mentioned in Section 1, we propose to measure the proximity between two signatures ς_1 and ς_2 with the semi-distance

$$\delta(\varsigma_1, \varsigma_2) \doteq \left(\int (\varsigma_1''(t) - \varsigma_2''(t))^2 \, dt \right)^{1/2}, \tag{22.1}$$

where $\varsigma''(t)$ is the tangential projection of the vector of second derivatives with respect to time of the signature-function ς, as this would account for the similarity (or dissimilarity) in tangential acceleration between ς_1 and ς_2. Moreover, this obviates the need for an important pre-processing of the recorded signatures, as the second order differentiation cancels out any location or size effect. In the sequel, we therefore assume that $\mathscr{S}$ belongs to Σ, the space of functions from $\mathbb{R}^+$ to $\mathbb{R}^2$, both of whose components are twice differentiable, dotted with the semi-distance δ.

22.4 Nonparametric functional signature recognition

Now, assume that we have a sample of i.i.d. replications of $(\mathscr{S},Z) \in \Sigma \times \{0,1\}$. To make it more explicit, we assume that we have a first sample of specimen signatures $(\varsigma_1, \varsigma_2, \ldots, \varsigma_n)$, replications of the genuine signature of a given person (those observations such that $Z = 1$ in the global sample), and a second sample of forgeries $(\varphi_1, \varphi_2, \ldots, \varphi_m)$ (ones such that $Z = 0$). Note that assuming we have access to forgeries is by no means constrictive : the fakes need not really mimic the true one, but could just be signatures of other persons in the database, or whatever. However, we can expect the procedure to be more efficient if it is trained on a set of "skilled" forgeries. Then, observing a signature ς, a Nadaraya-Watson-like estimator for the conditional probability $\pi(\varsigma)$ is given by

$$\hat{\pi}(\varsigma) = \frac{\sum_{j=1}^{m} K\left(\frac{\delta(\varsigma,\varphi_j)}{h}\right)}{\sum_{i=1}^{n} K\left(\frac{\delta(\varsigma,\varsigma_i)}{h}\right) + \sum_{j=1}^{m} K\left(\frac{\delta(\varsigma,\varphi_j)}{h}\right)}, \tag{22.2}$$

where K is a nonnegative kernel function, supported and decreasing on $[0,1]$, and h is the bandwidth, both of which being usual parameters in nonparametric kernel regression. See that $\hat{\pi}(\varsigma)$ is nothing else but the weighted average of the binary responses Z associated with all signatures of the global sample, with more weight (given through K) to the signatures close to ς, the notion of "closeness" being quantified by h. It directly follows that $\hat{\pi}(\varsigma)$ always belongs to $[0,1]$, and is close to 0 (respectively 1) when ς is very close (respectively distant) to all the genuine signatures. Note that we here define the case $\frac{0}{0}$ to be equal to 1, in the event the observed signature is very different, in terms of the considered closeness notion, to any signature of the database. The decision then directly follows by comparing $\hat{\pi}(s)$ with a given threshold, say c : if $\hat{\pi}(\varsigma) > c$, the observed signature is likely to be a fake and is therefore rejected. If $\hat{\pi}(\varsigma) \leq c$, the signature is accepted. The usual Bayes rule would set c to $1/2$, however, depending on the application, this threshold value can be adjusted to match the required standards.

5. From theory to practice

The above procedure has been implemented in R software, and tested on a freely available signature data set : the one used for the First International Signature Verification Competition (SVC2004), see Yeung et al (2004). In that database, each signature is represented as a discrete sequence of points (from 136 to 595 points, depending on the signature), each point being characterized by the X-coordinate, the Y-coordinate, the time stamp (cpu time) and the button status (pen up/pen down). A first task was thus to smooth those discrete representations of the signatures-functions, in order to be able to differentiate their components $\mathscr{X}(t)$ and $\mathscr{Y}(t)$ later on. To keep some uniformity between the signatures, we first rescaled the time stamp to between 0 and 1. Then, we used a Local Polynomial kernel smoother, with Gaussian kernel and bandwidth of type k-nearest neighbor with $k = 10$, admittedly selected quite subjectively, to estimate the first and second derivatives of both $\mathscr{X}(t)$ and $\mathscr{Y}(t)$.

Now, given that the tangential acceleration is defined as the projection of the vector of second derivatives onto the unit vector tangent to the curve (that is, the normalized vector of first derivatives), we estimate the tangential acceleration function by

$$\hat{\mathscr{S}}''(t) = (\hat{\mathscr{X}}''(t), \hat{\mathscr{Y}}''(t))^t \frac{(\hat{\mathscr{X}}'(t), \hat{\mathscr{Y}}'(t))}{\|(\hat{\mathscr{X}}'(t), \hat{\mathscr{Y}}'(t))\|}$$

where $(\hat{\mathscr{X}}'(t), \hat{\mathscr{Y}}'(t))$ and $(\hat{\mathscr{X}}''(t), \hat{\mathscr{Y}}''(t))$ are the previously mentioned kernel estimates of the first and second derivative vectors. Five tangential acceleration functions for genuine signatures, as well as a 'fake' tangential acceleration function, are represented in Figure 22.1 below for one user. The consistency of the tangential acceleration over the genuine signatures is clear, in contrast to what is shown for the forgery.

It is now easy to compute numerically the semi-distance (22.1) between any two signature-objects, and then to estimate the fake probability (22.2) for an observed signature ς. This was done using a Gaussian kernel and a bandwidth of type k-nearest neighbor, determined by least-squares cross-validation. Notably, the value k is seen to vary considerably from one user to another.

The database we used consisted of 100 sets of signatures data, each set containing 20 genuine signatures from one signature contributor and 20 skilled forgeries from at least four other contributors. For each user, we decided to split the 40 available signatures in two : 10 genuine signatures and 10 forgeries would be utilized as the training set, so supposedly the samples $(\varsigma_1, \varsigma_2, \ldots, \varsigma_n)$ and $(\varphi_1, \varphi_2, \ldots, \varphi_m)$ that we have in hand, with the other 20 (again, 10 genuine signatures and 10 forgeries) being our testing set. We ran the procedure over that testing set and computed the equal error rate (EER), that is, the false rejection rate plus the false acceptance rate, for each user. We observed important variations over the users, which renders the fact that some signatures are easier to reproduce than others - even in terms of tangential acceleration. For some users, the EER was 0, but for some others it was around 25%.

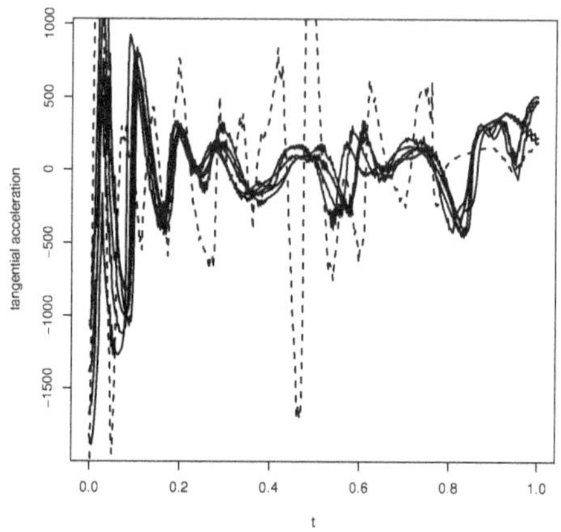

Fig. 22.1: Five 'genuine' tangential acceleration functions (plain line) and one 'fake' tangential acceleration function (dotted line)

On average, the EER was 9%, with a median of 5%, which is quite an encouraging result. Let us bear in mind that the proposed methodology has been applied to raw data (only a basic smoothing step has been carried out to estimate the derivatives of the functions of interest). Admittedly, an appropriate pre-processing of the data could dramatically improve the efficiency of the method proposed. What we have in mind for instance is the registration of the tangential acceleration functions, which would aim at aligning as close as possible the peaks and troughs observed in Figure 22.1 for example (see Ramsay (1998)). This would make different tangential acceleration functions of the same user still closer to one another, and therefore ease the recognition process. Note that other possibly useful pre-processing techniques are presented in Huang et al (2007). These ideas are left for future research.

22.5 Concluding remarks

In this work we propose an automatic signature recognition system, based on nonparametric functional regression ideas only. As opposed to currently used biometrical methodologies, often based on intricate and computationally intensive algorithms (neural networks, hidden Markov Chains, decision tress, etc.), this procedure is conceptually simple to understand, as the decision (fake or not) readily follows from the direct estimation of the probability of the observed signature being a forgery, and easy to implement, as kernel regression and related tools are now well understood and available in most statistical software. Besides, the method applied to raw data has shown pretty good results, while it is reasonable to think that an appropriate pre-processing of the data, like registration *inter alia*, would further improve the error rates observed so far.

Acknowledgements The author thanks Vincent Tu (UNSW) for interesting discussions.

References

1. Fan, J., Gijbels, I.: Local Polynomial Modelling and Its Applications. Chapman and Hall/CRC (1996)
2. Ferraty, F., Romain, Y.: Oxford handbook on functional data analysis (Eds). Oxford University Press (2011)
3. Ferraty, F., Vieu, P.: Nonparametric Functional Data Analysis: Theory and Practice. Springer (2006)
4. Huang, B.Q., Zhang, Y.B., Kechadi, M.T.: Preprocessing Techniques for Online Handwritting Recognition. Proceedings of the Seventh International Conference on Intelligent Systems Design and Applications, Rio de Janeiro (2007)
5. Impedovo, D., Pirlo, G.: Automatic signature verification : The state of the art. IEEE Trans. Syst. Man. Cybern. C, Appl. Rev. **38** (5), 609–635 (2008)
6. Impedovo, S., Pirlo, G., Modugno, R., Impedovo, D., Ferrante, A., Sarcinella, L., Stasolla, E.: Advancements in Handwritting Recognition. Manuscript, Università degli Studi di Bari (2010)
7. Ramsay, J.O.: Curve Registration. J. R. Stat. Soc. B **60**, 351–363 (1998)
8. Ramsay, J.O.: Functional Components of Variation in Handwriting, J. Am. Stat. Assoc. **95**, 9–15 (2000)
9. Ramsay, J.O., Silverman, B.W.: Functional data analysis. Springer (1997)
10. Wand, M.P., Jones, M.C.: Kernel Smoothing. Chapman and Hall/CRC (1995)
11. Yeung, D.T., Chang, H., Xiong, Y., George, S., Kashi, R., Matsumoto, T., Rigoll, G.: SVC2004: First International Signature Verification Competition, Proceedings of the International Conference on Biometric Authentication (ICBA), Hong Kong (2004)

Chapter 23
Longitudinal Functional Principal Component Analysis

Sonja Greven, Ciprian Crainiceanu, Brian Caffo, Daniel Reich

Abstract We introduce models for the analysis of functional data observed at multiple time points. The model can be viewed as the functional analog of the classical mixed effects model where random effects are replaced by random processes. Computational feasibility is assured by using principal component bases. The methodology is motivated by and applied to a diffusion tensor imaging (DTI) study on multiple sclerosis.

23.1 Introduction

Many studies now collect functional or imaging data at multiple visits or timepoints. In this paper we introduce a class of models and inferential methods for the analysis of longitudinal data where each repeated observation is functional.

Our motivating data set comes from a diffusion tensor imaging (DTI) study analyzing cross-sectional and longitudinal differences in brain connectivity in multiple sclerosis (MS) patients and controls. For each of the 112 subjects and each visit, we have functional anisotropy (FA) measurements along the corpus callosum tract in the brain. Figure 23.1 shows 2 example patients with 5 and 6 complete visits, respectively. Each visit's data for a subject is a finely sampled function, registered

Sonja Greven
Ludwig-Maximilians-Universität München, Munich, Germany, e-mail: sonja.greven@stat.uni-muenchen.de

Ciprian Crainiceanu
Johns Hopkins University, Baltimore, USA, e-mail: ccrainic@jhsph.edu

Brian Caffo
Johns Hopkins University, Baltimore, USA, e-mail: bcaffo@jhsph.edu

Daniel Reich
National Institutes of Health, Bethesda, USA, e-mail: daniel.reich@nih.gov

using 7 biological landmarks, with the argument of the function being the spatial distance along the tract.

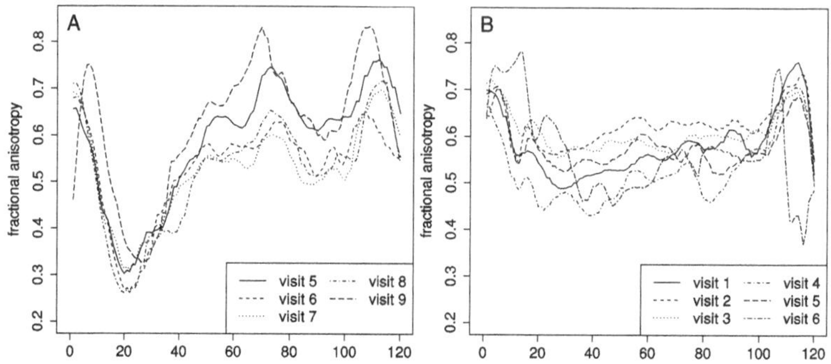

Fig. 23.1: Two example subjects (both MS patients) from the tractography data with 5 and 6 complete visits, respectively. Shown are the fractional anisotropy along the corpus callosum, measured at 120 landmark-registered sample points. Different visits for the same subject are indicated by line type and overlaid.

Longitudinal *scalar* data is commonly analyzed using the very flexible class of linear mixed models (Laird and Ware, 1982), which explicitly decompose the variation in the data into between- and within-subject variability. We propose a functional analog of linear mixed models by replacing random effects with random functional effects. We propose an estimation procedure that is based on principal components bases and extends functional principal component analysis (FPCA) to the longitudinal setting. Computation is very efficient, even for large data sets.

Our approach is different from functional mixed models based on the smoothing of fixed and random curves using splines or wavelets (Brumback and Rice, 1998; Guo, 2002; Morris and Carroll, 2006). In contrast to these methods focusing on the estimation of fixed and random curves, our approach is based on functional principal component analysis. In addition to the computational advantages, we are thus able to extract the main differences between subjects in their average profiles and in how their profiles evolve over time. Such a signal extraction, not possible using smoothing methods alone, allows the relation of subject-specific scores to other variables such as disease status or disease progression. Our approach can be seen as an extension of multilevel functional principal component analysis (Di et al., 2008). Our methods apply to longitudinal data where each observation is functional, and should thus not be confused with nonparametric methods for the longitudinal profiles of scalar variables (e.g. Yao et al., 2005).

23.2 The Longitudinal Functional Model and LFPCA

Consider first the functional analog of the popular random intercept-random slope model,

$$Y_{ij}(d) = \eta(d, T_{ij}) + X_{i,0}(d) + X_{i,1}(d)T_{ij} + U_{ij}(d) + \varepsilon_{ij}(d), \qquad (23.1)$$

where $Y_{ij}(\cdot)$ is a random function in $L_2[0,1]$ observed at a grid of values $d \in [0,1]$, and T_{ij} is the jth time-point for subject i, $i = 1,\ldots,I$, $j = 1,\ldots,J_i$. In this representation, $\eta(d, T_{ij})$ is a fixed main effect surface, $X_{i,0}(d)$ is the random functional intercept for subject i, $X_{i,1}(d)$ is the random functional slope for subject i, $U_{ij}(d)$ is the random subject- and visit-specific functional deviation, and $\varepsilon_{ij}(d)$ is random homoscedastic noise. We assume that $X_i(d) = \{X_{i,0}(d), X_{i,1}(d)\}$, $U_{ij}(d)$ and $\varepsilon_{ij}(d)$ are zero-mean, square-integrable, mutually uncorrelated random processes on $[0,1]$, and $\varepsilon_{ij}(d)$ is white noise measurement error with variance σ^2.

Model (23.1) can be generalized to a functional analog of a linear mixed model,

$$Y_{ij}(d) = \eta(d, Z_{ij}) + V_{ij}' X_i(d) + U_{ij}(d) + \varepsilon_{ij}(d), \qquad (23.2)$$

with vector-valued random process $X_i(d)$ and covariate vectors V_{ij} and Z_{ij}, but we will here focus on model (23.1) for simplicity.

To estimate model (23.1), we build on FPCA (e.g. Ramsay and Silverman, 2005) and extend multilevel FPCA (Di et al. 2008), using Mercer's theorem and the Karhunen-Loève expansion. We expand the covariance operators of the bivariate and univariate processes $X_i(d) = \{X_{i,0}(d), X_{i,1}(d)\}$ and $U_{ij}(d)$ as $K_X(d,d') = \sum_{k=1}^{\infty} \lambda_k \phi_k^X(d)\phi_k^X(d')'$ and $K_U(d,d') = \sum_{k=1}^{\infty} \nu_k \phi_k^U(d)\phi_k^U(d')$, where $\phi_k^X(d) = \{\phi_k^0(d), \phi_k^1(d)\}'$ and $\phi_k^U(d)$ are the eigenfunctions corresponding to the eigenvalues $\lambda_1 \geq \lambda_2 \geq \cdots \geq 0$, respectively $\nu_1 \geq \nu_2 \geq \cdots \geq 0$. The Karhunen-Loève expansions of the random processes are $X_i(d) = \sum_{k=1}^{\infty} \xi_{ik}\phi_k^X(d)$ and $U_{ij}(d) = \sum_{k=1}^{\infty} \zeta_{ijk}\phi_k^U(d)$, with principal components scores $\xi_{ik} = \int_0^1 X_{i,0}(s)\phi_k^0(s)ds + \int_0^1 X_{i,1}(s)\phi_k^1(s)ds$ and $\zeta_{ijk} = \int_0^1 U_{ij}(s)\phi_k^U(s)ds$ being uncorrelated mean zero random variables with variances λ_k and ν_k, respectively. The bivariate ϕ_k^X capture the potential correlation between random functional intercept and slope, which are allowed to co-vary between subjects. They allow the extraction of information on the main modes of variation with respect to both static and dynamic behavior of the functions. In practice, finite-dimensional approximations result in the following approximation to model (23.1),

$$Y_{ij}(d) = \eta(d, T_{ij}) + \sum_{k=1}^{N_X} \xi_{ik} V_{ij}' \phi_k^X(d) + \sum_{l=1}^{N_U} \zeta_{ijl}\phi_l^U(d) + \varepsilon_{ij}(d), \qquad (23.3)$$

where $V_{ij} = (1, T_{ij})'$, $\xi_{ik} \sim (0, \lambda_k)$, $\zeta_{ijl} \sim (0, \nu_l)$, $\varepsilon_{ij}(d) \sim (0, \sigma^2)$. $x_l \sim (0, a)$ denotes uncorrelated variables with mean 0 and variance a. Normality is not assumed but groups of variables are assumed uncorrelated, corresponding to our previous assumptions on the random processes. Model (23.3) is a linear mixed model. Model selection or testing for random effects could thus be used to choose N_X and N_U

(e.g. Greven and Kneib, 2010). We use the simpler and intuitive approach of choosing components up to a previously specified proportion of variance explained. We can show that the overall variance can be written as

$$\int_0^1 Var\{Y_{ij}(s)\}ds = \sum_{k=1}^{\infty} \lambda_k + \sum_{k=1}^{\infty} \nu_k + \sigma^2$$

if $\eta(d,T_{ij}) \equiv 0$ and the T_{ij} are random variables independent of all other variables with $E(T_{ij}) = 0$ and $Var(T_{ij}) = 1$.

23.3 Estimation and Simulation Results

We estimate model (23.3) as follows. The fixed effect mean surface $\eta(d,T)$ can be consistently estimated using a bivariate smoother such as penalized splines in d and T under a working independence assumption. We use the empirical covariances and the fact that $\text{Cov}\{Y_{ij}(d), Y_{ik}(d')\}$ can be expanded as

$$K_0(d,d') + T_{ik}K_{01}(d,d') + T_{ij}K_{01}(d',d) + T_{ij}T_{ik}K_1(d,d') + [K_U(d,d') + \sigma^2\delta_{dd'}]\delta_{jk},$$

to estimate the covariance operators based on linear regression. Here, $K_0(d,d')$, $K_1(d,d')$ and $K_{01}(d,d')$ denote the auto-covariance and cross-covariance functions for $X_{i,0}(d)$ and $X_{i,1}(d)$, and δ_{jk} is Kronecker's delta. We incorporated a bivariate smoothing step in the estimation of $K_U(d,d')$ and $K_X(d,d')$, which also allows estimation of σ^2. Eigenfunctions and variances can then be estimated from the spectral decomposition of the covariance functions. Estimation of the scores ξ_{ik} and ζ_{ijl} is based on best linear unbiased prediction in the linear mixed model (23.3). Matrix algebra involving Kronecker product, Woodbury formula etc. allows our implementation in an R function to be computationally very efficient, even for large data sets.

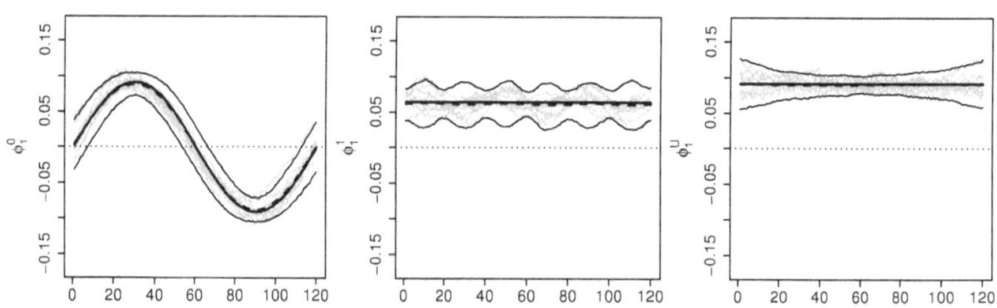

Fig. 23.2: The first true principal components (ϕ_1^0, ϕ_1^1) for X and ϕ_1^U for U (thick solid lines), the mean of the estimated functions (dashed), pointwise 5th and 95th percentiles of the estimated functions from 1000 simulations (thin solid), and estimated functions from 20 simulations (grey), without smoothing of covariances.

Our estimation approach performed well in extensive simulations, spanning different numbers of subjects and visits per subject, balanced and unbalanced designs, normal and non-normal scores, different eigenfunctions and mean functions. As an example, for one setting with 100 subjects, 4 unequally spaced time-points T_{ij} per subject and 120 sampling points d per curve, Figure 23.2 illustrates that eigenfunctions are well estimated; as are mean function, variances and scores (results not shown).

23.4 Application to the Tractography Data

Diffusion tensor imaging (DTI) is able to resolve individual functional tracts within the central nervous system white matter, a frequent target of multiple sclerosis (MS). DTI measures such as fractional anisotropy (FA) can be decreased or increased in MS due to lesions, loss of myelin and axon damage. A focus on single tracts can help in understanding the neuroanatomical basis of disability in MS. We are interested in differences between subjects both with respect to their mean tract profiles (static behavior) and the changes in their tract profiles over time (dynamic behavior).

Figure 23.3 exemplarily shows estimates for the first principal component (ϕ_1^0, ϕ_1^1). Positive scores correspond to a lower function with a particularly deep dip in the isthmus (at 20), but only to small changes over time. Estimated scores $\widehat{\xi}_{i1}$ are significantly higher in MS patients than controls. The patient group in particular seems to have a higher mean and a heavier right tail. This could be an indication of a mixture in this group of patients who are more or less affected by MS along this particular tract. Potential loading-based clustering into patient subgroups will be of interest in future work. Interestingly, FA for this component is not decreased uniformly along the tract, but only posterior to the genu (ca. 1-100), with the decrease being especially pronounced in the area of the isthmus (ca. 20). Our results thus identify the region of the corpus callosum (the isthmus) where MS seems to take its greatest toll. Other components indicate the ways in which that portion of the tract changes from one year to the next. In future work, we plan to examine whether these changes can portend disease course. This result could not have been obtained by using the average FA instead of our functional approach.

References

1. Greven, S., Crainiceanu, C.M., Caffo, B., Reich, D.: Longitudinal functional principal component analysis. Electron. J. Stat. **4**, 1022–1054 (2010)
2. Brumback, B.A., Rice, J.A.: Smoothing spline models for the analysis of nested and crossed samples of curves. J. Am. Stat. Assoc. **93**, 961–976 (1998)
3. Di, C.Z., Crainiceanu, C.M., Caffo, B.S., Punjabi, N.M.: Multilevel functional principal component analysis. Ann. Appl. Stat. **3**, 458–488 (2008)

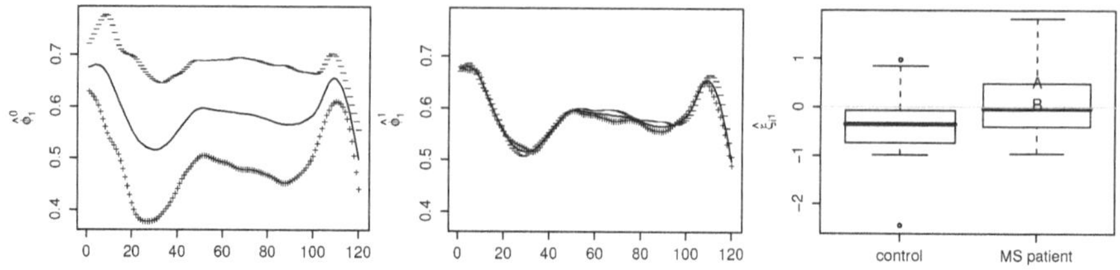

Fig. 23.3: The first estimated principal component (ϕ_1^0, ϕ_1^1) for the random intercept (left) and slope (middle) process X. Depicted are estimates for the overall mean $\eta(d)$ (solid line), and for $\eta(d)$ plus/minus $2\sqrt{\lambda_k}$ times the component. Boxplots on the right show estimates of the corresponding scores ξ_{ik} by case/control group. The two example patients shown in Figure 23.1 are indicated by A and B.

4. Greven, S., Kneib, T.: On the Behaviour of Marginal and Conditional AIC in Linear Mixed Models. Biometrika **97**, 773–789 (2010)
5. Guo, W.: Functional mixed effects models. Biometrics **58**: 121-128 (2002)
6. Laird, N., Ware, J.H.: Random-effects models for longitudinal data. Biometrics **38**, 963–974 (1982)
7. Morris, J.S., Carroll, R.J.: Wavelet-based functional mixed models. J. Roy. Stat. Soc. B **68**, 179–199 (2006)
8. Ramsay, J.O., Silverman, B.W.: Functional data analysis (Second Edition). Springer (2005)
9. Yao, F., Müller, H.G., Wang, J.L.: Functional data analysis for sparse longitudinal data. J. Am. Stat. Assoc. **100**: 577-590 (2005)

Chapter 24
Estimation and Testing for Geostatistical Functional Data

Oleksandr Gromenko, Piotr Kokoszka

Abstract We present procedures for the estimation of the mean function and the functional principal components of dependent spatially distributed functional data. We show how the new approaches improve on standard procedures, and discuss their application to significance tests.

24.1 Introduction

The data that motivates the research summarized in this note consist of curves $X(\mathbf{s}_k;t)$, $t \in [0,T]$, observed at spatial locations $\mathbf{s}_1, \mathbf{s}_2, \ldots, \mathbf{s}_N$. Such functional data structures are quite common, but typically the spatial dependence and the spatial distribution of the points $\mathbf{s}_k$ are not taken into account. A well–known example is the Canadian temperature and precipitation data used as a running example in Ramsay and Silverman (2005). The annual curves are available at 35 locations, some of which are quite close, and so the curves look very similar, others are very remote with notably different curves. Figure **??** shows the temperature curves together with the simple average and the average estimated by one of the methods proposed in this paper. Conceptually, the average temperature in Canada should be computed as the average over a fine regular grid spanning the whole country. In reality, there are only several dozen locations mostly in the inhabited southern strip. Computing an average over these locations will bias the estimate. Data at close by locations contribute similar information, and should get smaller weights than data at sparse locations. This is the fundamental principle of spatial statistics which however received only limited attention in the framework of functional data analysis. Another example of this type is the Australian rainfall data set, recently studied by Delaigle

Oleksandr Gromenko
Utah State University, Logan, USA, e-mail: agromenko@gmail.com

Piotr Kokoszka
Utah State University, Logan, USA, e-mail: Piotr.Kokoszka@usu.edu

and Hall (2010), which consists of daily rainfall measurements from 1840 to 1990 at 191 Australian weather stations. Many other environmental and geophysical data sets fall into this framework; examples are discussed in Gromenko et al. (2011), on which this note is based.

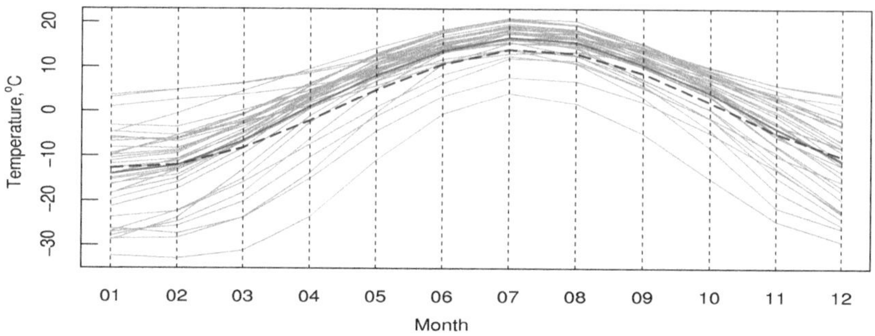

Fig. 24.1: Average annual temperature curves at 35 locations in Canada used as a running example in Ramsey and Silverman (2005). The continuous thick line is the simple average, the dashed line is an estimate that takes into account spatial locations and dependence.

Delicado et al. (2010) review recent contributions to the methodology for spatially distributed functional data; for geostatistical functional data, several approaches to kriging have been proposed. The focus of this note is the estimation of the mean function and of the functional principal components (FPC's). Accurate estimates of these quantities are required to develop many exploratory and inferential procedures of functional data. In order to define the mean function and the FPC's in an inferential setting, we assume that $\{X(\mathbf{s})\}$ is a random field taking values in $L^2 = L^2([0,1])$ which is is strictly stationary, i.e. for every shift $\mathbf{h}$,

$$(X(\mathbf{s}_1), X(\mathbf{s}_2), \ldots, X(\mathbf{s}_k)) \stackrel{d}{=} (X(\mathbf{s}_1+\mathbf{h}), X(\mathbf{s}_2+\mathbf{h}), \ldots, X(\mathbf{s}_k+\mathbf{h})), \qquad (24.1)$$

and square integrable in the sense that $E\|X(\mathbf{s})\|^2 < \infty$, where the norm is induced by the usual inner product in L^2. Under these conditions, the mean function $\mu(t) = EX(\mathbf{s};t)$ is well–defined. The FPC's also exist, and are defined as the eigenfunctions of the covariance operator

$$C(x) = E\left[\langle (X(\mathbf{s})-\mu), x\rangle (X(\mathbf{s})-\mu)\right], \quad x \in L^2.$$

We also assume that the field is also isotropic. A sufficient background in spatial statistics required to understand this note is presented in Chapters 2 and 3 of Gelfand et al. (2010).

For a sample of functions, $X_1, X_2, \ldots, X_N$, the sample mean is defined as $\bar{X}_N = N^{-1}\sum_{n=1}^{N} X_n$, and the sample covariance operator as

$$\widehat{C}(x) = N^{-1} \sum_{n=1}^{N} [\langle (X_n - \bar{X}_N), x \rangle (X_n - \bar{X}_N)], \quad x \in L^2.$$

The sample FPC's are typically computed as the eigenvalues of $\widehat{C}$. These are the estimates produced by several software packages, including the popular R package `fda`, see Ramsay et al. (2009). If the functions $X_k = X(\mathbf{s}_k)$ are spatially distributed, the sample mean and the sample FPC's need not be consistent, see Hörman and Kokoszka (2010). This happens if the spatial dependence is strong or if there are clusters of the points $\mathbf{s}_k$. We will demonstrate that in finite samples better estimators are available.

24.2 Estimation of the mean function

One approach to the estimation of mean function μ is to use the weighted average

$$\hat{\mu}_N = \sum_{n=1}^{N} w_n X(\mathbf{s}_n) \tag{24.2}$$

with the weights w_k minimizing $E \| \sum_{n=1}^{N} w_n X(\mathbf{s}_n) - \mu \|^2$ subject to the condition $\sum w_n = 1$. It can be shown that these weights satisfy the following system of $N+1$ linear equations:

$$\sum_{n=1}^{N} w_n = 1, \quad \sum_{k=1}^{N} w_k C_{kn} - r = 0, \quad n = 1, 2, \ldots, N, \tag{24.3}$$

where

$$C_{k\ell} = E[\langle X(\mathbf{s}_k) - \mu, X(\mathbf{s}_\ell) - \mu \rangle] \tag{24.4}$$

The estimation of the $C_{k\ell}$ is the central issue. Due to space constraints, we cannot provide all the details of the methods described below, we refer to Gromenko et al.(2011).

Method M1. This method postulates that at each time point t_j the scalar random field $X(\mathbf{s}; t_j)$ follows a parametric spatial model. The covariances $C_{k\ell}$ can be computed exactly by appropriately integrating the covariances of the models at each t_j, or approximately. This lead to two methods M1a (exact) and M1b (approximate).

Method M2. This method is based on the *functional* variogram

$$\begin{aligned} 2\gamma(\mathbf{s}_k, \mathbf{s}_\ell) &= E\|X(\mathbf{s}_k) - X(\mathbf{s}_\ell)\|^2 \\ &= 2E\|X(\mathbf{s}_k) - \mu\|^2 - 2E[\langle X(\mathbf{s}_k) - \mu, X(\mathbf{s}_\ell) - \mu \rangle] \\ &= 2E\|X(\mathbf{s}) - \mu\|^2 - 2C_{k\ell}. \end{aligned} \tag{24.5}$$

The variogram (24.5) can be estimated by its empirical counterparts, similarly as for scalar spatial data. A parametric model is fitted, and the $C_{k\ell}$ can then be computed using (24.5).

Method M3. This method uses a basis expansion of the functional data, it does not use the weighted sum (24.2). If the B_j form a basis, it estimates the inner products $\langle B_j, \mu \rangle$, and reconstructs an estimate of μ from them.

To compare the performance of these methods, Gromenko et al. (2011) simulated data at globally distributed points corresponding to the actual location of ionosonde stations; an ionosonde is a radar used to study the ionosphere. The quantity L is defined by

$$L = \frac{1}{R} \sum_{r=1}^{R} \int |\hat{\mu}_r(t) - \mu(t)| dt, \tag{24.6}$$

where R is the number of replications, was used to compare the methods. Figure 24.2 presents the results. It is seen that while methods M1 are the best, M2 is not significantly worse, and can be recommended, as it requires fitting only one variogram (the functional variogram (24.5)) rather than a separate variogram at every time point t_j. All methods that take spatial dependence into account are significantly better than the sample mean.

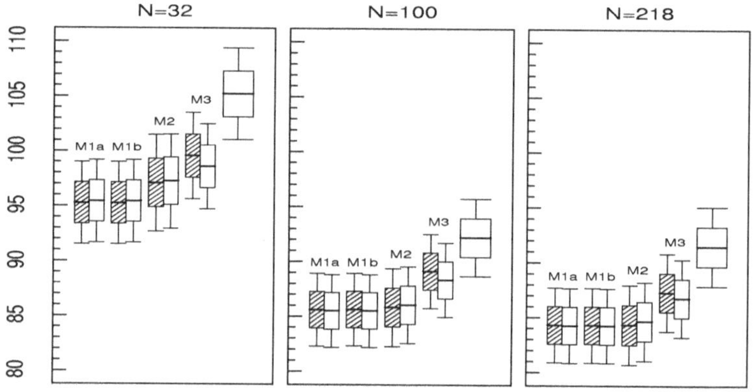

Fig. 24.2: Errors in the estimation of the mean function for sample sizes: 32, 100, 218. The dashed boxes are estimates using the Cressie Hawkins variogram, empty are for the Matheron variogram. The right–most box for each N corresponds to the simple average. The bold line inside each box plot represents the average value of L (24.6). The upper and lover sides of rectangles shows one standard deviation, and horizontal lines show two standard deviations.

24.3 Estimation of the functionalprincipal components

We now assume that the estimated mean function has been subtracted from the data, so in the following we set $EX(\mathbf{s}) = 0$. For the estimation of the FPC's, analogs of methods M2 and M3 can be developed. Extending Method M1 is also possible, but presents computational challenges because a parametric spatial model would need to be estimated for every pair (t_i, t_j). Evaluating the performance of such a method by simulations would take a prohibitively long time.

In both approaches, which we term CM2 and CM3, the FPC's are estimated by expansions of the form

$$v_j(t) = \sum_{\alpha=1}^{K} x_\alpha^{(j)} B_\alpha(t), \tag{24.7}$$

where the B_α are elements of an *orthonormal* basis.

Method CM2. Under the assumption of zero mean function, the covariance operator is the defined by $C(x) = E[\langle X(\mathbf{s}), x\rangle X(\mathbf{s})]$. It can be estimated by the weighted average

$$\widehat{C} = \sum_{k=1}^{N} w_k C_k, \tag{24.8}$$

where C_k is the operator defined by $C_k(x) = \langle X(\mathbf{s}_k), x\rangle X(\mathbf{s}_k)$. The weights w_k are computed by minimizing the Hilbert–Schmidt norm of the difference $\widehat{C} - C$ expanded into the basis $\{\langle\cdot, B_j\rangle\langle\cdot, B_k\rangle,\ 1 \le j,k \le K\}$, with suitably chosen K. A variogram in the space of Hilbert–Schmidt operators is suitably defined to fit a spatial dependence model. The orthonormality of the B_j plays a role in deriving the algorithm for the estimation.

Method CM3. The starting point is the expansion $X(\mathbf{s};t) \approx \sum_{j=1}^{K} \xi_j(\mathbf{s}) B_j(t)$, where, by the orthonormality of the B_j, the $\xi_j(\mathbf{s})$ form a stationary and isotropic mean zero spatial processes with observed values $\xi_j(\mathbf{s}_k) = \langle B_j, X(\mathbf{s}_k)\rangle$. Using the orthonormality of the B_j again, the estimation of C, can be reduced to the estimation of the means of the scalar spatial fields $\xi_i(\mathbf{s})\xi_j(\mathbf{s}),\ 1 \le i,j \le K$. The eigenfunctions of the estimated C can then be computed.

For the data generating processes designed to resemble the ionosonde data, methods CM2 and CM3 are fully comparable, but both are much better that the standard method which does not account for the spatial properties of the curves. Methods CM2 and CM3 have the same computational complexity.

24.4 Applications to inference for spatially distributed curves

Gromenko et al. (2011) developed a test of independence of two families of curves; there are curves $X(\mathbf{s}_k)$ and $Y(\mathbf{s}_k)$, and the problem is to test if the functional spatial fields X and Y are independent. The procedure requires estimation of the mean functions of the $X(\mathbf{s}_k)$ and the $Y(\mathbf{s}_k)$, as well as their FPC's. The problem is motivated by testing if decadal trends in the internal magnetic field of the Earth are correlated with the apparent long term trends in the ionospheric electron density. The test shows that the two families of curves are strongly dependent, but a highly significant conclusion is possible only after the spatial properties of the curves are taken into account. Using the estimators described in the previous sections, Gromenko and Kokoszka (2010) developed a test for the equality of means of the fields X and Y.

Acknowledgements This research was partially supported by NSF grants DMS-0804165 and DMS-0931948.

References

1. Delaigle, A., Hall, P.: Defining probability density function for a distribution of random functions. Ann. Stat. **38**, 1171–1193 (2010)
2. Delicado, P., Giraldo, R., Comas, C., Mateu, J.: Statistics for spatial functional data: some recent contributions. Environmetrics **21**, 224–239 (2010)
3. Gelfand, A. E., Diggle, P. J., Fuentes, M., Guttorp, P.: Handbook of Spatial Statistics. CRC Press (2010)
4. Gromenko, O., Kokoszka, P.: Testing the equality of mean functions of spatially distributed curves. Technical Report. Utah State University (2010)
5. Gromenko, O., Kokoszka, P., Zhu, L., Sojka, J.: Estimation problems for spatially distributed curves with application to testing the independence of ionospheric and magnetic field trends. Technical Report. Utah State University (2011)
6. Hörmann, S., Kokoszka, P.: Consistency of the mean and the principal components of spatially distributed functional data. Technical Report. Utah State University (2010)
7. Ramsay, J., Hooker, G., Graves, S.: Functional Data Analysis with R and MATLAB. Springer (2009)
8. Ramsay, J. O., Silverman, B. W.: Functional Data Analysis. Springer (2005)

Chapter 25
Structured Penalties for Generalized Functional Linear Models (GFLM)

Jaroslaw Harezlak, Timothy W. Randolph

Abstract GFLMs are often used to estimate the relationship between a predictor function and a response (e.g. a binary outcome). This manuscript provides an extension of a method recently proposed for functional linear models (FLM) - PEER (partially empirical eigenvectors for regression) to GFLM. The PEER approach to FLMs incorporates the structure of the predictor functions via a joint spectral decomposition of the predictor functions and a penalty operator into the estimation process via a generalized singular value decomposition. This approach avoids the more common two-stage smoothing basis approach to estimating a coefficient function. We apply our estimation method to a magnetic resonance spectroscopy data with binary outcomes.

25.1 Introduction

The coefficient function, β, in a GFLM represents the linear relationship between a transformed mean of the scalar response, y, and a predictor, x, formally written as $g(E[y]) = \int x(t)\beta(t)\,dt$, where $g(\cdot)$ is a so called link function. The problem typically involves a set of n responses $\{y_i\}_{i=1}^{n}$ corresponding to a set of observations $\{x_i\}_{i=1}^{n}$, each arising as a discretized sampling of an idealized function; i.e., $x_i \equiv (x_i(t_1),...,x_i(t_p))$, for some, $t_1,...,t_p$, of $[0,1]$. We assume the predictors have been sampled densely enough to capture a spatial predictor structure and thus $p >> n$.

Classical approaches (see for example, Crambes et.al., 2009 and Hall et.al., 2007) to the ill-posed problem of estimating β use either the eigenvectors determined by the predictors (e g. principal components regression - PCR) or methods based on

Jaroslaw Harezlak
Indiana University School of Medicine, Indianapolis, USA, e-mail: harezlak@iupui.edu

Timothy W. Randolph
Fred Hutchinson Cancer Research Center, Seattle, USA, e-mail: trandolp@fhcrc.org

a projection of the predictors onto a pre-specified basis and then obtaining an estimate from a generalized linear model formed by the transform coefficients. These methods, however, do not provide an analytically tractable way of incorporating the predictor's functional structure directly into the GFLM estimation process.

Here, we extend the framework developed in Randolph et al. (2011) which exploits the analytic properties of a joint eigen-decomposition for an operator pair—a penalty operator, L, and the operator determined by the predictor functions, X. More specifically, we exploit an eigenfunction basis whose functional structure is inherited by both L and X. As this basis is algebraically determined by the shared eigenproperties of both operators, it is neither strictly empirical (as with principal components) nor completely external to the problem (as in the case of B-spline regression models). Consequently, this approach avoids a separate fitting or smoothing step. We refer to this approach as PEER (partially empirical eigenvector regression) and here provide an overview of PEER as developed for FLMs and then describe the extension to GFLMs.

25.2 Overview of PEER

We consider estimates of the coefficient-function β arising from a squared-error loss with quadratic penalty. These may be expressed as

$$\tilde{\beta}_{\alpha,L} = \arg\min_{\beta}\{||y - X\beta||^2_{\mathbb{R}^n} + \alpha||L\beta||^2_{L^2}\}, \tag{25.1}$$

where L is a linear operator.

Within this classical formulation, PEER exploits the *joint* spectral properties of the operator pair (X,L). This perspective allows the estimation process to be guided by an informed construction of L. It succeeds when structure in the generalized singular vectors of the pair (X,L) is commensurate with the appropriate structure of β. How L imparts this structure via the GSVD is detailed in Randolph et al. (2011), and so the discussion here is restricted to providing the notation necessary for the GFLM setting.

A least-squares solution, $\hat{\beta}$, satisfies the normal equations $X'X\beta = X'y$. Estimates arise as minimizers, $\hat{\beta} = \arg\min_{\beta}||y - X\beta||^2$, but there are infinitely many such solutions and so regularization is required. The least-squares solution with a minimum norm is provided by the singular value decomposition (SVD): $X = UDV'$ where the left and right singular vectors, u_k and v_k, are the columns of U and V, respectively, and $D = \text{diag}\{\sigma_k\}_{k=1}^{p}$, with $\sigma_1 \geq \sigma_2 \geq \ldots \geq \sigma_r$ ($r = \text{rank(X)}$, $\sigma_r \approx 0$). The minimum-norm solution is $\hat{\beta}_+ = X^{\dagger}y = \sum_{\sigma_k \neq 0}(1/\sigma_k)u_k'y\,v_k$, where $X^{\dagger}$ denotes the Moore-Penrose inverse of X: $X^{\dagger} = VD^{\dagger}U'$, where $D^{\dagger} = \text{diag}\{1/\sigma_k \text{ if } \sigma_k \neq 0;\ 0 \text{ if } \sigma_k = 0\}$.

For functional data, however, $\hat{\beta}_+$ is an unstable estimate which motivates PCR estimate: $\tilde{\beta}_{\text{PCR}} = V_d D_d^{-1} U_d' y$ where $A_d \equiv \text{col}[a_1, ..., a_d]$ denotes the first d columns of a matrix A. Another classical way to obtain a more stable estimate in terms of the ordinary singular vectors is to impose a ridge penalty, $L = I$ (see Hoerl et.al., 1970) for which the minimizing solution to (25.1) is

$$\tilde{\beta}_{\alpha,\mathrm{R}} = (X'X + \alpha I)^{-1}X'y = \sum_{k=1}^{r} \left(\frac{\sigma_k^2}{\sigma_k^2 + \alpha} \right) \frac{1}{\sigma_k} u_k' y v_k, \tag{25.2}$$

For a given linear operator L and parameter $\alpha > 0$, the estimate in (25.1) takes the form

$$\tilde{\beta}_{\alpha,L} = (X'X + \alpha L'L)^{-1}X'y. \tag{25.3}$$

This cannot be expressed using the singular vectors of X alone, but the generalized singular value decomposition of the pair (X,L) provides a tractable and interpretable vector expansion.

We provide here a short description of the GSVD method. Additional details are available in the Randolph et al. (2011). It is assumed that X is an $n \times p$ matrix ($n \leq p$) of rank n, L is an $m \times p$ matrix ($m \leq p$) of rank m and the null spaces of X and L intersect trivially: $\mathrm{Null}(L) \cap \mathrm{Null}(X) = \{0\}$. This condition is needed to obtain a unique solution and is natural in our applications. It is not required, however, to implement the methods. We also assume that $n \leq m \leq p$, with $m + n \geq p$, and the rank of $Z := [X'L']'$ is at least p.

Then there exist orthogonal matrices U and V, a nonsingular matrix W and diagonal matrices S and M such that

$$\begin{aligned} X &= U\underline{S}W^{-1}, & \underline{S} &= \begin{bmatrix} 0 & S \end{bmatrix}, & S &= \mathrm{diag}\{\sigma_k\} \\ L &= V\underline{M}W^{-1}, & \underline{M} &= \begin{bmatrix} I & 0 \\ 0 & M \end{bmatrix}, & M &= \mathrm{diag}\{\mu_k\}. \end{aligned} \tag{25.4}$$

The diagonal entries of S and M are ordered as

$$\begin{matrix} 0 \leq \sigma_1 \leq \sigma_2 \leq \ldots \sigma_n \leq 1 \\ 1 \geq \mu_1 \geq \mu_2 \geq \ldots \mu_n \geq 0 \end{matrix} \quad \text{where} \quad \sigma_k^2 + \mu_k^2 = 1, \quad k = 1, \ldots, n. \tag{25.5}$$

Denote the columns of U, V and W by u_k, v_k and w_k, respectively. For the majority of matrices L the generalized singular vectors u_k and v_k are not the same as the ordinary singular vectors of X. One case when they are the same is for $L = I$.

The penalized estimate is a linear combination of the columns of W and the solution to the penalized regression in (25.1) can be expressed as

$$\tilde{\beta}_{\alpha,L} = \sum_{k=p-n+1}^{p} \left(\frac{\sigma_k^2}{\sigma_k^2 + \alpha \mu_k^2} \right) \frac{1}{\sigma_k} u_k' y w_k, \tag{25.6}$$

We refer to any $\tilde{\beta}_{\alpha,L}$ ($L \neq I$) as a PEER (partially empirical eigenvectors for regression) estimate. The utility of a penalty L depends on whether the true coefficient function shares structural properties with this GSVD basis. With regard to this, the importance of the parameter α may be reduced by a judicious choice of L (Varah, 1979) since the terms in (25.6) corresponding to the vectors $\{w_k : \mu_k = 0\}$ are independent of α.

25.2.1 Structured and targeted penalties

A *structured penalty* refers to a second term in (25.1) that involves an operator chosen to encourage certain functional properties in the estimate. Here we give examples of such penalties. If we begin with some knowledge about the subspace of functions in which the informative signal resides, then we can define a penalty based on it. For example, suppose $\beta \in \mathscr{Q} := \text{span}\{q_j\}_{j=1}^d$ for some $q_j \in L^2(\Omega)$. Set $Q = \sum_{j=1}^d q_j \otimes q_j$ and consider the orthogonal projection $P_{\mathscr{Q}} = QQ^\dagger$. Define $L_{\mathscr{Q}} = I - P_{\mathscr{Q}}$, then $\beta \in \text{Null}(L_{\mathscr{Q}})$ and $\tilde{\beta}_{\alpha,L_{\mathscr{Q}}}$ is unbiased.

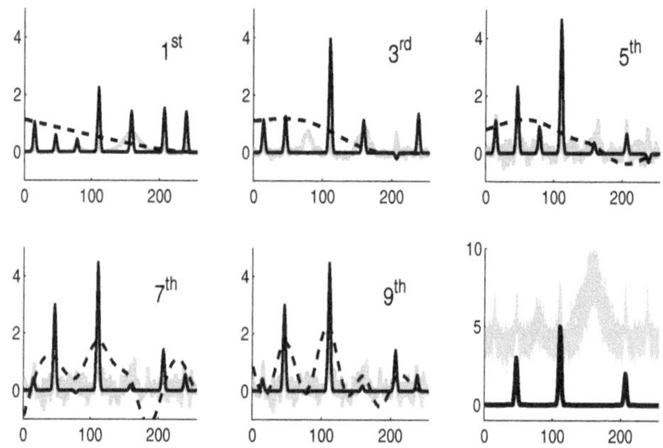

Fig. 25.1: Partial sums of penalized estimates. The first five odd-numbered partial sums from (25.6) for three penalties, L: 2nd-derivative (dotted black), ridge (solid gray), targeted (solid black); see text. The last panel exhibits β (solid black) and several predictors, x_i (light gray), from the simulation.

Figure 25.1 illustrates the estimation process with plots of some partial sums from equation (25.6) for three estimates. The ridge estimate is, naturally, dominated by the leading eigenvectors of X. The second-derivative penalized estimate is dominated first by low-frequency structure. The targeted PEER estimate shown here begins with the largest peaks corresponding the largest GSV components, but quickly converges to the informative features.

25.2.2 Analytical properties

For a general linear penalty operator L, the analytic form of the estimate and its basic properties of bias, variance and MSE are provided in Randolph et al. (2011). Any direct comparison between estimates using different penalty operators is confounded by the fact there is no simple connection between the generalized singular values/vectors and the ordinary singular values/vectors. Therefore, Randolph et al. (2011) first considered the case of targeted or projection-based penalties. Within this class, a parameterized family of estimates is comprised of *ordinary* singular values/vectors. Since the ridge and PCR estimates are contained in (or a limit of) this family, an analytical comparison with some targeted PEER estimates is possible.

25.3 Extension to GFLM

Generalization of PEER to the GFLM setting proceeds via replacement of the continuous responses $y_1, \ldots, y_n$ by responses coming from a general exponential family whose expectations $g(\mu_i)$ are linearly related to a functional predictor X_i. We specifically focus here on the binary responses and logistic regression setting. We replace the least squares criterion by a likelihood function appropriate for the member of the exponential family distribution and find the estimate of β by minimizing the following expression:

$$\tilde{\beta}_{\alpha,L} = \arg\min_\beta \{\sum_i l(g(y_i), X_i\beta) + \alpha||L\beta||^2_{L^2}\}, \tag{25.7}$$

where $l(\cdot)$ is the log-likelihood function. The fitting procedure for PEER in GFLM setting is a modification of an iteratively reweighted least squares (IRLS) method. In a similar spirit to the BLUP and REML estimation of the tuning parameter in the linear mixed model equivalent setting, we select the tuning parameter using the penalized quasi-likelihood (PQL) method associated with the generalized linear mixed models. REML criterion is preferred here, since it has been been shown to outperform the GCV method (see Reiss and Ogden, 2007).

25.4 Application to a magnetic resonance spectroscopy data

We apply the GFLM-PEER method to study the relationship of the magnetic resonance spectroscopy (MRS) data and neurocognitive impairment arising from the HIV Neuroimaging Consortium (HIVNC) study (see Harezlak et al., 2011 for the study description). In particular, we are interested in studying the relationship of the metabolite level concentrations in the brain and classification of the patients into

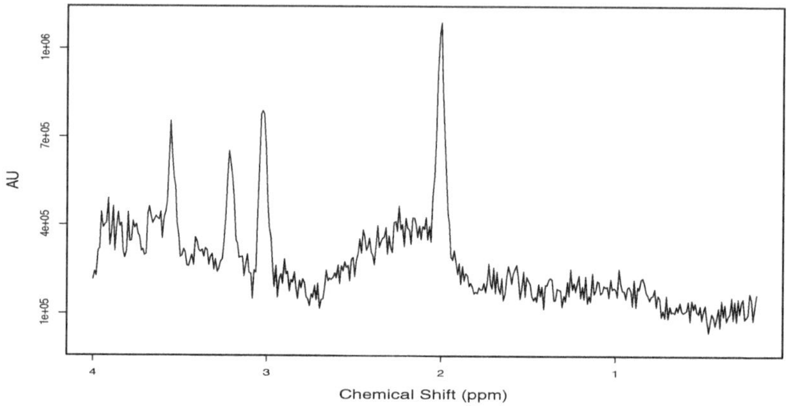

Fig. 25.2: A sample magnetic resonance spectroscopy (MRS) spectrum displaying brain metabolite levels in one frontal gray matter (FGM) voxel.

neurologically asymptomatic (NA) and neurocognitively impaired (NCI) groups. The predictor functions come in the form of spectra for each studied voxel in the brain (see Figure 25.2). Our method provides promising results when compared to the more established functional regression methods which do not take into account the external pure metabolite spectra profiles . We also obtain interpretable functional regression estimates that do not rely on a two–step procedure estimating the metabolite concentrations first and then using them as predictors in a logistic regression model.

25.5 Discussion

Estimation of the coefficient function β in a generalized functional linear model requires a regularizing constraint. When the data contain natural spatial structure (e.g., as derived from the physics of the problem), then the regularizing constraint should acknowledge this. In the FLM case, exploiting properties of the GSVD provided a new analytically-rigorous approach for incorporating spatial structure into functional linear models. In the GFLM case, we extend the IRLS procedure to take into account the penalty operator.

A PEER estimate is intrinsically based on GSVD factors. This fact guides the choice of appropriate penalties for use in both FLM and GFLM. Heuristically, the structure of the penalty's least-dominant singular vectors should be commensurate with the informative structure of β. The properties of an estimate are determined jointly by this structure and that in the set of predictors. The structure of the generalized singular functions provides a mechanism for using a priori knowledge in choosing a penalty operator allowing, for instance, one to target specific types of structure and/or avoid penalizing others. The effect a penalty has on the properties of the estimate is made clear by expanding the estimate in a series whose terms are the generalized singular vectors/values for X and L.

References

1. Crambes, C., Kneip, A., Sarda, P.: Smoothing spline estimators for functional linear regression. Ann. Stat. **37**, 35–72 (2009)
2. Hall, P., Horowitz, J.L.: Methodology and convergence rates for functional linear regression. Ann. Stat. **35**, 70–91 (2007)
3. Harezlak, J., Buchthal, S., Taylor, M., Schifitto, G., Zhong, J., Daar, E.: Persistence of HIV-associated cognitive impairment, inflammation and neuronal injury in the HAART era. AIDS, in press, doi: 10.1097/QAD.0b013e3283427da7 (2011)
4. Hoerl, A.E., Kennard, R.W.: Ridge regression: Biased estimation for nonorthogonal problems. Technometrics **12**, 55-67 (1970)
5. Randolph, T., Harezlak, J.,Feng, Z.: Structured penalties for functional linear models: partially empirical eigenvectors for regression. Unpublished manuscript (2011)
6. Reiss, P.T., Ogden, R.T.: Functional principal component regression and functional partial least squares. J. Am. Stat. Assoc. **102**, 984-986 (2007)
7. Varah, J.M.: A practical examination of some numerical methods for linear discrete ill-posed problems. SIAM Review **21** (1), 100-111 (1979)

Chapter 26
Consistency of the Mean and the Principal Components of Spatially Distributed Functional Data

Siegfried Hörmann, Piotr Kokoszka

Abstract This paper develops a framework for the estimation of the functional mean and the functional principal components when the functions form a random field. We establish conditions for the sample average (in space) to be a consistent estimator of the population mean function, and for the usual empirical covariance operator to be a consistent estimator of the population covariance operator.

26.1 Introduction

In this paper we study functional data observed at spatial locations. That is, the data consist of curves $X(\mathbf{s}_k;t)$, $t \in [0,T]$, observed at spatial points $\mathbf{s}_1, \mathbf{s}_2, \ldots, \mathbf{s}_N$. Such data structures are quite common, but often the spatial dependence and the spatial distribution of the points $\mathbf{s}_k$ are not taken into account. A well–known example is the Canadian temperature and precipitation data used as a running example in Ramsay and Silverman (2005). The annual curves are available at 35 locations, some of which are quite close, and so the curves look very similar, others are very remote with notably different curves. Ramsay and Silverman (2005) use the functional principal components and the functional linear model as exploratory tools.

Due to the importance of such data structures it is useful to investigate when the commonly used techniques designed for iid functional data retain their consistency for spatially distributed data, and when they fail. We establish conditions for consistency, or lack thereof, for the functional mean and the functional principal components. Our conditions combine the spatial dependence of the curves $X(\mathbf{s}_k;\cdot)$ and the distribution of the data locations $\mathbf{s}_k$.

Siegfried Hörmann
Université Libre de Bruxelles, Belgium, e-mail: shormann@ulb.ac.be

Piotr Kokoszka
Utah State University, USA, e-mail: piotr.kokoszka@usu.edu

While in time series analysis, the process is indexed by an equispaced scalar parameter, we need here a d-dimensional index space. For model building this makes a big difference since the dynamics and dependence of the process have to be described in "all directions", and the typical recurrence equations used in time series cannot be employed. The model building is further complicated by the fact that the index space is often continuous (geostatistical data). Rather than defining a random field $\{\xi(\mathbf{s});\ \mathbf{s}\in\mathbb{R}^d\}$ via specific model equations, dependence conditions are imposed, in terms of the decay of the covariances or using mixing conditions. Another feature peculiar to random field theory is the design of the sampling points; the distances between them play a fundamental role. Different asymptotics hold in the presense of clusters and for sparsely distributed points. At least three types of point distributions have been considered in the literature: When the region R_N where the points $\{\mathbf{s}_{i,N}; 1\leq i\leq N\}$ are sampled remains bounded, then we are in the so-called *infill domain sampling* case. Classical asymptotic results, like the law of large numbers or the central limit theorem will usually fail, see Lahiri (1996). The other extreme situation is described by the *increasing domain sampling*. Here a minimum separation between the sampling points $\{\mathbf{s}_{i,N}\}\in R_N$ for all i and N is required. We shall also explore the *nearly infill* situation studied e.g. by Lahiri (2003) and Park et al. (2009). In this case the domain of the sampling region becomes unbounded ($\mathrm{diam}(R_N)\to\infty$), but at the same time the number of sites in any given subregion tends to infinity.

26.2 Model and dependence assumptions

We assume $\{X(\mathbf{s}),\mathbf{s}\in\mathbb{R}^d\}$ is a random field taking values in $L^2=L^2([0,1])$, i.e. each $X(\mathbf{s})$ is a square integrable function defined on $[0,1]$. The value of this function at $t\in[0,1]$ is denoted by $X(\mathbf{s};t)$. With the usual inner product in L^2, the norm of $X(\mathbf{s})$ is

$$\|X(\mathbf{s})\|=\langle X(\mathbf{s}),X(\mathbf{s})\rangle^{1/2}=\left\{\int X^2(\mathbf{s};t)dt\right\}^{1/2}.$$

We assume that the spatial process $\{X(\mathbf{s}),\mathbf{s}\in\mathbb{R}^d\}$ is strictly stationary, i.e. for every $\mathbf{h}\in\mathbb{R}^d$,

$$\big(X(\mathbf{s}_1),X(\mathbf{s}_2),\dots,X(\mathbf{s}_k)\big)\overset{d}{=}\big(X(\mathbf{s}_1+\mathbf{h}),X(\mathbf{s}_2+\mathbf{h}),\dots,X(\mathbf{s}_k+\mathbf{h})\big). \tag{26.1}$$

We also assume that it is square integrable in the sense that

$$E\|X(\mathbf{s})\|^2<\infty. \tag{26.2}$$

Under (26.1) and (26.2), the common mean function is denoted by $\mu=EX(\mathbf{s})$. To develop an estimation framework for μ, we impose different assumptions on the decay of $E\langle X(\mathbf{s}_1)-\mu,X(\mathbf{s}_2)-\mu\rangle$, as the distance between $\mathbf{s}_1$ and $\mathbf{s}_2$ increases.

We shall use the distance function defined by the Euclidian norm in $\mathbb{R}^d$, denoted $\|\mathbf{s}_1 - \mathbf{s}_2\|_2$, but other distance functions can be used as well.

Assumption 1 *The spatial process* $\{X(\mathbf{s}), \mathbf{s} \in \mathbb{R}^d\}$ *is strictly stationary and square integrable, i.e. (26.1) and (26.2) hold. In addition,*

$$|E\langle X(\mathbf{s}_1) - \mu, X(\mathbf{s}_2) - \mu\rangle| \leq h(\|\mathbf{s}_1 - \mathbf{s}_2\|_2), \tag{26.3}$$

where $h : [0,\infty) \to [0,\infty)$ *with* $h(x) \searrow 0$, *as* $x \to \infty$.

The following example illustrates how Assumption 1 can be veryfied under strong mixing.

Example 1 *Let* $\{e_j\}$ *be the orthonormal basis obtained from the eigenfunctions of the covariance operator* $C(y) = E\langle X(\mathbf{s}) - \mu, y\rangle(X(\mathbf{s}) - \mu)$. *Then we obtain the Karhunen-Loève expansion of*

$$X(\mathbf{s}) - \mu = \sum_{j\geq 1} \xi_j(\mathbf{s}) e_j, \tag{26.4}$$

with $\xi_j(\mathbf{s}) = \langle X(\mathbf{s}) - \mu, e_j\rangle$, $j \geq 1$. *Suppose that the functional field* $\{X(\mathbf{s}), \mathbf{s} \in \mathbb{R}^d\}$ *is strictly stationary and* α-mixing. *That is*

$$\sup_{(A,B)\in\sigma(X(\mathbf{s}))\times\sigma(X(\mathbf{s}+\mathbf{h}))} |P(A)P(B) - P(A\cap B)| \leq \alpha(\mathbf{h}),$$

with $\alpha(\mathbf{h}) \to 0$ *if* $\|\mathbf{h}\|_2 \to \infty$. *Then the* $\xi_j(\mathbf{s})$ *inherit the* α*-mixing property, and thus (26.3) can be established using*

$$|E\langle X(\mathbf{s}_1) - \mu, X(\mathbf{s}_2) - \mu\rangle| \leq \sum_{j\geq 1} |E[\xi_j(\mathbf{s}_1)\xi_j(\mathbf{s}_2)]|$$

in combination with classical covariance inequalities for mixing scalar processes (e.g. those in Rio (1993)). We refer to Hörmann and Kokoszka (2011+) for details.

Assumption 1 is appropriate when studying estimation of the mean function. For the estimation of the covariance operator, we need to impose a different assumption. Recall that if z and y are elements in some Hilbert space H with norm $\|\cdot\|_H$, the operator $z \otimes y$, is defined by $z \otimes y(x) = \langle z, x\rangle y$. Further, if K is a linear operator in a separable Hilbert space H, then it is said to be Hilbert-Schmidt, if for some orthonormal basis $\{e_i\}$ of H

$$\|K\|_{\mathscr{H}}^2 := \sum_{i\geq 1} \|K(e_i)\|_H^2 < \infty.$$

Then $\|\cdot\|_{\mathscr{H}}$ defines a norm on the space of all operators satisfying this condition. The norm is independent of the choice of the basis. This space is again a Hilbert space with the inner product

$$\langle K_1, K_2\rangle_{\mathscr{H}} = \sum_{i\geq 1} \langle K_1(e_i), K_2(e_i)\rangle.$$

In the following assumption, we suppose that the mean of the functional field is zero. This is justified by notational convenience and because we deal with the consistent estimation of the mean function separately.

Assumption 2 *The spatial process $\{X(\mathbf{s}), \mathbf{s} \in \mathbb{R}^d\}$ is strictly stationary with zero mean and with 4 moments, i.e. $E\langle X(\mathbf{s}), x\rangle = 0$, $\forall x \in L^2$, and $E\|X(\mathbf{s})\|^4 < \infty$. In addition,*

$$\left|E\langle X(\mathbf{s}_1)\otimes X(\mathbf{s}_1) - C, X(\mathbf{s}_2)\otimes X(\mathbf{s}_2) - C\rangle_{\mathscr{H}}\right| \le H\left(\|\mathbf{s}_1 - \mathbf{s}_2\|_2\right), \tag{26.5}$$

where $H : [0,\infty) \to [0,\infty)$ with $H(x) \searrow 0$, as $x \to \infty$.

26.3 The sampling schemes

As already noted, for spatial processes assumptions on the distribution of the sampling points are as important as those on the covariance structure. To formalize the different sampling schemes introduced in the Introduction, we propose the following measure of "minimal dispersion" of some point cloud $\mathfrak{S}$:

$$I_\rho(\mathbf{s},\mathfrak{S}) = |\{\mathbf{y} \in \mathfrak{S} : \|\mathbf{s}-\mathbf{y}\|_2 \le \rho\}|/|\mathfrak{S}| \quad \text{and} \quad I_\rho(\mathfrak{S}) = \sup\left\{I_\rho(\mathbf{s},\mathfrak{S}),\ \mathbf{s} \in \mathfrak{S}\right\},$$

where $|\mathfrak{S}|$ denotes the number of elements of $\mathfrak{S}$. The quantity $I_\rho(\mathfrak{S})$ is the maximal fraction of $\mathfrak{S}$–points in a ball of radius ρ centered at an element of $\mathfrak{S}$. Notice that $1/|\mathfrak{S}| \le I_\rho(\mathfrak{S}) \le 1$. We call $\rho \mapsto I_\rho(\mathfrak{S})$ the *intensity function* of $\mathfrak{S}$.

Definition 1 *For a sampling scheme $\mathfrak{S}_N = \{\mathbf{s}_{i,N}; 1 \le i \le S_N\}$, $S_N \to \infty$, we consider the following conditions:*

(i) there is a $\rho > 0$ such that $\limsup_{N\to\infty} I_\rho(\mathfrak{S}_N) > 0$;
(ii) for some sequence $\rho_N \to \infty$ we have $I_{\rho_N}(\mathfrak{S}_N) \to 0$;
(iii) for any fixed $\rho > 0$ we have $S_N I_\rho(\mathfrak{S}_N) \to \infty$.

We call a deterministic sampling scheme $\mathfrak{S}_N = \{\mathbf{s}_{i,N}; 1 \le i \le S_N\}$

Type A *if (i) holds;*
Type B *if (ii) and (iii) hold;*
Type C *if (ii) holds, but there is a $\rho > 0$ such that $\limsup_{N\to\infty} S_N I_\rho(\mathfrak{S}_N) < \infty$.*

If the sampling scheme is stochastic we call it Type A, B *or* C *if relations (i), (ii) and (iii) hold with $I_\rho(\mathfrak{S}_N)$ replaced by $EI_\rho(\mathfrak{S}_N)$.*

Type A sampling is related to purely infill domain sampling which corresponds to $I_\rho(\mathfrak{S}_N) = 1$ for all $N \ge 1$, provided ρ is large enough. However, in contrast to the purely infill domain sampling, it still allows for a non-degenerate asymptotic theory for sparse enough subsamples (in the sense of Type B or C). A brief reflection shows that assumptions (i) and (ii) are mutually exclusive. Combining (ii) and (iii) implies that the points intensify (at least at certain spots) excluding the purely increasing domain sampling. Hence the Type B sampling corresponds to the nearly infill domain sampling. If only (ii) holds, but (iii) does not (Type C sampling) then the sampling scheme corresponds to purely increasing domain sampling.

26.4 Some selected results

Our first goal is to establish the consistency of the sample mean for functional spatial data. We consider Type B or Type C sampling and obtain rates of convergence. We consider here only a general setup. In Hörmann and Kokoszka (2011+) we have demonstrated that the obtained rates can be substantially improved in special cases. We also refer to Hörmann and Kokoszka (2011+) for the proofs of the following results.

For independent or weakly dependent functional observations X_k,

$$E\left\|\frac{1}{N}\sum_{k=1}^{N} X_k - \mu\right\|^2 = O\left(N^{-1}\right). \tag{26.6}$$

Proposition 1 below shows that for general functional spatial processes, the rate of consistency may be much slower than $O\left(N^{-1}\right)$; it is the maximum of $h(\rho_N)$ and $I_{\rho_N}(\mathfrak{S}_N)$ with ρ_N from (ii) of Definition 1. Intuitively, the sample mean is consistent if there is a sequence of increasing balls which contain a fraction of points which tends to zero, and the decay of the correlations compensates for the increasing radius of these balls.

Proposition 1 *Let Assumption 1 hold, and assume that $\mathfrak{S}_N$ defines a* non-random *design of Type A, B or C. Then for any $\rho_N > 0$,*

$$E\left\|\frac{1}{N}\sum_{k=1}^{N} X(\mathbf{s}_{k,N}) - \mu\right\|^2 \leq h(\rho_N) + h(0) I_{\rho_N}(\mathfrak{S}_N). \tag{26.7}$$

Hence, under the Type B or Type C non-random sampling, with ρ_N as in (ii) of Definition 1, the sample mean is consistent.

A question that needs to be addressed is whether the bound obtained in Proposition 1 is optimal. It is not surprising that (26.7) will not be *uniformly* optimal. This is because the assumptions in Proposition 1 are too general to give a precise rate for all the cases covered. In some sense, however, the rate (26.7) is optimal, as it is possible to construct examples which attain the bound (26.7). (See Example 5.2 in Hörmann and Kokoszka (2011+).)

Next we formulate the analogue of Proposition 1 establishing the rate of consistency of the empirical covariance operator

$$\hat{C}_N = \frac{1}{N}\sum_{k=1}^{N} X_k \otimes X_k.$$

(We assume that $EX_1 = 0$.)

Proposition 2 *Let Assumption 2 hold, and assume that $\mathfrak{S}_N$ defines a* non-random *design of Type A, B or C. Then for any $\rho_N > 0$*

$$E\left\|\widehat{C}_N - C\right\|_{\mathscr{H}}^2 \leq H(\rho_N) + H(0)I_{\rho_N}(\mathfrak{S}_N). \tag{26.8}$$

Hence under the Type B or Type C non-random sampling, with ρ_N as in (ii) of Definition 1, the empirical covariance operator is consistent.

The eigenvalues $\hat{\lambda}_{i,N}$ and eigenfunctions $\hat{e}_{i,N}$, $i \geq 1$, of the empirical covariance operator $\hat{C}_N$ are used to estimate the eigenvalues and eigenfunctions λ_i and e_i, respectively, of its population version $C = EX_1 \otimes X_1$. Using standard arguments the rates obtained in (26.8) can be transformed directly to the convergence rates of eigenvalues and eigenfunctions.

Corollary 1 *Assume that $\lambda_1 > \lambda_2 > \cdots > \lambda_{q+1}$ and let the assumptions of Proposition 2 hold. Define $\hat{c}_j = \operatorname{sign}\langle e_j, \hat{e}_{j,N}\rangle$. Then there is a constant κ depending only on the process $\{X(\mathbf{s},\cdot); \mathbf{s} \in \mathbb{R}^d\}$, such that*

$$\max_{1\leq i\leq q}\left\{E|\hat{\lambda}_{i,N} - \lambda_i|^2 + E\|\hat{e}_{i,N} - \hat{c}_i e_i\|^2\right\} \leq \kappa\left(H(\rho_N) + H(0)I_{\rho_N}(\mathfrak{S}_N)\right) \quad \forall N \geq 1.$$

Corollary 1 is important, as it shows when the functional principal components, defined by $Y_i = \langle X(\mathbf{s}), e_i\rangle$, can be consistently estimated. The introduction of the constants $\hat{c}_j$ is necessary, as the normalized eigenfunctions e_j (we assume $\|e_j\| = 1$) are only unique up to sign.

Our last result shows when the estimator $\hat{C}_N$ can be inconsistent.

Proposition 3 *Suppose representation (26.4) holds with stationary mean zero Gaussian processes ξ_j such that*

$$E[\xi_j(\mathbf{s})\xi_j(\mathbf{s}+\mathbf{h})] = \lambda_j\rho_j(h), \quad h = \|\mathbf{h}\|,$$

where each ρ_j is a continuous correlation function, and $\sum_j \lambda_j < \infty$. Assume the processes ξ_j and ξ_i are independent if $i \neq j$. If $\mathfrak{S}_N = \{\mathbf{s}_1, \mathbf{s}_2, \ldots, \mathbf{s}_n\} \subset \mathbb{R}^d$ with $\mathbf{s}_n \to \mathbf{0}$, then

$$\lim_{N\to\infty} E\|\widehat{C}_N - X(\mathbf{0}) \otimes X(\mathbf{0})\|_{\mathscr{H}}^2 = 0. \tag{26.9}$$

The proposition shows that the empirical operator approaches the random operator $X(\mathbf{0}) \otimes X(\mathbf{0})$. Thus it cannot be consistent.

Acknowledgements Research supported by the Banque National de Belgique and Communauté française de Belgique - Actions de Recherche Concertées

References

1. Hörmann, S., Kokoszka, P.: Consistency of the mean and the principal components of spatially distributed functional data. Submitted (2011)
2. Lahiri, S.N.: On the inconsistency of estimators based on spatial data under infill asymptotics. Sankhyā Ser. A, **58**, 403–417 (1996)
3. Lahiri, S.N.: Central limit theorems for weighted sums of a spatial process under a class of stochastic and fixed designs. Sankhyā Ser. A, **65**, 356–388 (2003)
4. Park, B.U., Kim, T.Y., Park, J-S., Hwang, S.Y.: Practically applicable central limit theorems for spatial statistics. Mathematical Geosciences, **41**, 555–569 (2009)
5. Ramsay, J.O. and Silverman, B.W.: Functional Data Analysis. Springer, New York (2005)
6. Rio, E.: Covariance inequalities for strongly mixing processes. Ann. Inst. H. Poincaré Probab. Statist., **29**, 587–597 (1993)

Chapter 27
Kernel Density Gradient Estimate

Ivana Horová, Kamila Vopatová

Abstract The aim of this contribution is to develop a method for a bandwidth matrix choice for kernel estimate of the first partial derivatives of the unknown density.

27.1 Kernel density estimator

Let a d-variate random sample $\mathbf{X}_1, \ldots, \mathbf{X}_n$ come from distribution with a density f. The kernel density estimator $\hat{f}$ is defined

$$\hat{f}(\mathbf{x}, H) = \frac{1}{n}\sum_{i=1}^{n} K_H(\mathbf{x} - \mathbf{X}_i) = \frac{1}{n}|H|^{-1/2}\sum_{i=1}^{n} K\big(H^{-1/2}(\mathbf{x} - \mathbf{X}_i)\big).$$

H is a symmetric positive definite $d \times d$ matrix called the bandwidth matrix, where $|H|$ stands for the determinant of H. The kernel function K is often taken to be a d-variate probability density function satisfying $\int_{\mathbb{R}^d} K(\mathbf{x})\mathrm{d}\mathbf{x} = 1$, $\int_{\mathbb{R}^d} \mathbf{x}K(\mathbf{x})\mathrm{d}\mathbf{x} = 0$, $\int \mathbf{x}\mathbf{x}^T K(\mathbf{x})\mathrm{d}\mathbf{x} = \beta_2 I_d$, I_d is an identity matrix and $\mathbf{x} = (x_1, \ldots, x_d)^T \in \mathbb{R}^d$ is a generic vector.

27.2 Kernel gradient estimator

Let $Df(\mathbf{x})$ denote a vector of the first partial derivatives, also referred as a gradient. Kernel gradient estimator is defined in a similar way as the kernel density estimate:

Ivana Horová
Masaryk University, Brno, Czech Republic, e-mail: horova@math.muni.cz

Kamila Vopatová
Masaryk University, Brno, Czech Republic, e-mail: vopatova@mail.muni.cz

$$\widehat{Df}(\mathbf{x},H)=\frac{1}{n}\sum_{i=1}^{n}DK_H(\mathbf{x}-\mathbf{X}_i),$$

where $DK_H(\mathbf{x})=|H|^{-1/2}H^{-1/2}DK\big(H^{-1/2}(\mathbf{x})\big)$ is a column vector of the first partial derivatives of the kernel function.

For a gradient estimator $\widehat{Df}(\mathbf{x},H)$ the Mean Integrated Square Error (MISE), the measure of the quality, is a matrix. Since performance based on scalar quantities rather than on matrices is easier, it is appropriate to apply a matrix norm. In accordance with the paper by Duong et al. (2008) the trace norm will be used and thus the trace of asymptotic mean square error (TAMISE) can be expressed as a sum of the trace of integrated variance (TIVar) and the trace of integrated square bias (TIBias^2). The resulting TAMISE-formula is of the form

$$\begin{aligned}\mathrm{TAMISE}\big[\widehat{Df}(\cdot,H)\big]&=\mathrm{TAMISE}(H)=\mathrm{TIVar}(H)+\mathrm{TIBias}^2(H)\\&=n^{-1}|H|^{-1/2}\mathrm{tr}\big[H^{-1}R(DK)\big]+\tfrac{1}{4}\beta_2(K)^2\,\mathrm{vech}^T H\Psi_6\mathrm{vech}H,\end{aligned}$$

where $R(g)=\int_{\mathbb{R}^d}g(\mathbf{x})g^T(\mathbf{x})d\mathbf{x}$ for any square integrable vector-valued function g, and vech is the vector half operator, so that $\mathrm{vech}H$ is a $d(d+1)/2\times 1$ vector of stacked columns of the lower triangular half of H. The term Ψ_6 involves partial derivatives up to the sixth order (for details see e.g. Duong et al. (2008), Vopatová et al. (2010)). Duong et al. (2008) proved the following proposition.

Proposition 27.1. *Let H_T be a bandwidth matrix minimizing* TAMISE(H)*, i.e.*

$$H_T=\arg\min\mathrm{TAMISE}(H).$$

Then $H_T=O\big(n^{-\frac{2}{d+6}}\big)J$, where J is a $d\times d$ matrix of ones.

The paper Horová et al. (2010) has been dealing with bandwidth matrix choice for bivariate density estimates. In this case assuming that the bandwidth matrix is diagonal there exist explicit solutions of the corresponding minimization problem $H_{opt}=\arg\min\mathrm{AMISE}(H)$. Then the following relation is valid:

$$\mathrm{AIVar}(H_{opt})=2\,\mathrm{AIBias}^2(H_{opt}).$$

Unfortunately, for $d>2$ there is not closed form expression for the optimal smoothing matrix. Nevertheless, the following theorem brings an analogous relation between TIVar and TIBias^2 without knowledge of the explicit form of H_T.

Theorem 27.1. *Let H_T be a minimizator of* TAMISE(H)*. Then*

$$\frac{d+2}{4}\mathrm{TIVar}(H_T)=\mathrm{TIBias}^2(H_T).$$

This equation can be rewritten as

$$|H_T|^{1/2}=\frac{d+2}{4}\frac{\mathrm{tr}\big[H_T^{-1}R(DK)\big]}{n\,\mathrm{TIBias}^2(H_T)}. \tag{27.1}$$

This relation is a basis of a method for bandwidth matrix choice we are going to present.

Corollary 27.1. *Let H_T be a minimizator of* TAMISE(H). *Then*

$$\begin{aligned}\mathrm{TAMISE}(H_T) &= \tfrac{d+6}{4}\,\mathrm{TIVar}(H_T)\\ &= \tfrac{d+6}{4}\,n^{-1}|H_T|^{-1/2}\,\mathrm{tr}\big[(H_T)^{-1}R(DK)\big],\end{aligned}$$

i.e.

$$\mathrm{TAMISE}(H_T) = O\big(n^{-\frac{4}{d+6}}\big).$$

This result corresponds to the result by Duong et al. (2008) and Chacón et al. (2009).

Let H_T be a diagonal matrix. Without lost of generality we can assume that there are constants c_i, $i=1,\dots,d$, $c_1=1$, such that

$$h_{i,T}^2 = c_i^2 h_{1,T}^2, \quad i=1,\dots,d.$$

The problem how to choose the constants c_i, $i=1,\dots,d$, or their suitable estimates will be treated later. Substitution of these entries in (1) and some computations lead to the interesting expression for $h_{1,T}$:

$$h_{1,T}^{d+6} = \frac{d+2}{n}|C|^{-1/2}\frac{\sum_{i=1}^{d}\frac{1}{c_i^2}\int\left(\frac{\partial K}{\partial x_i}\right)^2\mathrm{d}\mathbf{x}}{\beta_2(K)^2\,\mathrm{vech}^T C\Psi_6\,\mathrm{vech}\,C},$$

where $C=\mathrm{diag}\big(1,c_2^2,\dots,c_d^2\big)$. This expression generalizes our result for $d=2$ (see Vopatová et al. (2010)).

27.3 A proposed method

Our method is based on the equation given in Theorem. This approach consists in finding such a matrix H_T satisfying that equation. Since $\mathrm{TIBias}^2(H_T)$ depends on unknown partial derivatives of the density f, we use a suitable estimate of it

$$\begin{aligned}\widehat{\mathrm{TIBias}}^2(H_T) = \mathrm{tr}\Big[&\tfrac{1}{n^2}\textstyle\sum_{i,j=1}^{n}\int\big[(K_H*DK_H-DK_H)(\mathbf{x}-\mathbf{X}_i)\big]\\ &\times\big[(K_H*DK_H-DK_H)(\mathbf{x}-\mathbf{X}_j)\big]^T\mathrm{d}\mathbf{x}\Big].\end{aligned}$$

Let $\widehat{H}_T$ be a solution of the equation

$$|\widehat{H}_T|^{1/2} = \frac{d+2}{4n}\frac{\mathrm{tr}\big[\widehat{H}_T^{-1}R(DK)\big]}{\widehat{\mathrm{TIBias}}^2(\widehat{H}_T)} \qquad (27.2)$$

This equation represents a nonlinear equation for $d(d+1)/2$ unknowns entries of $\mathrm{vech}\widehat{H}_T$. In order to find these entries we need additional $d(d+1)/2-1$ equations for bandwidths $h_{ij,T}$.

Firstly, let us assume that $H_T = \mathrm{diag}(h_{1,T}^2, \dots, h_{d,T}^2)$. Then, $|H_T|^{1/2} = h_{1,T} \cdots h_{d,T}$ and the previous equation takes the form

$$\hat{h}_{1,T} \cdots \hat{h}_{d,T} = \frac{d+2}{4n} \frac{\mathrm{tr}\left[\widehat{H}_T^{-1} R(DK)\right]}{\widehat{\mathrm{TIBias}}^2(\widehat{H}_T)}.$$

In the previous paragraph it has been shown how the relation (1) could be satisfied in the case of diagonal matrix H_T. The problem now consists in finding an appropriate estimates of c_j, $j = 2, \dots, d$. To solve this problem we propose to use ideas of Scott (1992) and Duong et al. (2008). According to the Scott's rule the suitable estimates for bandwidths h_i, $i = 1, \dots, d$, for kernel density estimates are

$$\hat{h}_i = \hat{\sigma}_i n^{-1/(d+4)},$$

where $\hat{\sigma}_i$, $i = 1, \dots, d$, is an estimate of a sample standard deviation.

In the paper by Duong et al. (2008) the suitable estimates of bandwidths for kernel gradient estimators have been proposed:

$$h_{i,T}^2 = h_i^2 n^{\frac{4}{(d+4)(d+6)}} \left(\hat{\sigma}_i^2\right)^{\frac{4}{(d+4)(d+6)}}.$$

Combining previous ideas we obtain

$$\left(\frac{\hat{h}_{i,T}}{\hat{h}_{1,T}}\right)^2 = \left(\frac{\hat{\sigma}_i^2}{\hat{\sigma}_1^2}\right)^{\frac{(d+4)(d+6)+4}{(d+4)(d+6)}}.$$

It means that for $i = 2, \dots, d$

$$\hat{h}_{i,T}^2 = \hat{h}_{1,T}^2 \cdot c_i^2, \qquad c_i^2 = \left(\frac{\hat{\sigma}_i^2}{\hat{\sigma}_1^2}\right)^{\frac{(d+4)(d+6)+4}{(d+4)(d+6)}}.$$

Finally, we arrive at the relation

$$\hat{h}_{1,T}^{d+2} = \frac{d+2}{4n} |C|^{-1/2} \frac{\sum_{i=1}^d c_i^{-2} \int \left(\frac{\partial K}{\partial x_i}\right)^2 \mathrm{d}\mathbf{x}}{\widehat{\mathrm{TIBias}}^2(\widehat{H}_T)},$$

This equation can be solved by an appropriate numerical method.

Let us turn our attention to the full bandwidth matrix for the case $d = 2$. Let

$$H_T = \begin{pmatrix} h_1^2 = h_{11} & h_{12} \\ h_{12} & h_2^2 = h_{22} \end{pmatrix}$$

be a positive definite matrix. We adopt a similar idea as in the case of the diagonal matrix (see also Terrell (1990)). Let Σ be a sample covariance matrix and $\widehat{\Sigma}$ its estimate

$$\widehat{\Sigma} = \begin{pmatrix} \hat{\sigma}_{11}^2 & \hat{\sigma}_{12} \\ \hat{\sigma}_{12} & \hat{\sigma}_{22}^2 \end{pmatrix}.$$

In accordance with the expression for $h_{1,T}^{d+2}$ we can assume

$$\begin{aligned}
\hat{h}_{11,T} &= \hat{h}_{1,T}^2 = \left(\hat{\sigma}_{11}^2\right)^{13/12} n^{-1/4} \\
\hat{h}_{22,T} &= \hat{h}_{2,T}^2 = \left(\hat{\sigma}_{22}^2\right)^{13/12} n^{-1/4} \\
\hat{h}_{12,T}^2 &= \left(\hat{\sigma}_{12}^2\right)^{13/12} n^{-1/2} \\
\hat{h}_{12,T} &= \operatorname{sign} \hat{\sigma}_{12} |\hat{\sigma}_{12}|^{13/12} n^{-1/4}
\end{aligned}$$

Then

$$\begin{aligned}
|\widehat{H}_T| &= \hat{h}_{11}\hat{h}_{22} - \hat{h}_{12}^2 \\
&= \hat{h}_{11,T}^2 \left[\left(\hat{\sigma}_{11}^2 \hat{\sigma}_{22}^2\right)^{13/12} - \left(\hat{\sigma}_{12}^2\right)^{13/12} \right] / \left(\hat{\sigma}_{11}^4\right)^{13/12} \\
&= \hat{h}_{11,T}^2 \cdot S(\hat{\sigma}).
\end{aligned}$$

Hence

$$\hat{h}_{11,T} = \hat{h}_{1,T}^2 = \frac{\operatorname{tr}\left[\widehat{H}_T^{-1} R(DK)\right]}{\sqrt{S(\hat{\sigma})} \cdot \widehat{\text{TIBias}}^2 (\widehat{H}_T)}.$$

This is a nonlinear equation for the unknown $h_{11,T}$. This equation can be solved by an appropriate numerical method.

27.4 Simulations

In order to verify the quality of the proposed method, we conduct a short simulation study. As a quality criterion we used an average of the integrated Euclidean norm (IEN) of difference vector, i.e.

$$\overline{\text{IEN}}(H) = \operatorname{avg} \int_{\mathbb{R}^2} \|\widehat{Df}(\mathbf{x}, H) - Df(\mathbf{x})\|^2 \mathrm{d}\mathbf{x},$$

where the average is taken over simulated realizations.

Samples of the size $n = 100$ were drawn from densities listed in the following table ($X \sim N_2(\mu_1, \mu_2; \sigma_1^2, \sigma_2^2, \rho)$). Bandwidth matrices were selected for 100 random samples generated from each density.

(a) $X \sim N_2(0,0;1,1/4,0)$
(b) $X \sim N_2(0,0;1,1/4,2/3)$
(c) $\frac{1}{2}\mathcal{N}_2(1,0;4/9,4/9,0)+\frac{1}{2}\mathcal{N}_2(-1,0;4/9,4/9,0)$
(d) $\frac{1}{2}\mathcal{N}_2(-1,1;4/9,4/9,3/5)+\frac{1}{2}\mathcal{N}_2(1,-1;4/9,4/9,3/5)$
(e) $\frac{1}{3}\mathcal{N}_2(0,0;1,1,0)+\frac{1}{3}\mathcal{N}_2(0,4;1,4,0)+\frac{1}{3}\mathcal{N}_2(4,0;4,1,0)$
(f) $\frac{1}{5}\mathcal{N}_2(0,0;1,1,0)+\frac{1}{5}\mathcal{N}_2(1/2,1/2;4/9,4/9,0)$ $+\frac{3}{5}\mathcal{N}_2(13/12,13/12;25/81,25/81,0)$

Target densities.

The next table summarizes the results of $\overline{\text{IEN}}$ computed for the proposed bandwidth matrix H_{iter} and for the TAMISE-optimal bandwidth matrix H_T.

data	$\overline{\text{IEN}}(H_{iter})$	$\overline{\text{IEN}}(H_T)$
(a)	0.0801 (0.0306)	0.0866 (0.0263)
(b)	0.1868 (0.0705)	0.1998 (0.0605)
(c)	0.0599 (0.0109)	0.0500 (0.0115)
(d)	0.1709 (0.0242)	0.1049 (0.0269)
(e)	0.0055 (0.0013)	0.0049 (0.0010)
(f)	0.0889 (0.0294)	0.0907 (0.0265)

Average of IEN with a standard deviation.

Acknowledgements I. Horová has been supported by the Ministry of Education, Youth and Sports of the Czech Republic under the project MŠMT LC06024, K. Vopatová was funded by the University of Defence through a 'Institutional development project UO FEM – Economics Laboratory' project.

References

1. Chacón, J.E., Duong, T., Wand, M.P.: Asymptotics for general multivariate kernel density derivative estimators, Research Online (2009)
2. Duong, T., Cowling, A., Koch, I., Wand, M.P.: Feature significance for multivariate kernel density estimation. Comput. Stat. Data Anal. **52**, 4225–4242 (2008)
3. Horová, I., Koláček, J., Vopatová, K.: Visualization and Bandwidth Matrix Choice. To appear in Commun. Stat. Theory (2010)
4. Scott, D.W.: Multivariate density estimation: Theory, practice, and visualization. Wiley, New York (1992)
5. Terrell, G.R.: The maximal smoothing principle in density estimation. J. Ame. Stat. Assoc. **85** 470–477 (1990)
6. Vopatová, K., Horová, I., Koláček, J.: Bandwidth Matrix Choice for Bivariate Kernel Density Derivative. Proceedings of the 25th International Workshop on Statistical Modelling (Glasgow, UK), 561–564 (2010)

Chapter 28
A Backward Generalization of PCA for Exploration and Feature Extraction of Manifold-Valued Shapes

Sungkyu Jung

Abstract A generalized Principal Component Analysis (PCA) for manifold-valued shapes is discussed with forward and backward stepwise views of PCA. Finite and infinite dimensional shape spaces are briefly introduced. A backward extension of PCA for the shape spaces, called Principal Nested Spheres, is shown in more detail, which results in a non-geodesic decomposition of shape spaces, capturing more variation in lower dimension.

28.1 Introduction

Shapes of objects have long been investigated. A shape is often described as a mathematical property that is invariant under the similarity transformations of translation, scaling and rotation. Statistical shape analysis quantifies shapes as mathematical objects, understood as random quantities, and develops statistical procedures on the space of shapes. A traditional shape representation is obtained by locating landmarks on the object; see Dryden and Mardia (1998). On the other hand, modern shape representations are diverse; examples include the medial representations, (Siddiqi and Pizer (2008)), automatically generated landmarks and spherical harmonic representations (Gerig et al. (2004)) and the elastic curve representation (Srivastava et al. (2010)).

While these representations provide rich descriptions of the shape, the sample spaces of these representations, called shape spaces, naturally form non-Euclidean manifolds, the reason of which is either the invariance under the similarity transformations or the fact that the representation itself involves angles and directions. Due to the curvature involved in the geometry of the shape space, the usual methods using Euclidean geometry are not directly used, therefore a generalization of Euclidean method is needed to analyze the manifold-valued shapes.

Sungkyu Jung
University of Chapel Hill, Chapel Hill, USA, e-mail: sungkyu@email.unc.edu

We focus on a generalization of Principal Component Analysis (PCA), which is a widely used data exploration method in a variety of fields, for many purposes including dimension reduction and visualization of important data structures. Generalized PCA methods for manifold data can be viewed as forward or backward stepwise approaches (as proposed in Marron et al. (2010)). In the traditional forward view, PCA is constructed from lower dimension to higher dimension. In the backward point of view, PCA is constructed in reverse order from higher to lower dimensions. In other words, while PCA finds successive affine subspaces of dimensions, the forward PCA accretes the direction of great variance at each step and the backward PCA removes the direction of least variance. Both of these different views lead to the usual Euclidean PCA given by the eigen-decomposition of a covariance matrix, but lead to different methodologies for manifold data.

A usual approach in the manifold extensions of PCA uses the local approximation of manifold by a tangent space (see Dryden and Mardia (1998) and Fletcher et al. (2004)), which can be viewed as forward extensions. Recently, Huckemann et al. (2010) developed Geodesic PCA, that fits the first and second geodesic principal components without restriction of a mean, where a geodesic is the shortest path between two points on the manifold and is an analog of a line in Euclidean space. This is understood as a partially backward approach, since the advantage of the method comes from reverting the first and second steps of the successive procedure. Analysis of Principal Nested Spheres (Jung et al. (2010)) was developed to extend PCA in a non-geodesic (non-linear) way, which was possible by taking a full backward approach.

Note that we take the advantage of focusing on a specific geometry of manifolds. In particular, the sample space of shapes involves spherical geometry, on which we exemplify the backward generalization of PCA. We briefly introduce the geometry of finite dimensional shape space, due to Kendall's shape theory, and also of an infinite dimensional shape space of space curves. After understanding the geometry of the shape spaces, we revisit analysis of Principal Nested Spheres and introduce a natural framework to extend the method to the functional shape space of space curves.

28.2 Finite and infinite dimensional shape spaces

We briefly give basic background for finite dimensional or infinite dimensional shape spaces. Detailed introduction and discussions can be found at Dryden and Mardia (1998) for the finite dimensional shape space and Srivastava et al. (2010) for the infinite dimensional functional shape space.

Landmark-based shape space: The shape of an object with $k > 2$ geometric landmarks in $m \geq 2$ dimension is identified as a point in Kendall's shape space (Kendall (1984)). An object is represented by the corresponding configuration matrix X, which is a $k \times m$ matrix of Cartesian coordinates of landmarks. The preshape of the configuration is obtained by removing the effect of translation and scaling and is given by $Z = HX/\|HX\|$, where H is the $(k-1)\times k$ Helmert sub-matrix (p. 34 of Dryden and Mardia (1998)), which is a form of centering matrix. Provided that $\|HX\| > 0$, $Z \in S^k_m := S^{m(k-1)-1}$, where $S^d = \{x \in \mathbb{R}^{d+1} : \|x\|_2 = 1\}$ is the unit hypersphere of dimension d embedded in the Euclidean space $\mathbb{R}^{d+1}$

The shape of a configuration matrix X is represented by the equivalence set under rotation, $[Z] = \{Z\Gamma : \Gamma \in SO(m)\}$, where $SO(m)$ is the set of all $m\times m$ rotation matrices. The space of all possible shapes is then a non-Euclidean space called the shape space and is denoted by Σ^k_m. Since the shape space is not explicitly expressed, sometimes it is helpful to define Σ^k_m as a quotient of the preshape space, i.e. $\Sigma^k_m = S^k_m/SO(m)$. In practice, the shape space is often approximated by the preshape space through a Procrustean method. Sometimes strictly focusing on the shape space is preferred, but the computations are based on the metric of preshape space.

Square-root velocity representation of space curves: A space curve in $m \geq 1$ dimension can be used to represent a boundary of an object. Let $\beta \in L_2([0,1],\mathbb{R}^m) = \{f : \int_{[0,1]} \|f(t)\|^2 dt < \infty\}$ be a square integrable parameterized curve in $\mathbb{R}^m$ that represents a shape of an object. While this functional form is a straightforward extension of the finite dimensional landmarks, the invariance under similarity transformations is not yet obtained.

Srivastava et al. (2010) introduced the square-root velocity representation of space curves, which leads to a preshape space. Specifically, the square-root velocity function q of β is defined as $q(t) = \dot{\beta}(t)/\sqrt{\|\dot{\beta}(t)\|}$. The q function is invariant to scaling and translation of the original function β, and thus the space of q functions is the preshape space of the curves.

Since $\|q\|^2 = \int \langle q(t), q(t)\rangle dt = \int \langle \dot{\beta}(t), \dot{\beta}(t)\rangle dt/\|\dot{\beta}\| = 1$, the space of such q functions is a subset of the infinite dimensional unit sphere in L_2 space, i.e.

$$q \in S^\infty_m = \{f \in L_2([0,1],\mathbb{R}^m) : \|f\| = 1\}.$$

The shape space of curves is defined as a quotient of the preshape space $S^\infty_m/SO(m)$. Srivastava et al. also established an invariance under re-parameterization. Similar to the finite dimensional shape space, it is a common practice to approximate the shape space by the preshape space or to make use of metrics on the preshape space in defining a metric of the shape space.

28.3 Principal Nested Spheres

The shape spaces introduced in the previous section are quotient spaces of the preshape spheres S^d, where the dimension d being either finite or infinite. The analysis of Principal Nested Spheres (PNS), proposed in Jung et al. (2010), was first developed as a backward generalization of PCA for S^d, $d < \infty$. A natural extension of PNS to the infinite dimensional sphere is shown.

When the sample space is the finite dimensional unit d-sphere S^d (which is an approximation of the Kendall's shape space), PNS gives a decomposition of S^d that captures the non-geodesic variation in a lower dimensional sub-manifold. The decomposition sequentially provides the best k-dimensional approximation $\mathscr{A}_k$ of the data for each $k = 0,1,\ldots,d-1$. $\mathscr{A}_k$ is called the k-dimensional PNS, since it is essentially a sphere and is *nested* within (i.e. a sub-manifold of) the higher dimensional PNS. The sequence of PNS is then

$$\mathscr{A}_0 \subset \mathscr{A}_1 \subset \cdots \subset \mathscr{A}_{d-1} \subset S^d.$$

The analysis of PNS provides intuitive approximations of the directional or shape data for every dimension, captures the non-geodesic variation, and provides intuitive visualization of the major variability in terms of shape changes.

The procedure involves iterative reduction of the dimensionality of the data. We first fit a $d-1$ dimensional *subsphere* $\mathscr{A}_{d-1}$ of S^d that best approximates the data by a least squares criterion. A subsphere is defined by an axis $v \in S^d$ and radius $r \in (0,\pi/2]$ as $\mathscr{A}_{d-1} = \{x \in S^d : \rho(x,v) = r\}$, where ρ is a metric on S^d. Given $x_1,\ldots,x_n \in S^d$, the principal subsphere minimizes $\sum_{i=1}^n \rho(x_i,\mathscr{A}_{d-1})^2$. This principal subsphere is not necessarily a great sphere (i.e. a sphere with radius 1, analogous to the great circle for S^2), which makes the resulting decomposition non-geodesic. Each data point has an associated residual, which is a signed distance to its projection on $\mathscr{A}_{d-1}$. Then with the data projected onto the subsphere we continue to search for the best fitting $d-2$ dimensional subsphere. These steps are iterated to find lower dimensional principal *nested* spheres. For visualization and further analysis, we obtain a Principal Scores matrix of the data, essentially consisting of the residuals of each level and is a complete analog of Principal Component scores matrix.

Now for the infinite dimensional sphere, the same idea can be carried over. However, since the sample space is infinite dimensional the iterative reduction of dimensions starts not at the original $S^\infty := S^\infty_m$ but at some finite dimensional sphere S^N. Specifically, suppose we choose a set of countably many basis functions $\{\psi_1,\psi_2,\ldots\}$ of S^∞, where $\psi_i \in S^\infty$, $\|\psi_i\| = 1$ for each i, and $\langle\psi_i,\psi_j\rangle = 0$ for $i \neq j$, that spans $L_2([0,1],\mathbb{R}^m)$. For each $q \in S^\infty$, there exists a sequence of real numbers λ_j such that $q = \sum_{i=1}^\infty \lambda_j\psi_j$ satisfying $\sum_{j=1}^\infty \lambda_j^2 = 1$. Then the finite dimensional sphere S^N for some N is defined as $S^N = \{q \in S^\infty : q = \sum_{j=1}^N \lambda_j\psi_j, \sum_{j=1}^N \lambda_j^2 = 1\}$. In practice N can be chosen to be larger than the sample size, and the basis functions $\{\psi_j\}$ shall be chosen to contain most variation of random quantity.

Once we have the finite dimensional approximation S^N of S^∞, the dimension reduction becomes a complete analogue of the vector space version. The application of principal nested spheres for the shape space of curves is discussed in a working paper with J. S. Marron and A. Srivastava.

28.4 Conclusion

We briefly introduced two different forms of shape spaces, both related to the spherical geometry. By explicitly using the specific geometry, a backward generalization of PCA for the shape spaces, called the analysis of Principal Nested Spheres, is developed. An advantage of taking the backward viewpoint is that a non-geodesic extension of PCA is possible, and thus gives a more succinct representation than geodesic-based methods.

The backward viewpoint can also be exploited stirctly to the Kendall's shape space, leading to lower dimensional shape spaces successively, i.e. $\Sigma_m^k \supset \Sigma_m^{k-1} \supset \cdots \supset \Sigma_m^3$. For the planar shapes ($m = 2$), Jung et al. (2011) propose to use the successive reduction to predict the number of effective landmarks and to choose fewer landmarks while preserving the variance power.

Some sample spaces of shapes are not simply related to the spherical geometry. In particular, the sample space of the medial representations is a direct product of simple manifolds. A non-geodesic PCA for that types of manifolds can also be developed by taking a backward generalization, see Jung et al. (2010a, 2011a).

References

1. Dryden, I.L., Mardia, K.V.: Statistical shape analysis. Wiley Series in Probability and Statistics, John Wiley & Sons Ltd., Chichester (1998)
2. Fletcher, P.T., Joshi, S., Lu, C., Pizer, S. M.: Principal Geodesic Analysis for the study of nonlinear statistics of shape. IEEE Transactions on Medical Imaging **23** (8), 995–1005 (2004)
3. Gerig, G., Styner, M., Szekely, G.: Statistical shape models for segmentation and structural analysis. Proceedings of IEEE International Symposium on Biomedical Imaging (ISBI), vol. I, pp. 467-473 (2004)
4. Huckemann, S., Hotz, T., Munk, A.: Intrinsic shape analysis: Geodesic PCA for Riemannian manifolds modulo isometric lie group actions. Stat. Sinica **20** (1), 1–58 (2010)
5. Jung, S., Dryden, I.L., Marron, J.S.: Analysis of Principal Nested Spheres. Submitted to Biometrika (2010)
6. Jung, S., Liu, X., Pizer, S., Marron, J.S.: Generalized PCA via the backward stepwise approach in image analysis. In: Angeles, J. et al. (Eds.) Brain, Body and Machine: Proceedings of an International Symposium on the 25th Anniversary of McGill University Centre for Intelligent Machines, Advances in Intelligent and Soft Computing 83, pp. 11-123 (2010a)
7. Jung, S., Huckemann, S., Marron, J.S.: Principal Nested Nested Shape Spaces with an application to reduction of number of landmarks. Under preparation (2011)
8. Jung, S., Foskey, M., Marron, J.S.: Principal Arc Analysis for direct product manifold. To appear in Ann. Appl. Stat. (2011a)
9. Kendall, D.G.: Shape manifolds, procrustean metrics and complex projective spaces. B. Lond. Math. Soc. **16**, 81–121 (1984)
10. Marron, J.S., Jung, S., Dryden, I.L.: Speculation on the generality of the backward stepwise view of pca. In: Proceedings of MIR 2010: 11th ACM SIGMM International Conference on Multimedia Information Retrieval, Association for Computing Machinery, Inc., Danvers, MA, pp. 227-230 (2010)
11. Siddiqi, K., Pizer, S.M.: Medial Representations: Mathematics, Algorithms and Applications. Springer (2008)
12. Srivastava, A., Klassen, E., Joshi, S. H., Jermyn, I. H.: Shape Analysis of Elastic Curves in Euclidean Spaces. IEEE T. Pattern Anal. **99**, accepted for publication (2010)

Chapter 29
Multiple Functional Regression with both Discrete and Continuous Covariates

Hachem Kadri, Philippe Preux, Emmanuel Duflos, Stéphane Canu

Abstract In this paper we present a nonparametric method for extending functional regression methodology to the situation where more than one functional covariate is used to predict a functional response. Borrowing the idea from Kadri et al. (2010a), the method, which support mixed discrete and continuous explanatory variables, is based on estimating a function-valued function in reproducing kernel Hilbert spaces by virtue of positive operator-valued kernels.

29.1 Introduction

The analysis of interaction effects between continuous variables in multiple regression has received a significant amount of attention from the research community. In recent years, a large part of research has been focused on functional regression where continuous data are represented by real-valued functions rather than by discrete, finite dimensional vectors. This is often the case in functional data analysis (FDA) when observed data have been measured over a densely sampled grid. We refer the reader to Ramsay and Silverman (2002, 2005) and Ferraty and Vieu (2006) for more details on functional data analysis of densely sampled data

Hachem Kadri
INRIA Lille - Nord Europe/Ecole Centrale de Lille, Villeneuve d'Ascq, France, e-mail: hachem.kadri@inria.fr

Philippe Preux
INRIA Lille - Nord Europe/Université de Lille, Villeneuve d'Ascq, France, e-mail: philippe.preux@inria.fr

Emmanuel Duflos
INRIA Lille - Nord Europe/Ecole Centrale de Lille, Villeneuve d'Ascq, France, e-mail: emmanuel.duflos@inria.fr

Stéphane Canu
INSA de Rouen, St Etienne du Rouvray, France, e-mail: scanu@insa-rouen.fr

or fully observed trajectories. In this context, various functional regression models (Ramsay and Silverman, 2005) have been proposed according to the nature of explanatory (or covariate) and response variables, perhaps the most widely studied is the generalized functional linear model where covariates are functions and responses are scalars (Cardot et al., 1999,2003; James, 2002; Müller and Stadtmüller, 2005; Preda, 2007).

In this paper, we are interested in the case of regression models with a functional response. Two subcategories of such models have appeared in the FDA literature: covariates are scalars and responses are functions also known as "functional response model" (Faraway, 1997; Chiou et al., 2004); both covariates and responses are functions (Ramsay and Dalzell, 1991; He et al., 2000; Cuevas et al., 2002; Prchal and Sarda, 2007; Antoch et al., 2008). In this work, we pay particular attention to this latter situation which corresponds to extending multivariate linear regression model to the functional case where all the components involved in the model are functions. Unlike most of previous works which consider only one functional covariate variable, we wish to perform a regression analysis in which multiple functional covariates are used to predict a functional response. The methodology which is concerned with solving such task is referred to as a multiple functional regression.

Previous studies on multiple functional regression (Han et al., 2007; Matsui et al., 2009; Valderrama et al., 2010) assume a linear relationship between functional covariates and responses and model this relationship via a multiple functional linear regression model which generalizes the model in Ramsay and Dalzell (1991) to deal with more than one covariate variable. However, extensions to nonparametric models have not been considered. Nonparametric functional regression (Ferraty and Vieu, 2002,2003) is addressed mostly in the context of functional covariates and scalar responses. More recently, Lian (2007) and Kadri et al. (2010a) showed how function-valued reproducing kernel Hilbert spaces (RKHS) and operator-valued kernels can be used for the nonparametric estimation of the regression function when both covariates and responses are curves. Building on these works, we present in this paper a nonparametric multiple functional regression method where several functions would serve as predictors. Furthermore, we aim at extending this method to handle mixed discrete and functional explanatory variables. This should be helpful for situations where a subset of regressors are comprised of repeated observations of an outcome variable and the remaining are independent scalar or categorical variables. In Antoch et al. (2008) for example, the authors discuss the use of a functional linear regression model with a functional response to predict electricity consumption and mention that including the knowledge of special events such as festive days in the estimation procedure may improve the prediction.

The remainder of this paper is organized as follows. Section 2 reviews the multiple functional linear regression model and discusses its nonparametric extension. This section also describes the RKHS-based estimation procedure for the nonparametric multiple functional regression model. Section 3 concludes the paper.

29.2 Multiple functional regression

Before presenting our nonparametric multiple function regression procedure, we start this section with a brief overview of the multiple functional linear regression model (Matsui et al., 2009; Valderrama et al., 2010). This model extends functional linear regression with a functional response (Ramsay and Dalzell, 1991; Ramsay and Silverman, 2005) to deal with more than one covariate and seeks to explain a functional response variable $y(t)$ by several functional covariates $x_k(s)$. A multiple functional linear regression model is formulated as follows:

$$y_i(t) = \alpha(t) + \sum_{k=1}^{p} \int_{I_s} x_{ik}(s)\beta_k(s,t)ds + \varepsilon_i(t), \quad t \in I_t, \quad i = 1,\ldots,n, \tag{29.1}$$

where $\alpha(t)$ is the mean function, p is the number of functional covariates, n is the number of observations, $\beta_k(s,t)$ is the regression function for the k-th covariate and $\varepsilon_i(t)$ a random error function. To estimate the functional parameters of this model, one can consider the centered covariate and response variables to eliminate the functional intercept α. Then, $\beta_k(.,.)$ are approximated by a linear combination of basis functions and the corresponding real-valued basis coefficients can be estimated by minimizing a penalized least square criterion. Good candidates for the basis functions include the Fourier basis (Ramsay and Silverman, 2005) and the B-spline basis (Prchal and Sarda, 2007).

It is well known that parametric models suffer from the restriction that the input-output relationship has to be specified a priori. By allowing the data to model the relationships among variables, nonparametric models have emerged as a powerful approach for addressing this problem. In this context and from functional input-output data $(x_i(s), y_i(t))_{i=1}^n \in (\mathscr{G}_x)^p \times \mathscr{G}_y$ where $\mathscr{G}_x : I_s \longrightarrow \mathbb{R}$ and $\mathscr{G}_y : I_t \longrightarrow \mathbb{R}$, a nonparametric multiple functional regression model can be defined as follows:

$$y_i(t) = f(x_i(s)) + \varepsilon_i(t), \quad s \in I_s,\ t \in I_t, \quad i = 1,\ldots,n,$$

where f is a linear operator which perform the mapping between two spaces of functions. In this work, we consider a slightly modified model in which covariates could be a mixture of discrete and continuous variables. More precisely, we consider the following model

$$y_i(t) = f(x_i) + \varepsilon_i(t), \quad i = 1,\ldots,n, \tag{29.2}$$

where $x_i \in X$ is composed of two subsets x_i^d and $x_i^c(s)$. $x_i^d \in \mathbb{R}^k$ is a $k \times 1$ vector of discrete dependent or independent variables and $x_i^c(s)$ is a vector of p continuous functions, so each x_i contains k discrete values and p functional variables.

Our main interest in this paper is to design an efficient estimation procedure of the regression parameter f of the model (29.2). An estimate f^* of $f \in \mathscr{F}$ can be obtained by minimizing the following regularized empirical risk

$$f^* = \arg\min_{f\in\mathscr{F}} \sum_{i=1}^{n} \|y_i - f(x_i)\|^2_{\mathscr{G}_y} + \lambda \|f\|^2_{\mathscr{F}} \tag{29.3}$$

Borrowing the idea from Kadri et al. (2010a), we use function-valued reproducing kernel Hilbert spaces (RKHS) and operator-valued kernels to solve this minimization problem. Function-valued RKHS theory is the extension of the scalar-valued case to the functional response setting. In this context, Hilbert spaces of function-valued functions are constructed and basic properties of real RKHS are restated. Some examples of potential applications of these spaces can be found in Kadri et al. (2010b) and in the area of multi-task learning (discrete outputs) see Evgeniou et al. (2005). Function-valued RKHS theory is based on the *one-to-one correspondence* between reproducing kernel Hilbert spaces of function-valued functions and positive operator-valued kernels. We start by recalling some basic properties of such Spaces. We say that a Hilbert space $\mathscr{F}$ of functions $X \longrightarrow \mathscr{G}_y$ has the *reproducing property*, if $\forall x \in X$ the evaluation functional $f \longrightarrow f(x)$ is continuous. This continuity is equivalent to the continuity of the mapping $f \longrightarrow \langle f(x), g\rangle_{\mathscr{G}_y}$ for any $x \in X$ and $g \in \mathscr{G}_y$. By the Riesz representation theorem it follows that for a given $x \in X$ and for any choice of $g \in \mathscr{G}_y$, there exists an element $h_x^g \in \mathscr{F}$, s.t.

$$\forall f \in \mathscr{F} \quad \langle h_x^g, f\rangle_{\mathscr{F}} = \langle f(x), g\rangle_{\mathscr{G}_y}$$

We can therefore define the corresponding operator-valued kernel $K(.,.) \in \mathscr{L}(\mathscr{G}_y)$, where $\mathscr{L}(\mathscr{G}_y)$ denote the set of bounded linear operators from $\mathscr{G}_y$ to $\mathscr{G}_y$, such that

$$\langle K(x,z)g_1, g_2\rangle_{\mathscr{G}_y} = \langle h_x^{g_1}, h_z^{g_2}\rangle_{\mathscr{F}}$$

It follows that $\langle h_x^{g_1}(z), g_2\rangle_{\mathscr{G}_y} = \langle h_x^{g_1}, h_z^{g_2}\rangle_{\mathscr{F}} = \langle K(x,z)g_1, g_2\rangle_{\mathscr{G}_y}$ and thus we obtain the reproducing property

$$\langle K(x,.)g, f\rangle_{\mathscr{F}} = \langle f(x), g\rangle_{\mathscr{G}_y}$$

It is easy to see that $K(x,z)$ is a positive kernel as defined below:

Definition 29.1. We say that $K(x,z)$, satisfying $K(x,z) = K(z,x)^*$, is a positive operator-valued kernel if given an arbitrary finite set of points $\{(x_i, g_i)\}_{i=1,\dots,n} \in X \times \mathscr{G}_y$, the corresponding block matrix K with $K_{ij} = \langle K(x_i,x_j)g_i, g_j\rangle_{\mathscr{G}_y}$ is positive semi-definite.

Importantly, the converse is also true. Any positive operator-valued kernel $K(x,z)$ gives rise to an RKHS $\mathscr{F}_K$, which can be constructed by considering the space of function-valued functions f having the form $f(.) = \sum_{i=1}^{n} K(x_i,.)g_i$ and taking completion with respect to the inner product given by $\langle K(x,.)g_1, K(z,.)g_2\rangle_{\mathscr{F}} = \langle K(x,z)g_1, g_2\rangle_{\mathscr{G}_y}$.

The functional version of the Representer Theorem can be used to show that the solution of the minimization problem (29.3) is of the following form:

$$f^*(x) = \sum_{j=1}^{n} K(x, x_j) g_j \tag{29.4}$$

Substituting this form in (29.3) , we arrive at the following minimization over the scalar-valued functions g_i rather than the function-valued function f

$$\min_{g \in (\mathscr{G}_y)^n} \sum_{i=1}^{n} \|y_i - \sum_{j=1}^{n} K(x_i, x_j) g_j\|^2_{\mathscr{G}_y} + \lambda \sum_{i,j}^{n} \langle K(x_i, x_j) g_i, g_j \rangle_{\mathscr{G}_y} \tag{29.5}$$

This problem can be solved by choosing a suitable operator-valued kernel. Choosing K presents two major difficulties: we need to construct a function from an adequate operator, and which takes as arguments variables composed of scalars and functions. Lian (2007) considered the identity operator, while in Kadri et al. (2010) the authors showed that it will be more useful to choose other operators than identity that are able to take into account functional properties of the input and output spaces. They also introduced a functional extension of the Gaussian kernel based on the multiplication operator. Using this operator, their approach can be seen as a nonlinear extension of the functional linear concurrent model (Ramsay and Silverman, 2005). Motivated by extending the functional linear regression model with functional response, we consider in this work a kernel K constructed from the integral operator and having the following form:

$$(K(x_i, x_j) g)(t) = [k_{x^d}(x_i^d, x_j^d) + k_{x^c}(x_i^c, x_j^c)] \int k_y(s,t) g(s) ds \tag{29.6}$$

where k_{x^d} and k_{x^c} are scalar-valued kernels on $\mathbb{R}^k$ and $(\mathscr{G}_x)^p$ respectively and k_y the reproducing kernel of the space $\mathscr{G}_y$. Choosing k_{x^d} and k_y is not a problem. Among the large number of possible classical kernels k_{x^d} and k_y, we chose the Gaussian kernel. However, constructing k_{x^c} is slightly more delicate. One can use the inner product in $(\mathscr{G}_x)^p$ to construct a linear kernel. Also, extending real-valued functional kernels such as those in Rossi et Villa. (2006) to multiple functional inputs could be possible.

To solve the problem (29.5), we consider that $\mathscr{G}_y$ is a real-valued RKHS and k_y its reproducing kernel and then each function in this space can be approximated by a finite linear combination of kernels. So, the functions $g_i(.)$ can be approximated by $\sum_{l=1}^{m} \alpha_{il} k_y(t_l, .)$ and solving (29.5) returns to finding the corresponding real variables α_{il}. Under this framework and using matrix formulation, we find that the $nm \times 1$ vector α satisfies the system of linear equation

$$(\mathbf{K} + \lambda I)\alpha = Y \tag{29.7}$$

where the $nm \times 1$ vector Y is obtained by concatenating the columns of the matrix $(Y_{il})_{i \leq n,\, l \leq m}$ and $\mathbf{K}$ is the block operator kernel matrix $(\mathbf{K}_{ij})_{1 \leq i,j \leq n}$ where each $\mathbf{K}_{ij}$ is a $m \times m$ matrix.

29.3 Conclusion

We study the problem of multiple functional regression where several functional explanatory variables are used to predict a functional response. Using function-valued RKHS theory, we have proposed a nonparametric estimation procedure which support mixed discrete and continuous covariates. In future, we will illustrate our approach and evaluate its performance by experiments on simulated and real data.

Acknowledgements H. Kadri is supported by Junior Researcher Contract No. 4297 from the the Nord-Pas de Calais region.

References

1. Antoch, J., Prchal, L., De Rosa, M., Sarda, P.: Functional linear regression with functional response: application to prediction of electricity consumption. IWFOS'2008 Proceedings, Functional and operatorial statistics, Physica-Verlag, Springer (2008)
2. Cardot, H., Ferraty, F. and Sarda, P.: Functional linear model. Stat. Probab. Lett. **45**, 11–22 (199)
3. Cardot, H., Ferraty, F., Sarda, P.: Spline Estimators for the Functional Linear Model. Stat. Sinica **13**, 571–591 (2003)
4. Chiou, J.M., Müller, H.G., Wang, J.L.: Functional response models. Stat. Sinica **14**, 675–693 (2004)
5. Cuevas, A., Febrero, M., Fraiman, R.: Linear functional regression: the case of fixed design and functional response. Canad. J. Stat. **30**, 285–300 (2002)
6. Evgeniou, T., Micchelli, C. A., Pontil, M.: Learning multiple tasks with kernel methods. Journal of Machine Learning Research **6**, 615–637 (2005)
7. Faraway, J.: Regression analysis for a functional response. Technometrics **39**, 254–262 (1997)
8. Ferraty, F., Vieu, P.: The functional nonparametric model and applications to spectrometric data. Computation. Stat. **17**, 545–564 (2002)
9. Ferraty, F., Vieu, P.: Curves discrimination: a nonparametric functional approach. Comput. Stat. Data Anal. **44**, 161–173 (2003)
10. Ferraty, F., Vieu, P.: Nonparametric Functional Data Analysis. Springer, New York (2006)
11. Han, S.W., Serban, N., Rouse, B.W.: Novel Perspectives On Market Valuation of Firms Via Functional Regression. Technical report, Statistics group, Georgia Tech. (2007)
12. He, G., Müller, H.G., Wang, J.L.: Extending correlation and regression from multivariate to functional data. In: Puri, M.L. (ed.) Asymptotics in statistics and probability, VSP International Science Publishers, pp. 301-315 (2000)
13. James, G.: Generalized linear models with functional predictors. J. Royal Stat. Soc. B **64**, 411–432 (2002)
14. Kadri, H., Preux, P., Duflos, E., Canu, S., Davy, M.: Nonlinear functional regression: a functional RKHS approach. In: Proc. of the 13th Int'l Conf. on Artificial Intelligence and Statistics (AI & Stats). JMLR: W&CP 9, pp. 374-380 (2010a)
15. Kadri, H., Preux, P., Duflos, E., Canu, S., Davy, M.: Function-Valued Reproducting Kernel Hilbert Spaces and Applications. NIPS workshop on TKML (2010b)
16. Lian, H.: Nonlinear functional models for functional responses in reproducing kernel hilbert spaces. Canad. J. Stat. **35**, 597–606 (2007)
17. Matsui, H., Kawano, S., Konishi, S.: Regularized functional regression modeling for functional response and predictors. Journal of Math-for-industry, **1**, 17–25 (2009)

18. Müller, H.G., Stadtmüller, U.: Generalized functional linear models. Ann. Stat. **33**, 774–805 (2005)
19. Prchal, L., Sarda, P. (2007). Spline estimator for the functional linear regression with functional response. Preprint (2007)
20. Preda, C.: Regression models for functional data by reproducing kernel Hilbert spaces methods. J. Stat. Plan. Infer. **137**, 829–840 (2007)
21. Ramsay, J., Dalzell, C.J.: Some tools for functional data analysis. J. Roy. Stat. Soc. B **53**, 539–572 (1991)
22. Ramsay, J., Silverman, B.: Applied functional data analysis. Springer, New York (2002)
23. Ramsay, J., Silverman, B.: Functional data analysis (Second Edition). Springer, New York (2005)
24. Rossi, F., Villa, N.: Support vector machine for functional data classification. Neurocomputing **69** (7-9), 730–742 (2006)
25. Valderrama, M.J., Ocaña, F.A., Aguilera, A.M., Ocaña-Peinado, F.M.: Forecasting pollen concentration by a two-Step functional model. Biometrics **66**, 578–585 (2010)

Chapter 30
Combining Factor Models and Variable Selection in High-Dimensional Regression

Alois Kneip, Pascal Sarda

Abstract This presentation provides a summary of some of the results derived in Kneip and Sarda (2011). The basic motivation of the study is to combine the points of view of model selection and functional regression by using a factor approach. For highly correlated regressors the traditional assumption of a sparse vector of parameters is restrictive. We therefore propose to include principal components as additional explanatory variables in an augmented regression model.

30.1 Introduction

We consider a high dimensional linear regression model of the form

$$Y_i = \boldsymbol{\beta}^T \mathbf{X}_i + \varepsilon_i, \quad i = 1, \ldots, n, \tag{30.1}$$

where $(Y_i, \mathbf{X}_i)$, $i = 1, \ldots, n$, are i.i.d. random pairs with $Y_i \in \mathbb{R}$ and $\mathbf{X}_i = (X_{i1}, \ldots, X_{ip})^T \in \mathbb{R}^p$. We will assume without loss of generality that $\mathbb{E}(X_{ij}) = 0$ for all $j = 1, \ldots. p$. Furthermore, $\boldsymbol{\beta}$ is a vector of parameters in $\mathbb{R}^p$ and $(\varepsilon_i)_{i=1,\ldots,n}$ are centered i.i.d. real random variables independent with $\mathbf{X}_i$ with $Var(\varepsilon_i) = \sigma^2$. The dimension p of the vector of parameters is assumed to be typically larger than the sample size n.

Model (30.1) comprises two main situations which have been considered independently in two separate branches of statistical literature. On one side, there is the situation where $\mathbf{X}_i$ represents a (high dimensional) vector of different predictor variables. Another situation arises when the regressors are p discretizations (for example at different observations times) of a same curve. In this case model (30.1) represents a discrete version of an underlying continuous *functional linear regression model.*

Alois Kneip
Universität Bonn, Bonn, Germany, e-mail: akneip@uni-bonn.de

Pascal Sarda
Institut de Mathématiques de Toulouse, France, e-mail: sarda@math.univ-toulouse.fr

The first situation is studied in a large literature on model selection in high dimensional regression. The basic structural assumption can be described as follows: There is only a relatively small number of predictor variables with $|\beta_j| > 0$ which have a significant influence on the outcome Y. In other words, the set of nonzero coefficients is sparse, $S := \#\{j|\beta_j \neq 0\} \ll p$.

The most popular procedures to identify and estimate nonzero coefficients β_j are Lasso and the Dantzig selector. Some important references are Tibshirani (1996), Meinshausen and Bühlmann (2006), Zhao and Yu (2006), , Candes and Tao (2007) van de Geer (2008) Bickel et al. (2009). An important technical condition in this context is that correlations between explanatory variables are "sufficiently weak".

In sharp contrast, the setup considered in the literature on functional regression rests upon a very different type of structural assumptions. Some references are Ramsay and Dalzell (1991), Cardot et al. (1999), Cuevas et al. (2002), Yao et al. (2005), Cai and Hall (2006), Hall and Horowitz (2007), Cardot et al. (2007) and Crambes et al. (2009). We consider the simplest case that $X_{ij} = X_i(t_j)$ for random functions $X_i \in L^2([0,1])$ observed at an equidistant grid $t_j = \frac{j}{p}$. The main structural assumption on coefficients can then be subsumed as follows: $\beta_j := \frac{\beta(t_j)}{p}$, where $\beta(t) \in L^2([0,1])$ is a continuous slope function, and as $p \to \infty$, $\sum_j \beta_j X_{ij} = \sum_j \frac{\beta(t_j)}{p} X_i(t_j) \to \int_0^1 \beta(t) X_i(t) dt$.

Obviously, in this setup no variable $X_{ij} = X_i(t_j)$ corresponding to a specific observation at grid point t_j will possess a particulary high influence on Y, and there will exist a large number of small, but nonzero coefficients β_j of size proportional to $1/p$. Additionally, there are necessarily very strong correlations between explanatory variables $X_{ij} = X_i(t_j)$ and $X_{il} = X_i(t_l)$, $j \neq l$.

Further analysis then usually relies on the Karhunen-Loève decomposition which provides a decomposition of random functions in terms of functional principal components of the covariance operator of X_i. In the discretized case analyzed in this paper this amounts to consider an approximation of X_i by the principal components of the covariance matrix $\mathbf{\Sigma} = \mathbb{E}(\mathbf{X}_i \mathbf{X}_i^T)$. In practice, often a small number k of principal components will suffice to achieve a small L^2-error.

Based on this insight, the most frequently used approach in functional regression is to approximate $\mathbf{X}_i \approx \sum_{r=1}^k (\widehat{\boldsymbol{\psi}}_r^T \mathbf{X}_i) \widehat{\boldsymbol{\psi}}_r$ in terms of the first k estimated principal components $\widehat{\boldsymbol{\psi}}_1, \ldots, \widehat{\boldsymbol{\psi}}_k$, and to rely on the approximate model $Y_i \approx \sum_{r=1}^k \alpha_r \widehat{\boldsymbol{\psi}}_r^T \mathbf{X}_i + \varepsilon_i$. Here, k serves as smoothing parameter. The new coefficients $\boldsymbol{\alpha}$ are estimated by least squares, and $\widehat{\beta}_j = \sum_{r=1}^k \widehat{\alpha}_r \widehat{\psi}_{rj}$. Resulting rates of convergence are given in Hall and Horowitz (2007).

The above arguments show that a suitable regression analysis will have to take into account the underlying structure of the explanatory variables X_{ij}. The basic motivation of this paper now is to combine the points of view of the above branches of literature in order to develop a new approach for model adjustment and variable selection in the practically important situation of strongly correlated regressors.

30.2 The augmented model

In the following we assume that the vectors of regressors $\mathbf{X}_i \in \mathbb{R}^p$ can be decomposed in the form of a *factor model*

$$\mathbf{X}_i = \mathbf{W}_i + \mathbf{Z}_i, \; i = 1, \ldots, n, \tag{30.2}$$

where $\mathbf{W}_i$ and $\mathbf{Z}_i$ are two uncorrelated random vectors in $\mathbb{R}^p$. The random vector $\mathbf{W}_i$ is intended to describe high correlations of the X_{ij} while the components Z_{ij}, $j = 1, \ldots, p$ of $\mathbf{Z}_i$ are uncorrelated. This implies that the covariance matrix $\boldsymbol{\Sigma}$ of $\mathbf{X}_i$ adopts the decomposition

$$\boldsymbol{\Sigma} = \boldsymbol{\Gamma} + \boldsymbol{\Psi}, \tag{30.3}$$

where $\boldsymbol{\Gamma} = \mathbb{E}(\mathbf{W}_i \mathbf{W}_i^T)$, while $\boldsymbol{\Psi}$ is a diagonal matrix with diagonal entries $var(Z_{ij}) > 0$, $j = 1, \ldots, p$. Note that factor models can be found in any textbook on multivariate analysis and must be seen as one of the major tools in order to analyze samples of high dimensional vectors. Also recall that a standard factor model is additionally based on the assumption that a finite number k of factors suffices to approximate $\mathbf{W}_i$ precisely. This means that $\mathbf{W}_i = \sum_{r=1}^{k} (\boldsymbol{\psi}_r^T \mathbf{W}_i) \boldsymbol{\psi}_r$, where $\boldsymbol{\psi}_1, \ldots, \boldsymbol{\psi}_k$ are orthonormal eigenvectors corresponding to the k largest eigenvalues $\lambda_1 > \cdots > \lambda_k > 0$ of the standardized matrix $\frac{1}{p}\boldsymbol{\Gamma}$.

Each of the two components $\mathbf{W}_i$ and $\mathbf{Z}_i$ separately may possess a significant influence on a response variable Y_i. Indeed, if $\mathbf{W}_i$ and $\mathbf{Z}_i$ were known, a possibly substantial improvement of model (30.1) would consist in a regression of Y_i on the $2p$ variables $\mathbf{W}_i$ **and** $\mathbf{Z}_i$

$$Y_i = \sum_{j=1}^{p} \beta_j^* W_{ij} + \sum_{j=1}^{p} \beta_j Z_{ij} + \varepsilon_i, \quad i = 1, \ldots, n \tag{30.4}$$

with different sets of parameters β_j^* and β_j, $j = 1, \ldots, p$, for each contributor. We here again assume that ε_i, $i = 1, \ldots, n$ are centered i.i.d. random variables with $Var(\varepsilon_i) = \sigma^2$ which are independent of W_{ij} and Z_{ij}.

By definition, W_{ij} and Z_{ij} possess substantially different interpretations. Z_{ij} describes the part of X_{ij} which is *uncorrelated with all other variables.* A nonzero coefficient $\beta_j \neq 0$ then means that the variation of X_{ij} has a *specific* effect on Y_i. We will of course assume that such nonzero coefficients are **sparse**, $\sharp\{j|\beta_j \neq 0\} \leq S$ for some $S \ll p$.

In contrast, the variables W_{ij} are heavily correlated. It therefore does not make any sense to assume that for some $j \in \{1, \ldots, p\}$ any particular variable W_{ij} possesses a specific influence on the predictor variable. However, the term $\sum_{j=1}^{p} \beta_j^{**} W_{ij}$ may represent an important, *common* effect of all predictor variables. The vectors $\mathbf{W}_i$ can obviously be rewritten in terms of principal components.

Noting that $\beta_j Z_{ij} = \beta_j (X_{ij} - W_{ij})$ and $\mathbf{W}_i = \sum_{r=1}^{k} (\boldsymbol{\psi}_r^T \mathbf{W}_i) \boldsymbol{\psi}_r$, it is easily seen that there exist coefficients $\alpha_1, \ldots, \alpha_k$ such that (30.4) can be rewritten in the form of the following **augmented model**:

$$Y_i = \sum_{r=1}^{k} \alpha_r \xi_{ir} + \sum_{j=1}^{p} \beta_j X_{ij} + \varepsilon_i, \tag{30.5}$$

where $\xi_{ir} = \boldsymbol{\psi}_r^T \mathbf{W}_i / \sqrt{p\lambda_r}$. The use of ξ_{ir} instead of $\boldsymbol{\psi}_r^T \mathbf{W}_i$ is motivated by the fact that $Var(\boldsymbol{\psi}_r^T \mathbf{W}_i) = p\lambda_r$, $r = 1,\ldots,k$. Therefore the ξ_{ir} are standardized variables with $Var(\xi_{i1}) = \cdots = Var(\xi_{i1}) = 1$.

Obviously, the augmented model may be considered as a synthesis of the standard type of models proposed in the literature on functional regression and model selection. It generalizes the classical multivariate linear regression model (30.1). If a k-factor model holds exactly, i.e. $rank(\boldsymbol{\Gamma}) = k$, then the only substantial restriction of (30.4)- (30.5) consists in the assumption that Y_i depends *linearly* on W_i and Z_i. The analysis of Kneip and Sarda (2011) is somewhat more general and includes the case that a factor model only holds approximately. Also the problem of determining a suitable dimension k is considered.

30.3 Estimation

Assuming a sparse set of nonzero coefficients, the basic idea of our approach consists in applying Lasso in order to retrieve all nonzero parameters α_r, $r = 1,\ldots,k$, and β_j, $j = 1,\ldots,p$.

Since ξ_{ir} are latent, unobserved variables they are replaced by the predictors $\widehat{\xi}_{ir} = \widehat{\boldsymbol{\psi}}_r^T \mathbf{X}_i / \sqrt{p\widehat{\lambda}_r}$, where $\widehat{\lambda}_1 \geq \widehat{\lambda}_2 \geq \ldots$ are the eigenvalues of the standardized empirical covariance matrix $\frac{1}{p}\widehat{\boldsymbol{\Sigma}} = \frac{1}{n}\sum_{i=1}^n \mathbf{X}_i^T \mathbf{X}_i$, while $\widehat{\boldsymbol{\psi}}_1, \widehat{\boldsymbol{\psi}}_2, \ldots$ are associated orthonormal eigenvectors.

When replacing ξ_{ir} by $\widehat{\xi}_{ir}$ in (30.5), a direct application of model selection procedures does not seem to be adequate, since $\widehat{\xi}_{ir}$ and the predictor variables X_{ij} are heavily correlated. Therefore, instead of the originals vectors $\mathbf{X}_i$ we use the corresponding projections $\widetilde{\mathbf{X}}_i = \widehat{\mathbf{P}}_k \mathbf{X}_i$, where $\widehat{\mathbf{P}}_k = \mathbf{I}_p - \sum_{r=1}^k \widehat{\boldsymbol{\psi}}_r \widehat{\boldsymbol{\psi}}_r^T$..

For a pre-specified parameter $\rho > 0$ estimators $\widetilde{\boldsymbol{\alpha}} = (\widetilde{\alpha}_1,\ldots,\widetilde{\alpha}_k)^T$ and $\widetilde{\boldsymbol{\beta}} = (\widetilde{\beta}_1,\ldots,\widetilde{\beta}_p)^T$ are then obtained by minimizing

$$\frac{1}{n}\sum_{i=1}^{n}\left(Y_i - \boldsymbol{\alpha}^T \widehat{\boldsymbol{\xi}}_i - \boldsymbol{\beta}^T \widetilde{\mathbf{X}}_i\right)^2 + 2\rho\left(\sum_{r=1}^{k}|\alpha_r| + \sum_{j=1}^{p}|\beta_j|\right)$$

over all vectors $\boldsymbol{\alpha} \in \mathbb{R}^k$ and $\boldsymbol{\beta} \in \mathbb{R}^p$. Here, $\widehat{\boldsymbol{\xi}}_i = (\widehat{\xi}_{i1},\ldots,\widehat{\xi}_{ik})^T$.

In a final step a back-transformation is performed, and the final estimators of α_r and β_j are given by

$$\widehat{\beta}_j = \frac{\widetilde{\beta}_j}{\left(\frac{1}{n}\sum_{i=1}^{n}(\widehat{\mathbf{P}}_k\boldsymbol{X}_i)_j^2\right)^{1/2}},\ j=1,\ldots,p,$$

$$\widehat{\alpha}_r = \widetilde{\alpha}_r - \sqrt{p\widehat{\lambda}_r}\sum_{j=1}^{p}\widehat{\psi}_{rj}\widehat{\beta}_j,\ r=1,\ldots,k.$$

30.4 Theoretical properties of augmented model

A precise theoretical analysis based on finite sample inequalities can be found in Kneip and Sarda (2011). In this section we will confine ourselves to sum up some of the main results from the point of view of an asymptotic theory as $n,p \to \infty$ with $\frac{\log p}{n} \to 0$.

For a precise description of setup and necessary assumptions we again refer to Kneip and Sarda (2011). Qualitatively it is required that the tails of the distribution of the variables X_{ij}, W_{ij} and Z_{ij} decrease sufficiently fast, while the error terms ε_{ij} are normal. Furthermore, there has to exist a constant $D_1 > 0$ such that $\inf_{j=1,\ldots,p} Var(Z_{ij}) \geq D_1$ for all p. A third essential condition is that each principle component of $\frac{1}{p}\boldsymbol{\Gamma}$ explains a considerable proportion of the total variance of $\mathbf{W}_i$, for all $r=1,\ldots,k$ and all p we have $\frac{\lambda_r}{\frac{1}{p}\sum_{j=1}^{p}\mathbb{E}(W_{ij}^2)} \geq v$ for some $v > 0$.

The main underlying condition ensuring identifiability of the coefficients and allowing to derive consistency results is **sparseness** of β_j, $\sharp\{j|\beta_j \neq 0\} \leq S$ for some $S \ll p$. We hereby rely on recent results in the literature on model selection which also apply for the case $p > n$. Established theoretical results show that under some regularity conditions (validity of the "restricted eigenvalue conditions") model selection procedures *allow to identify sparse solutions* even if there are multiple vectors of coefficients satisfying the normal equations. For a discussion of these issues see Candès and Tao (2007) or Bickel et al. (2009).

If k and $S \geq \sharp\{j|\beta_j \neq 0\}$ are fixed, then under some regularity conditions it can be shown that as $n,p \to \infty$, $\log p/n \to 0$, we obtain with $\rho \sim \sqrt{\frac{\log p}{n}}$

$$\sum_{r=1}^{k}|\widehat{\alpha}_r - \alpha_r| = O_P(\sqrt{\frac{\log p}{n}})$$

$$\sum_{j=1}^{p}|\widehat{\beta}_j - \beta_j| = O_P(\sqrt{\frac{\log p}{n}}).$$

Moreover,

$$\frac{1}{n}\sum_{i=1}^{n}\left(\sum_{r=1}^{k}\xi_{ir}\widehat{\alpha}_r + \sum_{j=1}^{p}X_{ij}\widehat{\beta}_j - (\sum_{r=1}^{k}\xi_{ir}\alpha_r + \sum_{j=1}^{p}X_{ij}\beta_j)\right)^2 = O_P(\frac{\log p}{n} + \frac{1}{p})$$

Kneip and Sarda (2011) also present an extensive simulation study. It is shown that if $\mathbf{W}_i$ possesses a non-negligible influence on the the response variable, then variable selection based on the standard regression model (30.1) does not work at all, while the augmented model is able to yield sensible parameter estimates.

References

1. Bickel, P.J., Ritov, Y., Tsybakov, A.: Simulataneous analysis of Lasso and Dantzig selector. Ann. Stat., **37**, 1705–1732 (2009)
2. Cai, T., Hall, P.: Prediction in functional linear regression. Ann. Stat. **34**, 2159–2179 (2007)
3. Candès, E., Tao, T.: The Dantzig selector: statistical estimation when p is much larger than n. Ann. Stat. **35**, 2013–2351 (2007)
4. Cardot, H., Ferraty, F., Sarda, P.: Functional linear model. Stat. Probab. Lett. **45**, 11–22 (1999)
5. Cardot, H., Mas, A., Sarda, P.: CLT in functional linear regression models. Probab. Theor. Rel. **138**, 325–361 (2007)
6. Crambes, C., Kneip, A., Sarda, P.: Smoothing spline estimators for functional linear regression. Ann. Stat. **37**, 35–72 (2009)
7. Cuevas, A., Febrero, M., Fraiman, R.: Linear functional regression: the case of fixed design and functional response. Canad. J. Stat. **30**, 285–300 (2002)
8. Hall, P., Horowitz, J.L.: Methodology and convergence rates for functional linear regression. Ann. Stat. **35**, 70–91 (2007)
9. Kneip, A., Sarda, P.: Factor Models and Variable Selection in High Dimensional Regression Analysis. Revised manuscript (2011)
10. Meinshausen, N., Bühlmann, P.: High dimensional graphs and variable selection with the Lasso. Ann. Stat. **34**, 1436–1462 (2006)
11. Ramsay, J.O., Dalzell, C.J.: Some tools for functional data analysis (with discussion). J. Roy. Stat. Soc. B **53**, 539–572 (1991)
12. Tibshirani, R.: Regression shrinkage and selection via the Lasso. J. Roy. Stat. Soc. B **58**, 267–288 (1996)
13. van de Geer, S.: High-dimensional generalized linear models and the Lasso. Ann. Stat. **36**, 614–645 (2008)
14. Yao, F., Müller, H.-G., Wang, J.-L.: Functional regression analysis for longitudinal data. Ann. Stat. **37**, 2873–2903 (2005)
15. Zhao, P., Yu, B.: On model selection consistency of Lasso. J. Machine Learning Research **7**, 2541–2567 (2006)

Chapter 31
Factor Modeling for High Dimensional Time Series

Clifford Lam, Qiwei Yao, Neil Bathia

Abstract We briefly compare an econometric factor model and a statistical factor model, the latter being to capture the linear dynamic structure of the data $\mathbf{y}_t$ only. We propose a method for decomposing $\{\mathbf{y}_t\}$ into a common component and a white noise in the sense of a statistical factor model, together with an eye-ball test for finding the number of factors. Rates of convergence for various estimators are spelt out explicitly.

31.1 Introduction

A large panel of time series is a common endeavor in modern data analysis. In finance, understanding the dynamics of the returns of large number of assets is the key to asset pricing, portfolio allocation, and risk management. Environmental time series are often of a high dimension because of the large number of indices monitored across many different locations. Vector Autoregressive-Moving-Average model is useful for moderate dimension, but practically not viable when the dimension of the time series p is high, as the number of parameters involved is in the order of p^2. Therefore dimension-reduction is an important step in order to achieve an efficient and effective analysis of high-dimensional time series data. In relation to the dimension-reduction for independent observations, the added challenge here is to retain the dynamical structure of time series.

Clifford Lam
London School of Economics, London, UK, e-mail: C.Lam2@lse.ac.uk

Qiwei Yao
London School of Economics, London, UK, e-mail: Q.Yao@lse.ac.uk

Neil Bathia
London School of Economics, London, UK, e-mail: neilbathia@googlemail.com

Modeling by common factors is one of the most frequently used methods to achieve dimension-reduction in analyzing multiple time series. In fact it is constantly featured in econometrics literature. They are used to model different economic and financial phenomena, including asset pricing models (Ross 1976), yield curves (Chib and Ergashev 2009), macroeconomic behavior such as sector-effect or regional effect from disaggregated data (Quah and Sargent 1993, Forni and Reichlin 1998), macroeconomic forecasting (Stock and Watson 1998, 2002), consumer theory etc (Bai 2003).

The following econometric model represents a $p \times 1$ time series $\mathbf{y}_t$ as the sum of two unobservable parts:

$$\mathbf{y}_t = \mathbf{f}_t + \boldsymbol{\xi}_t,$$

where $\mathbf{f}_t$ is a factor term and is driven by r common factors with r smaller or much smaller than p, and $\boldsymbol{\xi}_t$ is an idiosyncratic term and consists of p idiosyncratic components. Since $\boldsymbol{\xi}_t$ is not necessarily a white noise, the identification and the inference for the above decomposition is inevitably challenging. For example, $\mathbf{f}_t$ and $\boldsymbol{\xi}_t$ are only asymptotically identifiable when p, the number of components $\mathbf{y}_t$, tends to ∞; see Chamberlain and Rothschild (1983). The generalized dynamic factor model proposed by Forni *et al.* (2000) is also of this form, which further allows components of $\{\boldsymbol{\xi}_t\}$ to be weakly correlated with each other, and $\mathbf{f}_t$ has dynamics driven by q common (dynamic) factors. See also Forni *et al.* (2004, 2005), Deistler *et al.* (2009), and Barigozzi *et al.* (2010) for further details, and Hallin and Liška (2007) for determining the number of factors.

The statistical factor model focuses on capturing the linear dynamic structure of $\mathbf{y}_t$:

$$\mathbf{y}_t = \mathbf{A}\mathbf{x}_t + \boldsymbol{\varepsilon}_t,$$

where $\mathbf{x}_t$ is an $r \times 1$ latent process with (unknown) $r < p$, $\mathbf{A}$ is a $p \times r$ unknown constant matrix, and $\boldsymbol{\varepsilon}_t \sim WN(\boldsymbol{\mu}_\varepsilon, \boldsymbol{\Sigma}_\varepsilon)$ is a vector white noise process. No linear combinations $\mathbf{c}'\mathbf{x}_t$ should be a white noise process as they should be absorbed in $\boldsymbol{\varepsilon}_t$. This setting can be traced back to Peña and Box (1987); see also its further development in dealing with cointegrated factors in Peña and Poncela (2006). With r much smaller than p, we achieve effective dimension reduction, where the serial dependence of $\{\mathbf{y}_t\}$ is driven by a much lower dimensional process $\{\mathbf{x}_t\}$. The fact that $\{\boldsymbol{\varepsilon}_t\}$ is white noise eases the identification of $\mathbf{A}$ and $\mathbf{x}_t$ tremendously. Although $(\mathbf{A}, \mathbf{x}_t)$ can be replaced by $(\mathbf{AH}, \mathbf{H}^{-1}\mathbf{x}_t)$ without changing the model, so that they are unidentifiable, it is easy to see that the r-dimensional linear space spanned by the columns of $\mathbf{A}$, denoted by $\mathscr{M}(\mathbf{A})$, is uniquely defined. In particular, for each p fixed, the model is identifiable in the sense that $\mathbf{A}\mathbf{x}_t$, called the common component, is uniquely defined. Furthermore, such model allows for estimation through simple eigenanalysis, as laid out in the next section.

31.2 Estimation Given r

Detailed assumptions for the statistical model are given in Lam, Yao and Bathia (2010) or Lam and Yao (2011). One important feature is that $\mathbf{\Sigma}_\varepsilon$ is allowed to have $O(1)$ elements, so that strong cross-sectional dependence of the noise is allowed, and as $p \to \infty$, the eigenvalues of $\mathbf{\Sigma}_\varepsilon$ can grow like p. This is more relevant in spatial data, where time series in a local neighborhood can have similar trend and noise series, so that cross-sectional dependence of the noise can be strong. Some real data examples with this feature will be presented in the talk.

Through the QR-decomposition $\mathbf{A} = \mathbf{QR}$, we can write the model as

$$\mathbf{y}_t = \mathbf{Q}\mathbf{f}_t + \boldsymbol{\varepsilon}_t,$$

where $\mathbf{Q}'\mathbf{Q} = \mathbf{I}$. Notice that

$$\mathbf{\Sigma}_\mathbf{y}(k) = \text{cov}(\mathbf{y}_{t+k}, \mathbf{y}_t) = \mathbf{Q}\mathbf{\Sigma}_\mathbf{f}(k)\mathbf{Q}' + \mathbf{Q}\mathbf{\Sigma}_{\mathbf{f},\boldsymbol{\varepsilon}}(k),$$

where $\mathbf{\Sigma}_\mathbf{f}(k) = \text{cov}(\mathbf{f}_{t+k}, \mathbf{f}_t)$ and $\mathbf{\Sigma}_{\mathbf{f},\boldsymbol{\varepsilon}}(k) = \text{cov}(\mathbf{f}_{t+k}, \boldsymbol{\varepsilon}_t)$. For $k_0 \geq 1$ given, define

$$\begin{aligned}\mathbf{L} &= \sum_{k=1}^{k_0} \mathbf{\Sigma}_\mathbf{y}(k)\mathbf{\Sigma}_\mathbf{y}(k)' \\ &= \mathbf{Q}\Big(\sum_{k=1}^{k_0} \{\mathbf{\Sigma}_\mathbf{f}(k)\mathbf{Q}' + \mathbf{\Sigma}_{\mathbf{f},\boldsymbol{\varepsilon}}(k)\}\{\mathbf{\Sigma}_\mathbf{f}(k)\mathbf{Q}' + \mathbf{\Sigma}_{\mathbf{f},\boldsymbol{\varepsilon}}(k)\}' \Big)\mathbf{Q}'.\end{aligned}$$

If we apply spectral decomposition to the term sandwiched by $\mathbf{Q}$ and $\mathbf{Q}'$, then we can write $\mathbf{L} = \mathbf{QUDU}'\mathbf{Q}'$, where $\mathbf{D}$ is a diagonal matrix, and $\mathbf{U}$ is $r \times r$ orthogonal. Hence the columns of $\mathbf{QU}$ are the eigenvectors of $\mathbf{L}$ corresponding to its r non-zero eigenvalues. We take $\mathbf{QU}$ as the $\mathbf{Q}$ to be used in our inference. A natural estimator of $\mathbf{Q}$ is then

$$\hat{\mathbf{Q}} = (\hat{\mathbf{q}}_1, \cdots, \hat{\mathbf{q}}_r),$$

where $\hat{\mathbf{q}}_i$ is the unit eigenvector corresponding to the i-th largest eigenvalue of $\hat{\mathbf{L}}$, which is a sample version of $\mathbf{L}$, with $\mathbf{\Sigma}_\mathbf{y}(k)$ replaced by the corresponding sample lag-k autocovariance matrix for $\mathbf{y}_t$. Consequently, we estimate the factors and the residuals respectively by

$$\hat{\mathbf{f}}_t = \hat{\mathbf{Q}}^T \mathbf{y}_t, \qquad \mathbf{e}_t = \mathbf{y}_t - \hat{\mathbf{Q}}\hat{\mathbf{f}}_t = (\mathbf{I}_p - \hat{\mathbf{Q}}\hat{\mathbf{Q}}^T)\mathbf{y}_t.$$

Some theories will be given in the talk with rate of convergence specified.

31.3 Determining r

The eigenvalues of $\hat{\mathbf{L}}$, denoted by $\hat{\lambda}_j$ for the j-th largest one, can help determine the number of factors r. In particular, if we have strong factors (see Lam, Yao and Bathia (2010) for the definition of the strength of factors), then the following holds:

$$\hat{\lambda}_{j+1}/\hat{\lambda}_j \asymp 1, j = 1, \cdots, r-1, \text{ and } \hat{\lambda}_{r+1}/\hat{\lambda}_r = O_P(n^{-1}).$$

The rate n^{-1} is non-standard, and is a result of defining $\mathbf{L}$ to include products of autocovariance matrices. This result suggests an eye-ball test of r, where a plot of the ratio of eigenvalues $\hat{\lambda}_{j+1}/\hat{\lambda}_j$ is made, and the first sharp drop in the plot indicates r.

More general results involving different strength of factors will be given in the talk, with simulation results and real data analyses presented as well.

References

1. Bai, J.: Inferential theory for factor models of large dimensions. Econometrica **71**, 135–171 (2003)
2. Bai, J., Ng, S.: Determining the number of factors in approximate factor models. Econometrica **70**, 191–221 (2002)
3. Bai, J., Ng, S.: Determining the number of primitive shocks in factor models. J. Bus. Econ. Stat. **25**, 52–60 (2007)
4. Barigozzi, M., Alessi, L., Capasso, M.: A robust criterion for determining the number of static factors in approximate factor models. European Central Bank Working Paper 903 (2010)
5. Bathia, N., Yao, Q., Ziegelmann, F.: Identifying the finite dimensionality of curve time series. Ann. Stat. to appear (2010)
6. Brillinger, D.R.: Time Series Analysis: Data Analysis and Theory (Second Edition). Holt, Rinehart & Winston, New York (1981)
7. Chamberlain, G.: Funds, factors, and diversification in arbitrage pricing models. Econometrica **51**, 1305–1323 (1983)
8. Chamberlain, G., Rothschild, M.: Arbitrage, factor structure, and mean-variance analysis on large asset markets. Econometrica **51**, 1281–1304 (1983)
9. Chib, S., Ergashev, B.: Analysis of multifactor affine yield curve models. J. Am. Stat. Assoc. **104**, 1324–1337 (2009)
10. Deistler, M., Anderson, B., Chen, W. and Filler, A. (2009). Generalized linear dynamic factor models – an approach via singular autoregressions. Eur. J. Control, Invited submission (2009)
11. Forni, M., Hallin, M., Lippi, M., Reichlin, L.: The generalized dynamic-factor model: identification and estimation. Rev. Econ. Stat. **82**, 540–554 (2000)
12. Forni, M., Hallin, M., Lippi, M., Reichlin, L.: The generalized dynamic-factor model: consistency and rates. J. Econometrics **119**, 231–255 (2004)
13. Forni, M., Hallin, M., Lippi, M., Reichlin, L.: The generalized dynamic-factor model: one-sided estimation and forecasting. J. Am. Stat. Assoc. **100**, 830-840 (2005)
14. Geweke, J.: The dynamic factor analysis of economic time series. In: Aigner, D.J. and Goldberger A.S. (eds.) Latent Variables in Socio-Economic Models, Chapter 19, Amsterdam, North-Holland (1977)
15. Hallin, M., Liska, R.: Determining the number of factors in the general dynamic factor model. J. Am. Stat. Assoc. **102**, 603–617 (2007)

16. Lam, C., Yao, Q., Bathia, N.: Estimation for latent factor models for high-dimensional time series. Manuscript, available at http://arxiv.org/abs/1004.2138 (2010)
17. Lam, C., Yao, Q.: Factor Modeling for High Dimensional Time Series. Under preparation (2011)
18. Pan, J., Peña, D., Polonik, W., Yao, Q.: Modelling multivariate volatilities via common factors. Available at http://stats.lse.ac.uk/q.yao/qyao.links/paper/pppy.pdf (2008)
19. Pan, J., Yao, Q.: Modelling multiple time series via common factors. Biometrika **95**, 356–379 (2008)
20. Péché, S.: Universality results for the largest eigenvalues of some sample covariance matrix ensembles. Probab. Theor. Rel. **143**, 481–516 (2009)
21. Peña, D., Box, E.P.: Identifying a simplifying structure in time series. J. Am. Stat. Assoc. **82**, 836–843 (1987)
22. Peña, D., Poncela, P.: Nonstationary dynamic factor analysis. J. Stat. Plan. Infer. **136**, 1237–1257 (2006)
23. Quah, D., Sargent, T.J.: A dynamic index model for large cross sections. In: J.H. Stock, J.H., Walton, M.W. (eds.), Business Cycles, Indicators and Forecasting NBER, Cahpter 7, pp. 285-309 (1993)
24. Ross, S.: The arbitrage theory of capital asset pricing. J. Fianance **13**, 341–360 (1976)
25. Sargent, T. J., Sims, C.A.: Business cycle modeling without pretending to have too much a priori economic theory. In: Sims, C. et al. (eds.) New methods in business cycle research, Minneapolis, Federal Reserve Bank of Minneapolis, pp. 45-108 (1977)
26. Stock, J.H., Watson, M.W.: Diffusion indexed. NBER Working Paper 6702 (1998)
27. Stock, J.H., Watson, M.W.: Macroeconomic forecasting using diffusion indices. J. Bus. Econ. Stat. **20**, 147–162 (2002)
28. Tiao, G.C., Tsay, R.S.: Model specification in multivariate time series (with discussion). J. Roy. Stat. Soc. B **51**, 157–213 (1989)
29. Wang, H.: Factor Profiling for Ultra High Dimensional Variable Selection. Available at SSRN: http://ssrn.com/abstract=1613452 (2010)
30. Wang, Y., Yao, Q., Li, P., Zou, J.: High dimensional volatility modeling and analysis for high-frequency financial data. Available at http://stats.lse.ac.uk/q.yao/qyao.links/paper/highfreq.pdf (2008)

Chapter 32
Depth for Sparse Functional Data

Sara López-Pintado, Ying Wei

Abstract The notions of depth for functional data provide a way of ordering curves from center-outward. These methods are designed for trajectories that are observed on a fine grid of equally spaced time points. However, in many applications the trajectories are observed on sparse irregularly spaced time points. We propose a model-based consistent procedure for estimating the depths when the curves are observed on sparse and unevenly spaced points.

32.1 Introduction

Functional data analysis is an exciting developing area in statistics. Many different statistical methods, such as principal components, analysis of variance, and linear regression, have been extended to functional data. The statistical analysis of curves can be significantly improved using robust estimators. New ideas of depth for functional data have been studied recently (see, e.g., Fraiman and Muniz, 2001, Cuevas et al., 2007, and López-Pintado and Romo, 2009. These concepts provide a way of ordering curves from center-outward and L-statistics can be defined for functional data. All these methods work well when the trajectories are observed on a fine grid and equally spaced time points (see Ramsay and Silverman, 2005). However, many times the trajectories are observed on irregularly spaced time points that vary a lot across trajectories. In this situation, some preliminary smoothing step like kernel smoothing, smoothing splines or local linear smoothing needs to be applied. When the number of observations for individual paths is small these methods do not perform well (Yao et al, 2005). In this paper we extend the ideas of band depth and modified band depth introduced in López-Pintado and Romo, 2009 to sparse func-

Sara López-Pintado
Columbia University, New York, USA, e-mail: sl2929@columbia.edu

Ying Wei
Columbia University, New York, USA, e-mail: yw2148@columbia.edu

tional data, where the data is onl! y observed on a set of sparse and unevenly spaced points.

The study is motivated by an early-life human growth research using the data from 1988 National Maternal and Infant Health Survey (NMIHS) and its 1991 Longitudinal Follow-up. The study included 2555 boys and 2510 girls national-wide who were born in the U.S. in the calendar year of 1988. Their heights and weights were taken sporadically only when they visited a hospital. Consequently, their growth paths were recorded on a set of sparse and irregularly-spaced time points and the number of measurements per subject is small. Moreover, among those subjects, low birth-weight infants($\leq$ 2500 g) were over-sampled, which constitute approximately 25% of the data. To understand the growth patterns of low birth-weight infants has been a long term research topic in epidemiology. The most informative growth pattern is represented by the underlying height process as a continuous function of age, since height growth is directly associated with individual growth hormone levels. T! he idea of depth provides a way of ordering the curves from the sample and a median growth curve can be defined. Moreover, a rank test based on depth can be used to test if boys born with normal weight have a different growth pattern in height than those who were born with low weight.

32.2 Method

32.2.1 Review on band depth and modified band depth

The concepts of band depth and modified band depth were introduced and analyzed in López-Pintado and Romo, 2009. These ideas provide a way of ordering curves from center-outward and classical order statistics can be generalized to functional data. We summarize next the band depth definitions.

Let $\mathscr{C}$ be the space of continuous real valued functions on the compact interval $[0,1]$ with probability measure $\mathscr{P}$ and let $x_1(t),\dots,x_n(t)$ be a random sample of i.i.d. curves drawn from the probability space $(\mathscr{C},\mathscr{P})$. Any k curves from the sample determine in $\mathbb{R}^2$ bands defined as

$$B = B\left(x_{i_1},\dots,x_{i_k}\right) = \left\{(t,y) \in I \times \mathbb{R} : \min_{r=1,\dots k} x_{i_r}(t) \leq y \leq \max_{r=1,\dots,k} x_{i_r}(t)\right\}.$$

For simplicity we will consider $k=2$ although all the results hold for any $kgeq2$. We denote the graph of x by $G(x) = \{(t,x(t)) : t \in I\} \subset \mathbb{R}^2$.

The band depth of a function x in $(\mathscr{C},\mathscr{P})$ was defined in López-Pintado and Romo, 2009, as $D(x;P) = \mathscr{P}(G(x) \subset B(X_1,X_2))$, where P is the probability distribution of the process X which generates the sample X_1,X_2. Alternatively, we can also express the band depth as

$$D(x;P) = \mathrm{E}\left[I(G(x) \subset B(X_1,X_2))\right]. \tag{32.1}$$

Let $x_1,...,x_n$ be an i.i.d. sample of functions with probability distribution P. The sample band-depth of x with respect to $x_1,...,x_n$ is

$$D_n(x) = \frac{\sum_{1\le i_1<i_2\le n} I(G(x) \subset B(x_{i_1},x_{i_2}))}{\binom{n}{2}}.$$

Essentially, the sample band depth is just the proportion of bands B defined by pairs of curves from the sample containing the graph of x.

An alternative and more flexible notion of depth is the modified band depth also introduced in López-Pintado and Romo, 2009). It is defined as $MD(x) = E_{X_1,X_2}[\lambda(A(x;X_1,X_2))]$, where $\lambda = \lambda_l/\lambda_l(I)$, λ_l is Lebesgue measure in $\mathbb{R}$ and

$$A(x;x_1,x_2) = \left\{ t \in I : \min_{r=1,2} x_r(t) \le x(t) \le \max_{r=1,2} x_r(t) \right\}.$$

Basically, $MD(x)$ measures how long is the curve x expected to be inside a stochastic band determined by two stochastic functions X_1 and X_2. The sample modified band-depth of any curve x with respect to an i.i.d. random sample $x_1,...,x_n$ is

$$MD_n(x) = \binom{n}{2}^{-1} \sum_{1\le i_1<i_2\le n} \lambda(A(x;x_{i_1},x_{i_2})).$$

.

The band depth and the modified band depth satisfy natural depth properties such as: invariance, monotonicity with respect to the deepest point, consistency and uniform consistency (see López-Pintado and Romo, 2009). Both of these notions of depth can be denoted as

$$D(x;P) = E_{X_1,X_2}[g(x;X_1,X_2)].$$

where $g(x;X_1,X_2) = I\{G(x) \subset B(X_1,X_2)\}$ for band depth, and $g(x;X_1,X_2) = \lambda(A(x;X_1,X_2))$ for the modified band depth.

In order to apply these notions of depth, the curves should be measured at a regular grid. In practice, this is rarely the case, usually each curve from the sample is observed at sparse and unevenly spaced time points. In the next section we extend the definition of (modified) band depth to this context of sparse functional data.

32.2.2 Adapted conditional depth for sparse data

Let $\mathscr{C}$ be the space of continuous real valued functions on the compact interval $[0,1]$ with probability measure $\mathscr{P}$, and $X, X_1, ..., X_n$ be a random sample of i.i.d. curves drawn from the probability space $(\mathscr{C}, \mathscr{P})$. Each curve is observed only on a set of random time points $T_i = (t_{i,1}, t_{i,2}, ..., t_{i,m_i})$, where m_i is total number observations of the ith curve. The individual measurement times $t_{i,j}$'s could be sparse and unevenly spaced. The proposed methods aim at estimating the depth of trajectory $X_i(t)$ while they are observed incompletely and sparsely.

We denote $w_{i,j}$ as the jth observation of the random function X_i at a random time $t_{i,j}$ with possible measurement errors $e_{i,j}$. Specifically, we assume that $w_{i,j}$ are obtained from the following model

$$w_{i,j} = X_i(t_{i,j}) + e_{i,j}, \tag{32.2}$$

where the error term $e_{i,j}$ are i.i.d. with $E(e_{i,j}) = 0$ and constant variance σ^2. Let $\pi(t) = (\pi_1(t), ..., \pi_k(t))$ be a k dimensional B-spline basis functions given the order of spline and a set of internal knots. Any smooth curves $X_i(t)$'s can be well approximated by a linear combination of $\pi(t)$ with appropriately chosen coefficients. Let β_i be the best fitting b-spline coefficients for $X_i(t)$, Model (32.2) can be approximated by the following nonparametric random coefficient model

$$w_{i,j} = \pi(t_{i,j})^\top \beta_i + e_{i,j}. \tag{32.3}$$

We further assume that the coefficients β_i are iid following certain multivariate distribution. Based on Model (32.3), we can predict the β_i's . Consequently, the trajectory $X_i(t)$ can also be predicted by $\widehat{X}_i(t) = \pi(t)^\top \widehat{\beta}_i$. One can show that it is the best linear predictor of $X_i(t)$. Finally, we estimate the depth of $X_i(t)$ based on the sample collection of $\widehat{X}_i(t)$'s.

References

1. Cuevas, A., Febrero, M., Fraiman, R.: Robust estimation and classification for functional data via projection-based depth notions. Computation. Stat. **22**, 481–496 (2007)
2. Fraiman, R., Muniz, G.: Trimmed means for functional data. TEST **10**, 419–440 (1999)
3. López-Pintado, S., Romo, J.: On the concept of depth for functional data. J. Am. Stat. Assoc. **104**, 486–503 (2009)
4. Ramsay, J. O., Silverman, B.: Functional data analysis (Second Edition). Springer Verlag, New York (2005)
5. Yao, F., Müller, H. G., Wang, J. L.: Functional data analysis for sparse longitudinal data. J. Am. Stat. Assoc. **100**, 577–590 (2005)

Chapter 33
Sparse Functional Linear Regression with Applications to Personalized Medicine

Ian W. McKeague, Min Qian

Abstract McKeague and Qian (2011) recently introduced a functional data-analytic approach to finding optimal treatment policies in the setting of personalized medicine based on genomic data. The policies are specified in terms of thresholds of gene expression at estimated loci along a chromosome. Methods for assessing the effectiveness of such treatment policies are described.

33.1 Introduction

Recent developments in high-throughput gene expression technology have the potential to advance the clinical prediction of disease origin, prognosis, and therapeutic response. This has opened up the possibility of tailoring treatments to individual patients by taking into account biomarker information extracted from biopsy tissue or sera. There is now an extensive literature on personalized medicine based on the analysis of genetic profiles, including gene expression; see, e.g., van 't Veer and Bernards (2008). Unfortunately, however, it is very difficult to establish the effectiveness of such treatment policies, not only from the practical point of view of designing clinical trials that can exploit the new understanding of the human genome, but also because the statistical methodology to assess the effectiveness of such treatment policies has received limited attention.

Consider a randomized clinical trial in which the goal is to develop an effective individualized treatment policy based on the gene expression profile of the patient. Our aim is to introduce policies determined by thresholds in gene expression at a small number of estimated genetic loci; such policies are easier to interpret and more feasible to implement (in bioassays, say) than policies based on a complete gene ex-

Ian W. McKeague
Columbia University, New York, USA, e-mail: im2131@columbia.edu

Min Qian
The University of Michigan, USA, e-mail: minqian@umich.edu

pression profile. Given gene expression profiles, treatment assignments, and clinical outcomes of all the patients in the trial, we then provide a way of simultaneously evaluating the effectiveness of such a treatment policy that optimizes interactions between gene expression and treatment, as well as locating genes that possibly account for such interactions. Here the effectiveness of a treatment policy is measured by the mean outcome when all patients follow the policy.

Early research in the area of individualized treatment concentrated on identifying qualitative interactions between treatment and patient pretreatment clinical variables, with treatments for various subsets of patients selected by hypothesis testing. In recent years, statistical methods for estimating a patient's risk category based on genetic information have been developed, and such estimates can help inform decisions as to the best treatment. This approach is only appropriate in settings where the best therapy (from among competing therapies) in each risk category is known. There is also an abundance of literature focusing on methods for identifying genes associated with clinical outcomes. These methods can be used to predict outcome under each competing treatment, so an individualized treatment policy can be formulated by choosing the treatment that achieves the best predicted outcome. Such treatment policies are generally less cost effective, though, since they are likely to involve genes that do not interact with treatment.

We take the point of view that gene expression profiles can be regarded as functional predictors with a continuous "time" index representing the genetic locus along a chromosome. A class of treatment policies that take interaction between treatment and an individual gene expression profile into account is then formulated in terms of a point impact linear regression model [McKeague and Sen (2010)], which specifies the functional predictor as having an impact on the outcome at a sparse collection of unknown time points (loci of genes). This type of model naturally leads to a class of treatment policies based on thresholding gene expression at such loci. The treatment policy we propose is then derived using a two step procedure that first estimates (by least squares) the effect of interaction on the outcome, and second, optimizes the interaction effect over competing treatments. This approach leads to a simple and interpretable treatment policy in terms of a sparse collection of genetic loci that best explain the interaction between gene expression and treatment.

33.2 Threshold-based point impact treatment policies

For simplicity we consider only two competing treatments (denoted by $A = \pm 1$), and a scalar outcome Y for which large values are desirable. We represent the gene expression profile in terms of a functional predictor $X = \{X(t), t \in [0,1]\}$ having sample paths in $\mathscr{X}$, where t indicates the locus along the genome or a specific chromosome. We will study threshold-based point impact (TPI) treatment policies $d : \mathscr{X} \to \{-1, 1\}$ that assign treatment solely on the basis of whether the value of X at a single locus exceeds a threshold, e.g., $d(X) = 1$ if $X(\theta) > c$, -1 otherwise, for some $\theta \in [0,1]$ and $c \in \mathbb{R}$. Our approach could be extended to allow multiple loci and thresholds, or to include additional pretreatment covariates, but we will avoid such complications for clarity of presentation.

A natural measure of the effectiveness of a general treatment policy d is the expected outcome, $P^d[Y]$, that would have resulted if d had been used to choose treatment for the entire study population; here P^d denotes the distribution of (X,A,Y) given $A = d(X)$, and $V(d) \equiv P^d[Y]$ is called the *value* of the policy. For simplicity we assume throughout that we have i.i.d. data on $(X,A,Y) \sim P$ from a randomized clinical trial (as when unbiased coin tosses are used to assign subjects to treatment groups), although our approach could readily be extended to observational studies in which the treatment assignment probability is unknown.

A treatment policy that maximizes the function $d \mapsto V(d)$ over all possible d is called *globally optimal*. It is easy to verify that any $d_0(x) \in \arg\max_{a\in\{-1,1\}} E(Y|X = x, A = a)$, $x \in \mathscr{X}$, is globally optimal, where E denotes expectation under P. Since A is independent of X and has mean zero,

$$E(Y|X,A) = E(Y|X) + T_0(X)A, \tag{33.1}$$

where $T_0(X) = E(Y|X,A = 1) - E(Y|X)$ is the *treatment effect*, so we can write $d_0(x) = \text{sign}(T_0(x))$ where $\text{sign}(0) \equiv 1$. Unfortunately, such a policy would be difficult to implement and estimate because it involves the nonparametric regression function $T_0(x)$ with an infinite dimensional predictor. For this reason, we restrict our attention to TPI policies, which are easier to implement and interpret.

We consider a regularization approach in which $T_0(x)$ is approximated in a way that allows direct estimation of the optimal TPI policy. More specifically, our proposed approach is to model the treatment effect in (33.1) by a point impact model in which $T_0(X)$ depends on X only through its value at a single locus: $T_0(X) = \alpha + \beta X(\theta)$, where the intercept α represents the main effect of treatment, and the slope β represents the interaction between treatment and the gene expression at locus θ. Under this model, the globally optimal policy d_0 is TPI, and we propose to replace the parameters (α, β, θ) by estimates $(\hat{\alpha}_n, \hat{\beta}_n, \hat{\theta}_n)$, leading to the TPI policy

$$\hat{d}_n(x) = \text{sign}(\hat{\alpha}_n + \hat{\beta}_n x(\hat{\theta}_n)), \; x \in \mathscr{X}. \tag{33.2}$$

To furnish such estimates, we need to introduce a model for the main effect of gene expression, $E(Y|X)$, and for simplicity we use a linear model: $E(Y|X) = \boldsymbol{U}^T\delta$, where $\boldsymbol{U} = (U_1, \ldots, U_J)^T$ is a given J-dimensional vector-valued function of X, and $\delta \in \mathbb{R}^J$ is a vector of parameters. For example, each U_j could be of the form $\int_0^1 \phi(t)X(t)\,dt$ for a given real-valued function ϕ, or depend only on the value of X at a pre-specified locus (e.g., representing a specific gene). This leads to the model for the full conditional mean function:

$$E(Y|X,A) = \boldsymbol{U}^T\delta + (\alpha + \beta X(\theta))A. \tag{33.3}$$

The parameters $\eta = (\delta, \alpha, \beta, \theta)$ can be estimated by least squares:

$$\hat{\eta}_n = (\hat{\delta}_n, \hat{\alpha}_n, \hat{\beta}_n, \hat{\theta}_n) = \arg\min_{\delta,\alpha,\beta,\theta} \mathbb{P}_n[Y - \boldsymbol{U}^T\delta - (\alpha + \beta X(\theta))A]^2, \tag{33.4}$$

where $\mathbb{P}_n$ is the empirical distribution for a sample of size n from the randomized clinical trial. It can be shown that the resulting estimated TPI policy $\hat{d}_n$ is asymptotically globally optimal in the sense that $V(\hat{d}_n)$ converges in probability to $V(d_0)$ as $n \to \infty$, provided the point impact model for the treatment effect is correctly specified. Moreover, this result holds even if the linear model used for the main effect of gene expression is misspecified; this robustness property is a consequence of $E(A|X) = 0$, implying that $(\hat{\alpha}_n, \hat{\beta}_n, \hat{\theta}_n)$ and $\hat{\delta}_n$ are asymptotically independent.

33.3 Assessing the estimated TPI policy

To assess the effectiveness of the estimated TPI policy $\hat{d}_n$, we use an estimator of the form $\hat{V}(\hat{d}_n)$, for various choices of estimators $\hat{V}(d)$ of $V(d)$. The estimators $\hat{V}$ are based on exploiting different aspects of the model (33.3). In each case, under suitable conditions it can be shown that the error involved in this assessment, $\hat{V}(\hat{d}_n) - V(\hat{d}_n)$, is asymptotically normal with mean zero and a variance that can be consistently estimated, see McKeague and Qian (2011). This leads to asymptotically valid Wald-type confidence intervals for $V(\hat{d}_n)$ in the sense that the probability of $V(\hat{d}_n)$ falling in the interval tends to the nominal coverage level. These intervals can then provide an attractive way of assessing the potential clinical effectiveness of $\hat{d}_n$ in the study population.

Three estimators of $V(d)$ are studied in detail by McKeague and Qian (2011). Before we define these estimators, note that $V(d)$ is identifiable in terms of the randomization probability $p(a|x)$ (earlier assumed to be $1/2$ for simplicity) on the basis of the identity:

$$V(d) = P^d[Y] = \int Y dP^d = \int Y \frac{dP^d}{dP} dP = E\Big[W(X,A;d)Y\Big], \tag{33.5}$$

where the weight $W(X,A;d) = 1_{A=d(X)}/p(A|X)$ is a version of the Radon–Nikodym derivative dP^d/dP, and we have used the assumption that $p(a|X) > 0$ almost surely for each value of a. Also note that $EW(X,A;d) = 1$, but in an empirical version of (33.5) it is preferable to normalize the observed weights by their sample mean.

This leads to the inverse probability of treatment weighted (IPTW) estimator:

$$\hat{V}_I(d) = \frac{\mathbb{P}_n[W(X,A;d)Y]}{\mathbb{P}_n[W(X,A;d)]}.$$

Clearly, in our randomized trial setting where $p(a|x)$ is known, V_I is consistent (by the law of large numbers). Alternatively, the randomization probability can be eliminated by re-expressing (33.5) as

$$V(d) = E\big[1_{d(X)=1}E(Y|X,A=1) + 1_{d(X)=-1}E(Y|X,A=-1)\big]. \tag{33.6}$$

The value can then be estimated by plugging-in the estimator of $E(Y|X,A)$ specified by the model (33.3), resulting in the so-called G-computation estimator (Robins, 1986):

$$\hat{V}_G(d) = \mathbb{P}_n\big[\boldsymbol{U}^T\hat{\delta}_n + (21_{d(X)=1} - 1)(\hat{\alpha}_n + \hat{\beta}_n X(\hat{\theta}_n))\big].$$

This estimator does not involve $p(a|x)$ and is consistent if the model (33.3) is correctly specified, since $\hat{\eta}_n$ is a consistent estimator of η_0. Moreover, since $E(A|X) = 0$, $\hat{V}_G$ is consistent even if the main effect of X in (33.3) is misspecified (as long as $T_0(X)$ is correctly specified and there is a constant term in $\boldsymbol{U}$). Although $\hat{V}_G(d)$ has a more restricted scope than the other estimators, it is more stable in the sense that, instead of using a weighted average of Y, it uses a weighted average of a model-based estimator of the conditional expectation of Y.

The third estimator is a version of an estimator derived by Murphy et al. (2001) based on semiparametric efficiency theory. This estimator utilizes both the randomization probability and a plug-in estimator of $E(Y|X,A)$ specified by the model:

$$\hat{V}_L(d) = \hat{V}_G(d) + \mathbb{P}_n[W(X,A;d)\hat{R}_n],$$

where $\hat{R}_n = Y - \boldsymbol{U}^T\hat{\delta}_n - (\hat{\alpha}_n + \hat{\beta}_n X(\hat{\theta}_n))A$ is the residual from fitting model (33.3). This estimator is derived by projecting the score function $W(X,A;d)(Y - V(d))$ of the IPTW estimator onto the orthogonal complement of the tangent space for the treatment assignment probability. It is consistent (given knowledge of $p(a|x)$) even if (33.3) is misspecified, and it is locally efficient at model (33.3).

We have restricted attention to a single-stage decision problem. However, time-varying treatments are common, and are needed, e.g., for individuals with a chronic disease who experience a waxing and waning course of illness. The goal then is to construct a policy that tailors the type and dosage of treatment through time according to the individual's changing health status. There is a thriving statistical literature in this area, but, to our knowledge, high-dimensional/functional predictors have not been considered. In future work it would be interesting to extend our approach to this multi-stage setting.

Acknowledgements Ian McKeague's research was supported by NSF grant DMS-0806088. Min Qian's research was supported by NIH grants R01 MH080015 and P50 DA10075 (PI: Susan Murphy).

References

1. McKeague, I. W., Qian, M.: Evaluation of treatment policies via sparse functional linear regression. Submitted to J. Am. Stat. Assoc. (2011)
2. McKeague, I. W., Sen, B.: Fractals with point impact in functional linear regression. Ann. Stat. **38**, 2559–2586 (2010)
3. Murphy, S. A., van der Laan, M. J., Robins, J. M. and CPPRG: Marginal mean models for dynamic regimes. J. Am. Stat. Assoc. **96**, 1410–1423 (2001)

4. van 't Veer, L.J., Bernards, R.: Enabling personalized cancer medicine through analysis of gene expression patterns. Nature **452**, 564–570 (2008)
5. Robins, J.M.: A new approach to causal inference in mortality studies with sustained exposure periods — application to control of the healthy worker survivor effect. Math. Modelling **7**, 1393–1512 (1986)

Chapter 34
Estimation of Functional Coefficients in Partial Differential Equations

Jose C. S. de Miranda

Abstract We present a methodology to estimate, up to a constant factor, functional parameters in PDE models of the following type: $f\frac{\partial^2 u}{\partial t^2}+g\frac{\partial u}{\partial t}+hu=\frac{\partial}{\partial x}\left[\mathscr{K}\frac{\partial u}{\partial x}\right]$. The parameters f,g,h and $\mathscr{K}$ depend solely on x. We assume we know N functions $v_1(x,t),\ldots,v_N(x,t)$ that satisfy, for each i, $1\leq i\leq N$, $v_i=u_i+\varepsilon_i$, where u_i is a solution of the PDE and ε_i is uncorrelated zero mean noise which satisfies a frequency domain condition.

34.1 Introduction

Dynamical systems where the state space is a function space can, in some instances, be described by partial differential equations. This is the case of the classical modelling of the dynamical behaviour of heat transfer and wave propagation whose mathematical modelling is successfully performed using partial differential equations. The parameters that define these dynamical systems are in many cases the coefficients that appear in their corresponding PDEs. In case of linear homogeneous, i.e., without forcing factors, PDEs, given a sufficiently large set of linearly independent solutions, in a deterministic setting, it is possible, at least theoretically, to exactly find these coefficients up to some multiplicative common factor. However, in a random setting, where these solutions are subject to noise, the determination of the coefficients becomes a dynamical systems' inference problem with functional data. This work is in the intersection of two areas: Inverse Problems and Functional Data Analysis. For IP see Isakov (2006) and Kirsch (1996). A comprehensive exposition of FDA is found in Ferraty and Vieu (2006), Ramsay and Silverman (2004 and 2005) and references therein. See also Dabo-Niang and Ferraty (2008) for the collection of IWFOS 2008 works.

Jose C. S. de Miranda
University of São Paulo, São Paulo, Brazil, e-mail: simon@ime.usp.br

The estimation of functional coefficients associated to systems of ODEs can be found in Ramsay (2007). In Huttunen (2007), the estimation of the non-homogeneous heat equation is studied and extensions to more general equations are indicated. The focus is on fast numeric algorithms.

In this work we present a methodology to estimate the functional coefficients, $f(x)$, $g(x)$, $h(x)$, and $\mathscr{K}(x)$ that appear in the following PDE:

$$f\frac{\partial^2 u}{\partial t^2}+g\frac{\partial u}{\partial t}+hu=\frac{\partial}{\partial x}\left[\mathscr{K}\frac{\partial u}{\partial x}\right]. \tag{34.1}$$

This PDE has as its particular cases both heat and wave equations with functional coefficients which are used for the modelling of heat and wave phenomena in non-homogeneous space-varying conditions. We assume we have at our disposal N functions $v_1(x,t),...,v_N(x,t)$ that satisfy, for each i, $1 \le i \le N$, $v_i = u_i + \varepsilon_i$, where u_i is a solution of the PDE and ε_i is uncorrelated zero mean noise which satisfies a frequency domain condition. We have perfect knowledge of the functions v_i but we do not know neither the solutions u_i nor the realizations of the random noise ε_i. The variables x and t will belong to the intervals $[0,L]$ and $[0,T]$ respectively. Based on these data, we will construct consistent estimators $\hat{f}$, $\hat{g}$, $\hat{h}$, $\hat{\mathscr{K}}$, of f, g, h, and $\mathscr{K}$. More precisely, we will construct consistent estimators of the equivalent class to which the vector $(f,g,h,\mathscr{K})$ belongs under the equivalence relation defined by $(f_1,g_1,h_1,\mathscr{K}_1) \equiv (f_2,g_2,h_2,\mathscr{K}_2) \iff \exists c \in \mathbb{R}\ (f_1,g_1,h_1,\mathscr{K}_1) = c(f_2,g_2,h_2,\mathscr{K}_2)$. We will denote this class by $\overline{(f,g,h,\mathscr{K})}$ and call it *shape*. Whenever $(f,g,h,\mathscr{K}) \neq 0$, a standard representative of shape is $sng(\mathscr{K}(0))\frac{(f,g,h,\mathscr{K})}{||(f,g,h,\mathscr{K})||_2}$, where $||(f,g,h,\mathscr{K})||_2 = \sqrt{||f||_2^2+||g||_2^2+||h||_2^2+||\mathscr{K}||_2^2}$, and $sng(a)=1$ if $a>0$ and $sng(a)=-1$ if $a<0$. If we have additional information about $(f,g,h,\mathscr{K})$, for example by means of the value y of a functional J applied to it, such as $J[(f,g,h,\mathscr{K})] = \int_0^L \mathscr{K}(x)dx = y \in \mathbb{R}$, then we will also be able to estimate $(f,g,h,\mathscr{K})$.

This extended abstract is organized as follows: in section 2 we construct the estimators for the functional parameters of the PDE; in section 3 we present the main result concerning the properties of these estimators; and, in section 4, we conclude this work with some final remarks.

34.2 Estimator construction

For ease of presentation we will suppose that N, the number of observed solutions, i.e., v_i, is a multiple of four, $N=4n$ say, and we will divide the functions v_i in four sets with n elements, $I_1=\{v_1,...,v_n\}$, $I_2=\{v_{n+1},...,v_{2n}\}$, $I_3=\{v_{2n+1},...,v_{3n}\}$, and $I_4=\{v_{3n+1},...,v_{4n}\}$, for example. Let λ, and μ be real numbers in the interval $(0,1)$.

Partial integration with respect to the second variable over the interval $[\lambda T - \mu\lambda t, \lambda T + \mu(1-\lambda)t]$ followed by integration with respect to t over $[0,T]$ and integration with respect to $(\lambda,\mu) \in (0,1)^2$ of equation (1) furnishes

$$f(x)\left(\frac{1}{T^3}\int_0^T t^2u(x,t)dt-\frac{1}{T^2}\int_0^T tu(x,t)dt+\frac{1}{6T}\int_0^T u(x,t)dt\right)+$$
$$g(x)\left(\int_0^T u(x,t)\mathscr{R}(t)dt\right)+$$
$$h(x)\int_0^T u(x,t)\mathscr{S}(t)dt=\frac{\partial}{\partial x}\left[\mathscr{K}(x)\frac{\partial}{\partial x}\int_0^T u(x,t)\mathscr{S}(t)dt\right] \tag{34.2}$$

where $\mathscr{R}(t)=\int_0^1(1-\mu)^2\left(\frac{(1\wedge\frac{t}{T-\mu T})^2-(0\vee\frac{t-\mu T}{T-\mu T})^2}{2}+\frac{t}{T}-(1\wedge\frac{t}{T-\mu T})\right)d\mu$

and $\mathscr{S}(t)=\int_0^T\int_0^1\mu(1-\mu)^2\left(\frac{\lambda^2}{6}(3-2\lambda)\Big|_{\lambda=0\vee\frac{t-\mu z}{T-\mu z}}^{\lambda=1\wedge\frac{t}{T-\mu z}}\right)d\mu dz.$

Denoting $\bar{u}^j=\frac{\sum_{i\in I_j}u_i}{n}$, due to linearity of the PIDE above, we can write, for $1\le j\le 4$,

$$f(x)\left(\frac{1}{T^3}\int_0^T t^2\bar{u}^j(x,t)dt-\frac{1}{T^2}\int_0^T t\bar{u}^j(x,t)dt+\frac{1}{6T}\int_0^T \bar{u}^j(x,t)dt\right)+$$
$$g(x)\left(\int_0^T \bar{u}^j(x,t)\mathscr{R}(t)dt\right)+h(x)\int_0^T \bar{u}^j(x,t)\mathscr{S}(t)dt=\frac{\partial}{\partial x}\left[\mathscr{K}(x)\frac{\partial}{\partial x}\int_0^T \bar{u}^j(x,t)\mathscr{S}(t)dt\right] \tag{34.3}$$

Let

$$\alpha(\bar{u}^j,x)=\alpha_j(x)=\left(\frac{1}{T^3}\int_0^T t^2\bar{u}^j(x,t)dt-\frac{1}{T^2}\int_0^T t\bar{u}^j(x,t)dt+\frac{1}{6T}\int_0^T \bar{u}^j(x,t)dt\right), \tag{34.4}$$

$$\beta(\bar{u}^j,x)=\beta_j(x)=\left(\int_0^T \bar{u}^j(x,t)\mathscr{R}(t)dt\right),\ \text{and}\ \gamma(\bar{u}^j,x)=\gamma_j(x)=\int_0^T \bar{u}^j(x,t)\mathscr{S}(t)dt. \tag{34.5}$$

We can thus write the system of four equations

$$\alpha_j(x)f(x)+\beta_j(x)g(x)+\gamma_j(x)h(x)=\frac{\partial}{\partial x}\left[\mathscr{K}(x)\frac{\partial}{\partial x}\gamma_j(x)\right]=\mathscr{K}'(x)\gamma_j'(x)+\mathscr{K}(x)\gamma_j''(x) \tag{34.6}$$

for $1\le j\le 4$.

Now, denote $\alpha=(\alpha_1,\alpha_2,\alpha_3,\alpha_4)$, $\beta=(\beta_1,\beta_2,\beta_3,\beta_4)$, $\gamma=(\gamma_1,\gamma_2,\gamma_3,\gamma_4)$, $\gamma'=(\gamma_1',\gamma_2',\gamma_3',\gamma_4')$, $\gamma''=(\gamma_1'',\gamma_2'',\gamma_3'',\gamma_4'')$, ${}^*\alpha=(\alpha_1,\alpha_2,\alpha_3)$, ${}^*\beta=(\beta_1,\beta_2,\beta_3)$, ${}^*\gamma=(\gamma_1,\gamma_2,\gamma_3)$. Solving the system leads to the following:

$$\mathscr{K}(z)=\mathscr{K}(0)exp\left(-\int_0^z\frac{det(\alpha,\beta,\gamma,\gamma'')}{det(\alpha,\beta,\gamma,\gamma')}dx\right) \tag{34.7}$$

$$f=\mathscr{K}\frac{det({}^*\gamma''-\frac{det(\alpha,\beta,\gamma,\gamma'')}{det(\alpha,\beta,\gamma,\gamma')}{}^*\gamma',\ {}^*\beta,\ {}^*\gamma)}{det({}^*\alpha,\ {}^*\beta.\ {}^*\gamma)} \tag{34.8}$$

$$g=\mathscr{K}\frac{det({}^*\alpha,\ {}^*\gamma''-\frac{det(\alpha,\beta,\gamma,\gamma'')}{det(\alpha,\beta,\gamma,\gamma')}{}^*\gamma',\ {}^*\gamma)}{det({}^*\alpha,\ {}^*\beta.\ {}^*\gamma)} \tag{34.9}$$

$$h=\mathscr{K}\frac{det({}^*\alpha,\ {}^*\beta,\ {}^*\gamma''-\frac{det(\alpha,\beta,\gamma,\gamma'')}{det(\alpha,\beta,\gamma,\gamma')}{}^*\gamma')}{det({}^*\alpha,\ {}^*\beta,\ {}^*\gamma)} \tag{34.10}$$

More briefly:
$\mathcal{K} = \kappa(\alpha,\beta,\gamma) = \kappa^*(\bar{u}^1,\bar{u}^2,\bar{u}^3,\bar{u}^4)$, $f = F(\alpha,\beta,\gamma) = F^*(\bar{u}^1,\bar{u}^2,\bar{u}^3,\bar{u}^4)$,
$g = G(\alpha,\beta,\gamma) = G^*(\bar{u}^1,\bar{u}^2,\bar{u}^3,\bar{u}^4)$, and $h = H(\alpha,\beta,\gamma) = H^*(\bar{u}^1,\bar{u}^2,\bar{u}^3,\bar{u}^4)$.

Denoting $\bar{v}^j = \frac{\sum_{i\in I_j} v_i}{n}$, for $1 \le j \le 4$, define $\hat{\alpha}_j(x), \hat{\beta}_j(x)$ and $\hat{\gamma}_j(x)$ by substituting $\bar{u}^j$ with $\bar{v}^j$ in equations (4) and (5), and use the previous notations concerning vectors.

Now, we define our estimators for the functional coefficients as
$\hat{\mathcal{K}} = \kappa(\hat{\alpha},\hat{\beta},\hat{\gamma}) = \kappa^*(\bar{v}^1,\bar{v}^2,\bar{v}^3,\bar{v}^4)$,
$\hat{f} = F(\hat{\alpha},\hat{\beta},\hat{\gamma}) = F^*(\bar{v}^1,\bar{v}^2,\bar{v}^3,\bar{v}^4)$,
$\hat{g} = G(\hat{\alpha},\hat{\beta},\hat{\gamma}) = G^*(\bar{v}^1,\bar{v}^2,\bar{v}^3,\bar{v}^4)$,
$\hat{h} = H(\hat{\alpha},\hat{\beta},\hat{\gamma}) = H^*(\bar{v}^1,\bar{v}^2,\bar{v}^3,\bar{v}^4)$.

Once again, we observe that the model (1) is unidentifiable as can be clearly seen by the presence of the constant $\mathcal{K}(0)$ in (7), and, consequently, in (8), (9) and (10). However, if, for example, we know that $\int_0^L \mathcal{K}(x)dx = y$, then, from (7) we have that $\mathcal{K}(0) = y\left[\int_0^L exp\left(-\int_0^z \frac{det(\alpha,\beta,\gamma,\gamma'')}{det(\alpha,\beta,\gamma,\gamma')}dx\right)dz\right]^{-1}$, and identifiability is guaranteed.

34.3 Main results

Let us assume noise has a representation in the frequency domain. Using complex notation we write

$$\varepsilon_i(x,t) = \sum_{l,m\in\mathbb{Z}} a_{i,l,m} e^{\iota\left(\frac{2\pi l x}{L} + \frac{2\pi m t}{T}\right)}. \tag{34.11}$$

From here on we will assume that noise satisfies the following conditions:

1. for all i,l and m, $\mathbb{E}a_{i,l,m} = 0$;
2. for all i,j,l,m,p and q, such that $(i,l,m) \neq (j,p,q)$, $\mathbb{C}ov(a_{i,l,m},a_{j,p,q}) = 0$;
3. $\frac{1}{n^2}\sum_{i=1}^n \sum_{l,m\in\mathbb{Z}} l^4\mathbb{E}(a_{i,l,m}^2) \to 0$ as $n \to \infty$.

Clearly, condition [3] is fulfilled in case noise is identically distributed (which means that, in the frequency domain, for all i and j and for all l and m we have $a_{i,l,m} = a_{j,l,m}$) and satisfies $\sum_{l,m\in\mathbb{Z}} l^4\mathbb{E}(a_{i,l,m}^2) < \infty$. Condition [2] imposes uncorrelatedness of both inter and intra noise Fourier coefficients. In case we have pairwise independent noise, i.e. $\varepsilon_i \perp \varepsilon_j$, for $i \neq j$, we still need the fulfillment of $\mathbb{C}ov(a_{i,l,m},a_{i,p,q}) = 0$ in order to obey [2]; this is also the case if we have independent noise.

Observe that Theorem 1, Corollary 1 and Theorem 2 concern shape estimation. We may assume an arbitrary non zero value for $\hat{\mathcal{K}}(0) = \mathcal{K}(0)$ or consider the standard representatives in the class, for pairing the functions. Using the notation established thus far, we can state the following:

Theorem 34.1. *Let the functional coefficients f,g,h and $\mathcal{K}$ that appear in (1) be such that $(f,g,h,\mathcal{K}) \neq (0,0,0,0)$ and, for all $x \in [0,L]$, $det(\alpha,\beta,\gamma,\gamma') \neq 0 \neq det(^*\alpha,\ ^*\beta.\ ^*\gamma)$. Assume that noise satisfies conditions 1, 2 and 3 stated* *above. Then, the mean integrated squared errors* $\mathbb{E}(\|\hat{\mathscr{K}} - \mathscr{K}\|_2^2)$, $\mathbb{E}(\|\hat{f} - f\|_2^2)$, $\mathbb{E}(\|\hat{g} - g\|_2^2)$, *and* $\mathbb{E}(\|\hat{h} - h\|_2^2)$ *go to zero as N goes to infinity.*

This result is independent of the value of $\mathscr{K}(0)$. The condition $det(\alpha,\beta,\gamma,\gamma') \neq 0$ implies that $\{\bar{u}_1,\bar{u}_2,\bar{u}_3,\bar{u}_4\}$ is linearly independent.

Corollary 34.1. *Under the same assumptions of Theorem 1, the estimators* $\hat{f}$, $\hat{g}$, $\hat{h}$, *and* $\hat{\mathscr{K}}$, *are consistent.*

We write $\phi'(p)(w)$ for Hadamard derivative of ϕ at p calculated at the vector w. Using the functional delta method, see van der Vaart (1996) and Kosorok (2006), we can also prove the following convergence in distribution results

Theorem 34.2. *Under the hypothesis of Theorem 1 and the additional assumption that noise is i.i.d. like* ε, *we have*
$\sqrt{N}(\hat{f} - f) \rightsquigarrow T_f(\nu)$, $\sqrt{N}(\hat{g} - g) \rightsquigarrow T_g(\nu)$, $\sqrt{N}(\hat{h} - h) \rightsquigarrow T_h(\nu)$, *and* $\sqrt{N}(\hat{\mathscr{K}} - \mathscr{K}) \rightsquigarrow T_{\mathscr{K}}(\nu)$, *where*
$T_f(\nu) = F^{*\prime}(\bar{u}^1,\bar{u}^2,\bar{u}^3,\bar{u}^4)(\nu_1,\nu_2,\nu_3,\nu_4)$,
$T_g(\nu) = G^{*\prime}(\bar{u}^1,\bar{u}^2,\bar{u}^3,\bar{u}^4)(\nu_1,\nu_2,\nu_3,\nu_4)$,
$T_h(\nu) = H^{*\prime}(\bar{u}^1,\bar{u}^2,\bar{u}^3,\bar{u}^4)(\nu_1,\nu_2,\nu_3,\nu_4)$,
and $T_{\mathscr{K}}(\nu) = \kappa^{*\prime}(\bar{u}^1,\bar{u}^2,\bar{u}^3,\bar{u}^4)(\nu_1,\nu_2,\nu_3,\nu_4)$.
Here $\{\nu_1,\nu_2,\nu_3,\nu_4\}$ *are i.i.d. like* ν *and* ν *has spectral representation given by* $\nu(x,t) = \sum_{l,m\in\mathbb{Z}} b_{l,m} e^{\iota(\frac{2\pi lx}{L} + \frac{2\pi mt}{T})}$ *with uncorrelated coefficients* $b_{l,m} \sim \mathscr{N}(0,\sigma_{l,m}^2)$, *where* $\sigma_{l,m}^2$ *is the common variance of* $a_{i,l,m}$.

34.4 Final remarks

We observe that the methodology presented here, i.e., the integration of the PDE in a conveniently chosen region, the analytic solution of the new integrated PIDE for the functional coefficients and, finally, the substitution of the true PDE solutions with averages of the measured solutions in the analytical expressions for the functional coefficients, can be applied for a larger class of PDEs that include PDEs of the type: $\sum_{i=0}^{m} f_i(x)\frac{\partial^i u}{\partial t^i}(x,t) = \frac{\partial}{\partial x}\left[\mathscr{K}\frac{\partial u}{\partial x}\right]$ and others.

We can define other estimators for f,g,h, and $\mathscr{K}$ by first smoothing our functional data. This can be done, for example, by simply smoothing v_i using a low pass filter and then using the smoothed versions of v_i in the estimators' equations. We expect this procedure to yield good results in case of high frequency noise, more specifically, in case Fourier coefficients of noise are completely negligible in all frequencies from zero till a threshold frequency where all the Fourier coefficients and the energy of all the solutions u_i are negligible for frequencies greater than this threshold frequency.

The conditions that we have assumed regarding the frequency domain behaviour of the noise can be relaxed, more specifically, there may be some moderate correlation among Fourier coefficients, both inter and intra noise, and the consistency of the estimators will still be maintained.

Observe that, although the estimators $\hat{f}, \hat{g}, \hat{h}, \hat{\mathscr{K}}$, are consistent, they are, in principle, dependent on the choice of the sets I_1, I_2, I_3, I_4. However, we can symmetrize this estimator to get one which is independent of this choice by taking the average of the estimators based on all possible partitions of $\{1,...,N\}$ in four sets with the same size. Although in a practical situation we will not be able to calculate all these $\frac{N!}{(n!)^4 4!}$ estimators for large N, we will still considerably reduce this "choice" effect by choosing a random subset of these partitions with a reasonable number of elements and then taking the average of the estimators associated to these partitions. Clearly, this symmetrization or proxy-symmetrization must be subjected to fulfillment of Theorem 1 hypothesis in order for us to have final consistent estimators.

Acknowledgements The author thanks OLSJC.

References

1. Dabo-Niang, S., Ferraty, F.: Functional and Operatorial Statistics (eds.) Contributions to statistics, Physica Verlag, Heidelberg (2008)
2. de Miranda, J.C.S.: Functional parameter estimation in Partial Differential Equations. Preprint (2009).
3. Ferraty, F., Vieu, P.: Nonparametric functional data analysis. Springer, New York (2006)
4. Huttunen, J. M. J., Kaipio, J. P.: Approximation error analysis in nonlinear state estimation with an application to state-space identification. Inverse Probl. **23**, 2141–2157 (2007)
5. Isakov, V.: Inverse problems for partial differential equations. Springer Science + Business Media, Inc. (2006)
6. Kirsch, A.: An introduction to the mathematical theory of inverse problems. Springer Verlag, New York (1996)
7. Kosorok, M.R.: Introduction to Empirical Processes and Semiparametric Inference. Springer Science + Business Media Inc. (2006)
8. Ramsay, J. O., Hooker, G., Cao, J., Campbell, D.: Parameter estimation for differential equations (with discussion). J. Roy. Stat. Soc. B **69**, 741–796 (2007).
9. Ramsay, J. O., Silverman, B. W.: Applied Functional Data Analysis. Springer Verlag, New York (2002)
10. Ramsay, J. O., Silverman, B. W.: Functional data analysis (Second Edition). Springer Verlag, New York (2005)
11. van der Vaart, A. W., Wellner, J. A.: Weak Convergence and Empirical Processes With Applications to Statistics. Springer Verlag, New York (1996)

Chapter 35
Functional Varying Coefficient Models

Hans-Georg Müller, Damla Şentürk

Abstract Functional varying coefficient models provide a versatile and flexible analysis tool for relating longitudinal responses to longitudinal predictors. Two key innovations are: Representing the varying coefficient functions through auto- and cross-covariances of the underlying stochastic processes; and including history effects through a smooth history index function. This presentation is a review of the paper Şentürk and Müller (2010).

35.1 Introduction

We consider functional data as independent and identically distributed realizations of a stochastic process. A basic problem is that these realizations are rarely directly observed. Commonly, one instead has available measurements of the trajectories that are obtained at discrete time points. These measurements are usually contaminated by measurement errors. In many longitudinal studies, the number of available measurements per subjects is quite small and often also the distribution of the measurement times is irregular and may be assumed to follow a random distribution for each subject. Then the times where measurements are obtained are considered to be sparse and irregular. Often one observes two or more random trajectories per subject and is interested in relating these trajectories to each other. In the following, we consider predictor trajectories X and response trajectories Y, and aim at regressing functional responses Y on functional predictors X, where $X,Y \in L^2([0,T])$ for a $T > 0$.

For an overview of functional data analysis, see Ramsay and Silverman (2005) and Ferraty and Vieu (2006). For the purpose of regressing trajectories Y on trajec-

Hans-Georg Müller
University of California, Davis, USA, e-mail: mueller@wald.ucdavis.edu

Damla Şentürk
Penn State University, University Park, USA, e-mail: dsenturk@stat.psu.edu

tories X, various functional regression models have been proposed. These include the functional linear model (Ramsay and Dalzell, 1991; Cuevas et al., 2002; Cardot et al., 2003, Yao et al., 2005a) or the functional additive model (Yao and Müller, 2008). The functional linear regression model with functional predictors and functional responses is given by

$$E(Y(t)|X) = \mu_Y(t) + \int (X(s) - \mu_X(s))\beta(s,t)\,ds,$$

where $\mu_Y(t) = EY(t)$, $\mu_X(s) = EX(s)$ are the mean functions and $\beta(\cdot,\cdot)$ is a smooth regression parameter function (surface).

A problem with this model is that its implicit time relations are often not realistic: Values of the predictor trajectories $X(s)$ influence current responses $Y(t)$, for $s > t$. This can be perceived as the future influencing the past, if the two functions share the same time axis, while it is not an issue if the two random functions are not time-related (see, e.g., Yao et al., 2008, where such a situation is discussed for gene expression trajectories). Thus the model might be hard to interpret when both X and Y are recorded concurrently, as is typically the case in longitudinal studies. In such cases one might prefer models that respect time order; compare the discussion in Malfait and Ramsay (2003) and Müller and Zhang (2005).

A well-established model that passes the time respecting criterion is the varying coefficient model

$$E(Y(t)|X) = \beta_0(t) + \beta_1(t)X(t),$$

which reduces the regression relation to a linear relation at current times t. This model has been proposed in Cleveland et al. (1991) and has been widely studied, among others by Wu and Chiang (2000), Huang et al. (2004) and Fan and Zhang (2008). The following is a review of proposals in Şentürk and Müller (2010), where more details can be found.

35.2 Varying coefficient models with history index

Current implementations of varying coefficient models have a number of shortcomings: One is that effects of time-lagged predictors $X(s)$ for $s < t$ are not included, while it is likely that in many longitudinal relationships there will be lingering effects of predictor values at preceding times. An example is the effect of calorie intake on body mass index. Secondly, available implementations do not work well for sparse longitudinal data – there may not be enough data available in local windows around t to reasonably estimate the regression coefficient functions $\beta_0(t), \beta_1(t)$ by the usual fitting of parametric regression models within a local window around t.

Thirdly, if predictors are measured with noise, as is common in longitudinal studies, these estimates are biased, as has been demonstrated in Şentürk and Müller (2008).

The above discussion motivates alternative approaches for varying coefficient modeling. An extension of the varying coefficient model is

$$E\{Y(t)|X\} = \beta_0(t) + \beta_1(t) \int_0^{\Delta} \gamma(u)X(t-u)du,$$

for a suitable lag $\Delta > 0$. Here γ is a *history index function*, while β_1 quantifies the effect of the recent history of the predictor function on the response and β_0 is an intercept function. Thus, the varying coefficient function β_1 represents the magnitude of the history influence which may vary over time. All functions are assumed to be smooth. For identifiability, it is opportune to require $\int_0^{\Delta} \gamma^2(u)du = 1$ and $\gamma(0) > 0$.

Since the history index function γ is time-invariant, the varying coefficient model with history index model separates history and time effects, which aids in its interpretability. A particular challenge arises in the case of longitudinal data, characterized by sparse and irregular observations per trajectory. The importance of such data for functional data analysis was discussed in James et al. (2000) and Yao et al. (2005b). While in traditional functional data analysis, one assumes samples of fully observed functions, measured without noise at arbitrarily dense grid points, a more realistic assumption is additive measurement errors $(\varepsilon_{ij}, \varepsilon_{ij})$. For longitudinal data, in addition one needs to deal with sparse and irregular designs, with measurements of X and Y given by

$$X_{ij} = X(S_{ij}) + \varepsilon_{ij}, \quad Y_{ij} = Y(T_{ij}) + \varepsilon_{ij},$$

where S_{ij}, T_{ij} are random times and their numbers N_{Xi}, N_{Yi} for the i-th subject are random and finite, e.g., bounded or Poisson distributed.

35.3 Functional approach for the ordinary varying coefficient model

When $\gamma(\cdot)$ is known, the history index model

$$E\{Y(t)|X\} = \beta_0(t) + \beta_1(t) \int_0^{\Delta} \gamma(u)X(t-u)du$$

reduces to an ordinary varying coefficient model. With centered predictors and responses $Y^C(t) = Y(t) - \mu_Y(t)$ and $X^C(t) = X(t) - \mu_X(t)$, one may rewrite this model as

$$E\{Y^C(t)|X(t)\} = \beta_1(t)X^C(t),$$

with $\beta_0(t) = \mu_Y(t) - \beta_1(t)\mu_X(t)$, β_0, β_1 "smooth". For longitudinal data, fitting this simpler model is already problematic.

In the functional approach, one targets the covariances

$$G_{XY}(s,t) = \text{cov}(X(s),Y(t)),\ G_{XX}(s,t) = \text{cov}\{X(s),X(t)\},\ G_{YY}(s,t) = \text{cov}\{Y(s),Y(t)\}.$$

The key is that these covariances can be consistently estimated for both longitudinal and dense designs, along with mean functions μ_X, μ_Y. To obtain all of these quantities, one may simply pool all data and the products of observations coming from the same subject and then judiciously apply smoothing methods. This approach has been described in Yao et al. (2005b) and is referred to as *PACE* – principal analysis by conditional expectation. A Matlab implementation is available at `http://anson.ucdavis.edu/~mueller/data/pace.html`. The covariance estimation steps in PACE (for the estimation of covariances G_{XX}, G_{YY}, G_{XY}) eliminate the effect of the noise contamination of the observations, which otherwise can be a difficult to address problem, since predictors are contaminated and presmoothing predictor trajectories is not an option for sparse longitudinal designs. PACE addresses the problems associated with the sparseness of the design through borrowing strength across subjects.

For the ordinary varying coefficient model, simple calculations lead to

$$\begin{aligned}\beta_1(t) &= \text{argmin}_\theta\{E(Y^C(t) - \theta X^C(t))^2\} = \frac{G_{XY}(t,t)}{G_{XX}(t,t)}\\ \beta_0(t) &= \mu_Y(t) - \beta_1(t)\mu_X(t).\end{aligned}$$

Therefore it suffices to substitute estimates $\hat{G}_{XX}$, $\hat{G}_{XY}$, $\hat{\mu}_Y$ that one obtains from PACE. The properties of these estimates have been well studied, and are then inherited by the resulting varying coefficient function estimates $\hat{\beta}_0$, $\hat{\beta}_1$, for both dense and sparse longitudinal designs with noisy predictors.

35.4 Fitting the history index model

Writing

$$\begin{aligned}E\{Y^C(t)|X^C(s), s \in [t-\Delta, t]\} &= \beta_1(t)\int_0^\Delta \gamma(s)X^C(t-s)ds\\ &= \int_0^\Delta \alpha_t(s)X^C(t-s)ds\end{aligned}$$

with new regression parameter functions $\alpha_t(s) = \beta_1(t)\gamma(s)$, one finds from $\int_0^\Delta \gamma^2(s)\,ds = 1$ that for each fixed time point t,

$$\gamma(s) = \frac{\alpha_t(s)}{\{\int_0^\Delta \alpha_t^2(s)\,ds\}^{1/2}}.$$

Although not strictly necessary, it turns out practically advantageous to average over finitely many time points, so that

$$\gamma(s) = \frac{\sum_{r=1}^{R} \alpha_{t_r}(s)}{[\int_0^{\Delta} \{\sum_{r=1}^{R} \alpha_{t_r}(s)\}^2 ds]^{1/2}},$$

for a small integer R.

Once the history index function γ is recovered, the history index model essentially reduces to an ordinary varying coefficient model and the previously described procedure applies. Representing functions α_t in the eigenbasis $\{\phi_{tm}, m = 1,2,\ldots\}$ of the processes $Z_t(s) = X^C(t-s)$, $s \in [0,\Delta]$, with auto-covariance function

$$G_t(s_1,s_2) = G_{XX}(t-s_1,t-s_2) = \sum_m \rho_{tm}\phi_{tm}(s_1)\phi_{tm}(s_2) \text{ for } s_1,s_2 \in [0,\Delta],$$

with eigenvalues ρ_{tm} and minimizing the expected squared deviation $E\{Y^C(t) - \int_0^{\Delta} \alpha_t(s)X^C(t-s)ds\}^2$ leads to the minimizer

$$\alpha_t^*(s) = \frac{1}{\rho_{tm}} \int_0^{\Delta} G_{XY}(t-s,t)\phi_{tm}(s)\,ds.$$

Due to the relationship with G_{XX}, eigenfunctions and eigenvalues $\{\phi_{tm}, \rho_{tm}\, m = 1,2,\ldots\}$ can be easily computed, once G_{XX} has been estimated. Consistency properties of $\hat{G}_{XX}, \hat{G}_{XY}$ are then inherited by $\hat{\alpha}_t^*$ and thus by the resulting estimate $\hat{\gamma}$ of the history index function.

A difficulty for the longitudinal case is the evaluation of $\widetilde{X}(t) = \int_0^{\Delta} \gamma(s)X^C(t-s)ds$, which is needed to transform the history index model into the varying coefficient model

$$E\{Y(t)|X\} = \beta_0(t) + \beta_1(t)\int_0^{\Delta} \gamma(u)X(t-u)du,$$

as the integrals defining $\widetilde{X}$ cannot be evaluated, even if γ is known, due to the sparseness of the data. Fortunately, the varying coefficient function β_1 in this model is found to have the representation

$$\beta_1(t) = G_{XY}(t,t)/\int_0^{\Delta} \gamma(s)G_{XX}(t-s,t)ds,$$

and it then suffices to substitute consistent estimates for G_{XX}, G_{XY}, γ, which are available also for the longitudinal case.

For the implementation of these methods, one needs to choose a number of auxiliary parameters. These include the lag parameter Δ, which determines the history domain and can be chosen by minimizing a suitable prediction error criterion. For selecting smoothing bandwidths to estimate the cross-sectional quantities $\mu_X, \mu_Y, G_{XX}, G_{XY}$, generalized cross-validation is an option and for the number of included components in the representation of the history index function, one may use fraction of variance explained or AIC-type criteria.

One may show under suitable regularity conditions that for longitudinal designs the estimators for the model functions β_0, β_1, γ are consistent.

References

1. Cardot, H., Ferraty, F., Sarda, P.: Spline estimators for the functional linear model. Stat. Sinica **13**, 571–591 (2003)
2. Chiang, C. T., Rice, J. A., Wu, C. O.: Smoothing spline estimation for varying coefficient models with repeatedly measured dependent variables. J. Am. Stat. Assoc. **96**, 605–619 (2001)
3. Cleveland, W. S., Grosse, E., Shyu, W. M.: Local regression models. In: Chambers, J. M., Hastie, T. J. (eds.) Statistical Models in S, pp. 309-376, Wadsworth & Brooks, Pacific Grove.
4. Cuevas, A., Febrero, M., Fraiman, R.: Linear functional regression: the case of fixed design and functional response. Canad. J. Stat. **33**, 285–300 (2002)
5. Ferraty, F., Vieu, P.: Nonparametric Functional Data Analysis. Springer, New York (2006)
6. Huang, J. Z., Wu, C. O., Zhou, L.: Polynomial spline estimation and inference for varying coefficient models with longitudinal data. Stat. Sinica **14**, 763–788 (2004)
7. James, G., Hastie, T. J., Sugar, C. A.: Principal component models for sparse functional data. Biometrika **87**, 587–602 (2000)
8. Malfait, N., Ramsay, J. O.: The historical functional linear model. Canad. J. Stat. **31**, 115–128 (2003)
9. Müller, H.G., Yao, F.: Functional additive models. J. Am. Stat. Assoc. **103**, 1534–1544 (2008)
10. Müller, H. G., Zhang, Y.: Time-varying functional regression for predicting remaining lifetime distributions from longitudinal trajectories. Biometrics **61**, 1064–1075 (2005)
11. Ramsay, J. O., Silverman, B. W.: Functional Data Analysis (Second Edition). Springer, New York (2005)
12. Ramsay, J. O., Dalzell, C. J.: Some tools for functional data analysis. J. Roy. Stat. Soc. B **53**, 539–572 (1991)
13. Şentürk, D., Müller, H. G.: Generalized varying coefficient models for longitudinal data. Biometrika **95**, 653–666 (2008)
14. Şentürk, D., Müller, H.G.: Functional varying coefficient models for longitudinal data. J. Am. Stat. Assoc. **105**, 1256–1264 (2010)
15. Wu, C. O., Chiang, C. T.: Kernel smoothing on varying coefficient models with longitudinal dependent variable. Stat. Sinica **10**, 433–456 (2000)
16. Yao, F., Müller, H. G., Wang, J. L.: Functional linear regression analysis for longitudinal data. Ann. Stat. **33**, 2873–2903 (2005a)
17. Yao, F., Müller, H.G., Wang, J.L.: Functional data analysis for sparse longitudinal data. J. Am. Stat. Assoc. **100**, 577–590 (2005b)

Chapter 36
Applications of Funtional Data Analysis to Material Science

S. Naya, M. Francisco-Fernández, J. Tarrío-Saavedra, J. López-Beceiro, R. Artiaga

Abstract Thermogravimetric curves (TG) and statistical functional nonparametric methods are used to classify (supervised) 7 wood species and to measure the influence of adding silica micro and nano-particles on the thermal degradation of an epoxy resin. A technique based on the Bayes rule and the Nadaraya-Watson regression estimator and a functional ANOVA for a one way treatment are applied respectively.

36.1 Introduction

Many problems in the material science field can be tackled using statistical functional data analysis (FDA) methodologies. In this work, we present two of these problems and describe the functional techniques used to analyze them. The first application refers to the classification of different wood species. The identification of wood is one of the most difficult tasks to perform related with the technology of this material. Wood identification analysis is typical in the furniture industries and the wood panel production. Often, the performed analysis has a non-uniform accuracy because of the operator. Therefore, the implementation of quantitative mod-

Salvador Naya
University of A Coruña, Spain, e-mail: salva@udc.es

Mario Francisco-Fernández
University of A Coruña, Spain, e-mail: mariofr@udc.es

Javier Tarrío-Saavedra
University of A Coruña, Spain, e-mail: jtarrio@udc.es

Jorge López-Beceiro
University of A Coruña, Spain, e-mail: jlopezb@udc.es

Ramón Artiaga
University of A Coruña, Spain, e-mail: rartiaga@udc.es

els and automatic recognition methods of wood samples are justified and useful. Wood samples are mainly classified based on the results of two techniques: image- and spectrum-based processing systems. In this work, the thermograms obtained by thermogravimetric analysis (TG) are used as a discriminant characteristic (Prime et al., 2009). The wood degradation in an inert atmosphere is dominated by the degradation behavior of its three main components (cellulose, lignin, and hemicellulose). The proportion of each wood component varies depending on the species (Tarrío et al., 2010b) and influences in the shape of the TG curves. Therefore, TG analysis becomes an interesting option to discriminate between classes of timber. These curves can be processed in a relatively simple way with functional analysis (Ferraty and Vieu, 2006).

The second problem analyzed in this work is that of performing an experimental design to evaluate the effect of the addition of fumed silica on the thermal degradation of an epoxy resin. The fumed silica epoxy-resin composites are prepared and characterized by TG and a one-way functional ANOVA method is applied. This procedure allows to test the possible differences in responses according to the treatments used, considering that the data are functions or curves. In our case, the silica content in each sample,with three levels (0, 10 and 20 wt%, weight percentage of fumed silica) is chosen as the treatment factor. Five experiments or replicates for each level are considered, which gives a balanced design. The response variable is functional and represents the mass of material depending on the temperature at which it is subjected to. Our analysis allows to answer questions like, can it be said that the thermal stability of the material increases or decreases with statistical significance? (Tarrío et al., 2010a).

36.2 Materials testing and data collecting

In the first problem, tests for 7 different wood species (European beech or *Fagus sylvatica*, European oak or *Quercus robur*, chestnut or *Castanea sativa*, *Eucalyptus globulus*, jatobá or *Hymenaea courbaril*, scots pine or *Pinus silvestris* and insignis pine or *Pinus radiata*) are carried out. Seven samples per each specie are tested. With respect to the second experiment, an epoxy resin matrix based on the diglycidyl ether of trimethylolpropane, Triepox GA, manufactured by Gairesa, and 1,3-benzenedimethanamine, is used. The fumed silica has been provided by Ferroatlntica I+D, Spain. Its average particle size is 0.15 μm. The samples are prepared for contents of 0, 10, and 20 wt% of fumed silica. In the two cases, tests are performed on a SDT 2960 TA Instruments thermo balance. This apparatus provides TG curves used in this study. A heating ramp of 20°C min^{-1} is applied in the range from 20 to 600°C at a rate of 50 mL min^{-1} of N_2.

36.3 Statistical methods

Different statistical procedures for functional data were applied to each one of the problems analyzed here and described in the Introduction section. Regarding the problem of classifying wood species, two nonparametric kernel functional discriminant methods (Ferraty and Vieu, 2006) and two approaches based on the boosting algorithm were applied to construct a classification rule to discriminate between the studied species. Each TG curve was classified as belonging to the specie to which the highest posterior probability is obtained, using leave-one-out cross validation.

The functional Nadaraya-Watson kernel nonparametric method (K-NPFDA), shown in (36.1), is used to carry out a supervised classification. Given a new TG curve, $x = x(t)$, obtained from a material to classify, the estimator of the posterior probability of belonging to a class g, with $g \in \{0, 1, \ldots, G\}$, is given by:

$$\hat{r}_h^{(g)}(x) = \frac{\sum_{i=1}^{n} I_{\{Y_i = g\}} K\left(\frac{\|x - X_i\|}{h}\right)}{\sum_{i=1}^{n} K\left(\frac{\|x - X_i\|}{h}\right)}, \tag{36.1}$$

where the observed TG curves, $X_i = X_i(t)$, are a sample of explanatory variables, while the response sample consists of the observations Y_i of a discrete random variable taking values in the set $\{0,1,\ldots,G\}$, the different classes. The parameter h is the bandwidth or smoothing parameter and $\|\cdot\|$ denotes the L_2 distance between the curves.

After a careful examination of the TG curves, further processing of the data has been found useful for standardizing them: $\tilde{f}(x) = \alpha f(x) + \beta$ with

$$\alpha = \frac{\sqrt{b-a}}{\sqrt{\int_a^b \left(f(t) - \frac{1}{b-a}\int_a^b f(s)ds\right)^2 dt}}\ ,\ \beta = 1 - \frac{\int_a^b f(t)dt}{\sqrt{b-a}\sqrt{\int_a^b \left(f(t) - \frac{1}{b-a}\int_a^b f(s)ds\right)^2 dt}},$$

in order to achieve

$$\frac{1}{b-a}\int_a^b \tilde{f}(t)dt = 0\ ,\ \frac{1}{b-a}\int_a^b \left(\tilde{f}(t) - \frac{1}{b-a}\int_a^b \tilde{f}(s)ds\right)^2 dt = 1$$

This transformation should act on the mean and variance to improve the discriminant power of the curves.

In our research, the Gaussian kernel, K, is used and the smoothing parameter, h, was selected according to a cross-validation method.

Additionally, a k Nearest Neighbors version of the kernel estimator (named KNN-NPFDA) is considered (Ferraty and Vieu, 2006), as well as two additional nonparametric methods based on the boosting algorithm, the B and the B-PCA methods.

As for the problem of studying the effect of adding fumed silica on the thermal degradation of an epoxy resin, when the data are functional (as in this case), an alternative to the classical Analysis of Variance (ANOVA) is the named functional ANOVA (FANOVA) (Cuevas et al., 2004). The covariates are factors while the response is functional. This technique, compared with the classical one, has the advantage of using all the information in the curves, instead of some specific values on them.

Following the nomenclature of (Cuevas et al., 2004), each functional datum can be written as $X_{ij}(t)$, where t usually represents time, with $t \in [a,b]$, i is the subscript that indicates the level of factor and j the replication number ($j = 1,2,\ldots,n_i$ and $i = 1,2,\ldots,k$). Variables $X_{ij}(t)$ can be considered as k independent samples of trajectories drawn from L_2-processes X_i, $i = 1,\ldots,k$.

The mean for each level is given by $E(X_i(t)) = m_i(t)$, while the covariance between two specific values of a curve, $Cov(X_i(s), X_i(t))$, can be easily estimated assuming stationarity.

We want to test:

$$H_0 : m_1 = m_2 = \ldots = m_k \tag{36.2}$$

The statistic implemented by (Cuevas et al., 2004) to test (36.2) is as follows:

$$V_n = \sum_{i<j} n_i \left\| \overline{X}_{i\cdot} - \overline{X}_{j\cdot} \right\|^2 \tag{36.3}$$

The use of (36.3) avoids the requirement of the hypothesis of homoscedasticity in the usual ANOVA. Under some assumptions (see Cuevas et al., 2004), it can be proved that the asymptotic distribution of V_n, under H_0, coincides with that of the statistic

$$V = \sum_{i<j}^{k} \left\| Z_i(t) - C_{ij} \cdot Z_j(t) \right\|^2, \tag{36.4}$$

where $C_{ij} = (p_i/p_j)^{1/2}$ (with p_i the limit of n_i/n) and $Z_1(t)$, ..., $Z_j(t)$ are independent Gaussian processes with mean zero and covariance $Cov(X_i(s), X_i(t))$.

To apply the test, if the n_i are large enough, hypothesis H_0 is rejected, at a level α, whenever $V_n > V_\alpha$ where $P_{H_0}(V > V_\alpha) = \alpha$. In a practical situation, the distribution of V under the null hypothesis can be approximated by applying a parametric bootstrap and the Monte Carlo method. This allows estimation of the value of the α-quantile, V_α. In this case, the use of the parametric bootstrap is justified because the distribution of V is a complicated function of k Gaussian processes.

36.4 Results and discussion

Figure 1 shows the two datasets. Starting with the wood species classification problem, in Table 36.1, the probabilities of correct classification and the temperature ranges for which they are maxima are shown. They have been calculated to classify

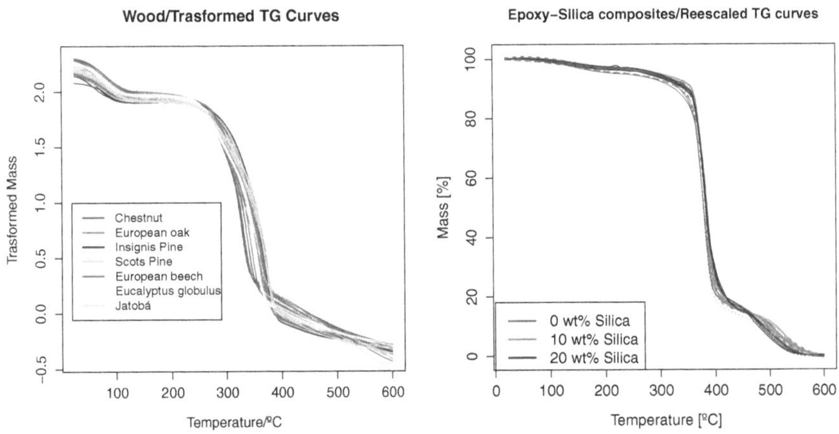

Fig. 36.1: Left panel: Transformed TG curves corresponding to the 7 wood species. Right panel: reescaled TG curves corresponding to epoxy-silica composites and adjusted with a penalized *b*-spline basis with 80 elements.

among 7 different species. This is the result of evaluating the probabilities at 1280 intervals of eight different sizes, from 50 to 400ºC.

The overall probabilities of correct classification when one wants to discriminate between the 7 species of wood is very high. Especially interesting are the results obtained using the methods K-NPFDA and B (7 elements in the basis and deph of tree equal to 3). In the first case we get a probability of correct classification of 0.88 for the interval 192.5-292.5 (see Table 36.1). This may succeed because of the different hemicellulose content and may be due to differences in hemicellulose degradation depending on the species, but we must do more experiments to prove it. The optimal interval (217.5-417.5ºC) obtained by the method B includes the degradation processes of the hemicellulose, cellulose and lignin getting a slightly higher probability of correct classification.

As for the problem of measuring the influence of adding fumed silica on the thermal degradation of an epoxy resin, the FANOVA test described in Section 3 is applied to the rescaled TG curves. The null hypothesis to be tested is $H_0 : m_1 = m_2 = m_3$, where m_i is the mean of the functional data within each of the three levels studied belonging to the factor *amount of fumed silica.*

The result of the application of this procedure in the case of the rescaled TG curves is the following: $V_n = 34470.8$ and $V_\alpha = 1235.671$. Therefore, $V_n >> V_\alpha$. The test is highly significant. At least one of the functional means is different from the others. The thermal stability of epoxy resin, which forms part of the composite material, undergoes a highly significant statistical increment with the addition of an increased amount of fumed silica (it supports higher temperatures before degrading). This is the indicator of an interaction between the epoxy resin and fumed silica. But, what levels are really different? Since only three groups are considered

	7 Groups Classification	
Methods	Optimal Interval (°C)	Prediction Prob.
K-NPFDA	192.5-292.5	0.88
Knn-NPFDA	182.5-282.5	0.88
B	217.5-417.5	0.90
B-PCA	195.0-345.0	0.83

Table 36.1: Prediction probabilities and optimal intervals obtained by each classification method. The TGA data were tested with 7 classes.

Groups compared	$V_{0.05}$	$V_{0.015}$	V_n	Result
0wt%–10wt%	1506.59	2252.66	13654.93	Significant
0wt%–20wt%	1439.34	2135.89	16199.65	Significant
10wt%–20wt%	774.42	1121.88	4616.23	Significant

Table 36.2: Pairwise comparisons using the functional ANOVA test with TG curves.

here, three pairwise comparisons, using the same functional ANOVA method, can be used as a first attempt to tackle this problem. To correct the problem of multiple testing, a Bonferroni correction is used. The idea behind this approach is to consider a new significance level, $\alpha_{\text{Bonf}} = \alpha/J$, J being the number of groups to be compared ($J = 3$, in our case), and compute individual tests using this new level. Table 36.2 shows the results of all pairwise comparisons with the functional ANOVA test using $\alpha = 0.05$ and $\alpha_{\text{Bonf}} = 0.05/3 \approx 0.015$. As it can be observed, according to these tests, the three groups are significant different.

36.5 New research lines

FDA methods have many applications in material science research. Some of them to be studied by us in the near future are, for example, classifying different types of bitumen and other materials for industrial applications and reverse engineering or using functional nonparametric methods applied to thermal and rheological functional data.

Acknowledgements This research has been partially supported by the Spanish Ministry of Science and Innovation, Grant MTM2008-00166 (ERDF included).

References

1. Cuevas, A., Febrero, M., Fraiman, R.: An anova test for functional data. Comput. Stat. Data An. **47**, 111–122 (2004)
2. Ferraty, F., Vieu, P.: Nonparametric Functional Data Analysis: Theory and Practice. Springer, New York (2006)
3. Tarrío-Saavedra, J., Naya, S., Francisco-Fernández, M., Artiaga, R., López-Beceiro, J.: Application of functional ANOVA to the study of thermal stability of micro-nano silica epoxy composites. Chemometrics Intell. Lab. Syst. **105**, 114–124 (2011)
4. Prime, R.B., Bair, H.E., Vyazovkin, S., Gallagher, P.K., Riga, A.: Thermogravimetric analysis (TGA). In: Menczel, J.D, Prime, R.B. (eds.) Thermal analysis of polymers. Fundamentals and applications. Wiley, San José (2009)
5. Tarrío-Saavedra, J., Naya, S., Francisco-Fernández, M., López-Beceiro, J., Artiaga, R.: Functional nonparametric classification of wood species from thermal data. J. Therm. Anal. Calorim., DOI: 10.1007/s10973-010-1157-2 (2010)

Chapter 37
On the Properties of Functional Depth

Alicia Nieto-Reyes

Abstract The properties that a functional depth should satisfy are proposed and a functional depth that fulfills them is defined. Before, the properties of the multidimensional depth were sought when dealing with functional depths.

37.1 Introduction

The notion of depth was introduced in Mahalanobis (1936). The main objective being to give a sense of order in a multidimensional data set, or distribution.

Subsequently, many definitions of multidimensional depth have being proposed, the most well-known are in Tukey (1975), Oja (1983), Liu (1990), ... When introducing this last definition of depth some properties were proposed to justify it as a multidimensional data depth. They were later adapted in Zuo and Serfling (2000) as the key properties required for any statistical depth function.

Thus, denoting by $\mathscr{P}$ the class of distributions on the Borel sets of $\mathbb{R}^p$ and by P_X the distribution of a general random vector X, in Zuo and Serfling (2000) the bounded and nonnegative mapping $D(\cdot,\cdot) : \mathbb{R}^p \times \mathscr{P} \longrightarrow \mathbb{R}$ is called a *statistical depth function* if it satisfies the following properties:

1. $D(Ax+b, P_{AX+b}) = D(x, P_X)$ holds for any $\mathbb{R}^p$-valued random vector X, any $p \times p$ nonsingular matrix A and any $b \in \mathbb{R}^p$.
2. $D(\theta, P) = \sup_{x \in \mathbb{R}^p} D(x, P)$ holds for any $P \in \mathscr{P}$ having a center of symmetry θ.
3. For any $P \in \mathscr{P}$ having deepest point θ, $D(x,P) \leq D(\theta + \alpha(x-\theta), P)$ holds for $\alpha \in [0,1]$.
4. $D(x,P) \to 0$ as $||x|| \to \infty$, for each $P \in \mathscr{P}$.

Alicia Nieto-Reyes
Universidad de Cantabria, Spain, e-mail: alicia.nieto@unican.es

Note that the above properties are affine invariance, maximality at center, monotonicity relative to the deepest point and vanishing at infinity and that they were given for multidimensional spaces.

As it is explained in Dabo-Niang and Ferraty (2008), with the progress in technology there has being an increase in functional data. This has resulted in the need of using depth for functional data. Thus, some definitions have appeared in Fraiman and Muniz (2001), Cuevas et al. (2007), López Pintado and Romo (2006) and (2009), Cuesta-Albertos and Nieto-Reyes (2008) and (2010) and Cuevas and Fraiman (2009).

Most of these depths are defined as an extension of the multidimensional ones. Probably, for this reason, the properties the authors have looked for has being the ones stated in Zuo and Serfling (2000).

In Section 2 we show the lack of adequacy of most of the properties in Zuo and Serfling (2000) for the functional setting while proposing the properties that a functional depth should fulfill. In Section 3 we give a new definition of functional depth directly thought for functional spaces, which meet these properties. Finally, in Section 4 we end with a conclusion where the layout of the talk is exposed.

37.2 Properties of functional depth

Let $\mathfrak{F}$ be a functional metric space, $\mathscr{P}$ a class of distributions on it, $P \in \mathscr{P}$ and P_X the distribution of a general random function X. Let us denote by $d(\cdot,\cdot)$ the associate distance to $\mathfrak{F}$. Thus, we propose that the mapping $D(\cdot,\cdot) : \mathfrak{F} \times \mathscr{P} \longrightarrow \mathbb{R}$ should fulfill the following properties:

1. $D(f(x), P_{f(X)}) = D(x, P_X)$ with $x \in \mathfrak{F}$ and $f : \mathfrak{F} \to \mathfrak{F}$ such that $d(f(x), f(y)) = a \cdot d(x,y) + b$ with $x, y \in \mathfrak{F}$, a, b constants in $\mathbb{R}$ and $a > 0$.

Note that when working in a functional space, the affine invariance should be translated into an invariance with respect to the functions that preserve the relative distances among the elements of the space.

2. $D(\theta, P) = \sup_{x \in \mathfrak{F}} D(x, P)$ holds for any $P \in \mathscr{P}$ having θ as center of halfspace symmetry.

The second property is analogous to the one for multidimensional spaces but there the symmetry being halfspace is not specified. The reason we have specified it is because it is more general than other types of symmetry like central and angular, and we consider it necessary for obtaining the idea of order in functional spaces we look for when using a functional depth.

3. For any $P \in \mathscr{P}$ having deepest point θ, $D(x,P) \leq D(y,P)$ holds for any $x, y \in \mathfrak{F}$ such that y is contained in the close band formed by x and θ.

Note that this is a more general property than the one asked for multidimensional depths because in the multidimensional depth the values of y have a particular form with respect to x. Here, it is only the set containing y which is defined, and not the form in the set.

4. For each $P \in \mathscr{P}$, $D(x,P) \to 0$ as $|x(v)| \to \infty$ for almost every $v \in I$, where I is the domain of definition of the elements of $\mathfrak{F}$.

The main reason for changing $\|x\| \to \infty$ by $|x(v)| \to \infty$ for almost every $v \in I$, is that in cases like the following the depth of the functions should not tend to zero: $x_n(t) = 1/\delta_n$ if $t \in [0, \delta_n)$ and zero otherwise, where $\{\delta_n\}_n \subset \mathbb{R}^+$ with $\lim_n \delta_n = 0$.

5. $\sup_{x \in \mathfrak{F}} |D(x,P_n) - D(x,P)| \to 0$ almost surely $[P]$, where P_n is a sequence of empirical distributions computed on a random sample taken from P.

Although this property was not stated as a property that a multidimensional depth should fulfill, it has being usually proved by some authors for the existent depths. We have written it here because of its major importance due to the main aim of a depth is to order a set of data.

6. *"The regions were the functions are close by have less weight in computing the depth than those where they are farther."*
   That is, denoting by E the expected value, by P_n is the empirical distribution and by $\mathfrak{C}$ the support of P_n, if there exists a J contained in the functions domain, I, such that $\{\max d(x(J), y(J)) : x, y \in \mathfrak{C}\} << E[d(x(J^c), y(J^c)) : x, y \in \mathfrak{C}]$, with $J^c = I - J$, then, $D(x,P_n) \approx D(x(J^c), P_n(J^c))$, where $x(J)$ refers to the curve x restricted to the domain J.

This last property is a philosophical one but of great importance due to it is quite likely to happen with real data that there are regions in which the curves may be quite similar. In fact, the previous known functional depths have this common problem. Particularly, we find this kind of curves when working with the functional version of microarray data. The main reason for introducing this property is that a curve can be in the middle, or near to, of the curves cloud in the interval(s) in which the curves are similar and far from the curves cloud in the interval(s) in which the curves differ. Thus, a curve satisfying this would be consider to have a high depth while it should have a low depth. Analogously, a curve that is far from the center area in the interval(s) in which the curves are alike and in the center or quite near to it in the rest, would be consider to have a low depth while it should have a high depth. Therefore a property that a functional depth should have is to take into account the intervals in which the curves are similar. A possible solution for that could be to consider only the intervals in which the distances among the curves were larger than a constant; but this is an ad hoc solution and so, not the one we should look for. A solution is found with the definition given in the following section.

Note that we have not said that a functional depth should be bounded and non-negative. Although these two properties are natural for the idea of depth and so, they are satisfied by the known functional depths and by the one we propose, they are not necessary for giving an idea of order.

37.3 A well-behave functional depth

The functional depth we propose consists in defining a layer formed by the most distance functions (or curves). Then, subsequent layers are defined inside the previous ones by the subsequent more distance curves. See the left plot on the Figure above, where the first layer is in blue, the second in red and the third in green.

Thus, the depth of a curve is given by the number of layers to which the curve belongs. Therefore, what makes this depth behave well is the fact that the regions in which the curves are close have less weight in computing the distances among curves than other regions. In addition, this definition of depth is robust in the sense that it does not matter how far away is a curve, or a region of the curve, from the curves cloud as it is in the same layer as if it were closer. Particularly, we can say that

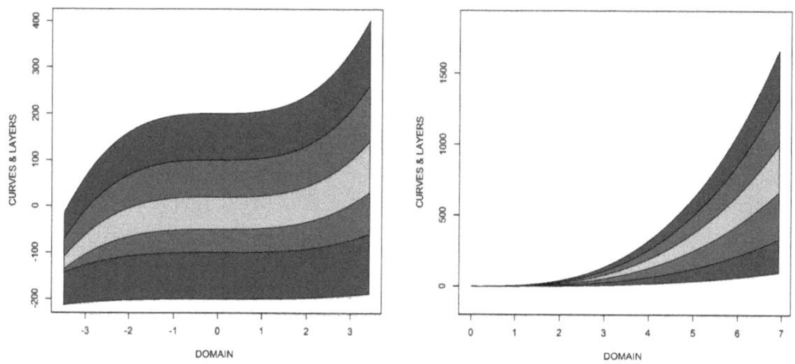

Fig. 37.1: Representation of six curves and their corresponding layers. In the left plot the images of the curves are different through all the domain and in the right tend to be equal in part of it.

this depth behaves well as it satisfies the properties of functional depth introduced in Section 2. Note that instead of distances we could work with areas, obtaining the same behavior for the depth.

Definition of depth given a set of curves
As above, let $\mathfrak{F}$ be a functional metric space and P a probability in it. Let us denote by $d(\cdot,\cdot)$ the associate distance to $\mathfrak{F}$. To compute the depth of a curve $C \in \mathfrak{F}$ with respect to a set of curves $\mathfrak{C} := \{C_1, \dots, C_n\} \subset \mathfrak{F}$, we use the empirical distribution. Let us now introduce the notation for the layers.
$\mathfrak{L}_1 := \{F \in \mathfrak{F} : d(F,C_i) \geq d(C_1,C_2)$ for an $i = 1,2$ with $C_1, C_2 \in \mathfrak{B}_1\}$
$\mathfrak{L}_k := \{F \in \mathfrak{F} - \cup_{l=1}^{k-1} \mathfrak{L}_l : d(F,C_i) \geq d(C_1,C_2)$ for an $i = 1,2$ with $C_1, C_2 \in \mathfrak{B}_k\}, k \geq 2$,
where
$\mathfrak{B}_1 := \{\arg\max d(C_i,C_j) / C_i, C_j \in \mathfrak{C}\}$,

$\mathfrak{B}_k := \{\arg\max d(C_i,C_j)/C_i,C_j \in \mathfrak{C} - \cup_{l=1}^{k-1}\mathfrak{B}_l\}$, for $k = 2,...,k_0$.
k_0 is such that $\mathfrak{C} - \cup_{l=1}^{k_0}\mathfrak{B}_l$ contains only one curve or is equal to $\mathfrak{B}_{k_0+1}$. In addition, let us denote
$\mathfrak{M} := \{C \in \mathfrak{C} : \min_{T\in S_C} d(C,T) \leq \min_{V\in R_C} d(\mathfrak{B}_k - C,V)$ with k s.t. $C \in \mathfrak{B}_k\}$, where $S_C = \mathfrak{C} - (\cup_{l=1}^{k-1}\mathfrak{L}_l \cup C)$ and $R_C = (\mathfrak{C} - \cup_{l=1}^{k}\mathfrak{L}_l) \cup C$. Given a curve C in $\mathfrak{F}$ there exist a k such that C is in the closure of $\mathfrak{L}_k$ minus $\mathfrak{B}_k$. Then, the depth of a curve C in $\bar{\mathfrak{L}}_{k+1} - \mathfrak{B}_{k+1}$ with respect to P is $D(C,P) := P(D_C)$ where

$$D_C := \begin{cases} \cup_{l=1}^{k}\mathfrak{L}_l & \text{if } C \in \mathfrak{M} \\ \cup_{l=1}^{k-1}\mathfrak{L}_l \cup C & \text{if } C \notin \mathfrak{M}. \end{cases}$$

Thus, the deepest curve/s among $C_1,\ldots,C_n$ is/are in $\mathfrak{B}_{k_0+1}$ and $\mathfrak{B}_k$ is empty for $k > k_0+1$.

Regarding the applications, when the cardinal of $\mathfrak{B}_{k_0+1}$ is larger than one and we want the deepest curve to be only one but do not need it to be in $\mathfrak{C}$, we can take as deepest curve the mean of the curves in $\mathfrak{B}_{k_0+1}$ (assuming it is in $\mathfrak{F}$). Note that, in fact, any curve of $\mathfrak{F}$ whose image is between the images of the curves in $\mathfrak{B}_{k_0+1}$ has maximum depth.

37.4 Conclusions

The presentation will be devoted to, departing from the properties for a multidimensional depth, propose the adequate ones for a functional depth. Particularly, we will study these properties with the existing functional depths. In addition, the behavior of the previous new functional depth will be studied and shown that it satisfies the above proposed properties for a functional depth. Finally, some applications will be shown where the importance of the proposed properties and depth can be observed.

Acknowledgements Thank you to J.A. Cuesta-Albertos for his help and useful comments.

References

1. Cuesta-Albertos J.A., Nieto-Reyes, A.: The random Tukey depth. Comput. Stat. Data An. **52** (11), 4979–4988 (2008)
2. Cuesta-Albertos J.A., Nieto-Reyes, A.: Functional Classification and the Random Tukey Depth. Practical Issues. In: Borgelt, C., Gonzalez-Rodriguez, G., Trutschnig, W., Lubiano, M.A., Gil, M.A., Grzegorzewski, P., Hryniewicz, O. (eds.) Combining Soft Computing and Statistical Methods in Data Analysis, pp. 121-126. Springer (2008)
3. Cuesta-Albertos J.A, Nieto-Reyes, A.: A random functional depth. In: Dabo-Niang, S., Ferraty, F. (eds.) Functional and Operational Statistics, pp. 121-126. Springer (2010)
4. Cuevas, A., Febrero-Bande, M, Fraiman, R.: Robust estimation and classification for functional data via projection-based depth notions. Computation. Stat. **22** (3), 481–496 (2007)

5. Cuevas, A, Fraiman, R.: On depth measures and dual statistics. A methodology for dealing with general data. J. Multivariate Anal. **100** (4), 753–766 (2009)
6. Dabo-Niang, S, Ferraty, F.: Functional and Operational Statistics. Springer (2008)
7. Fraiman, R, Muniz, G.: Trimmed means for functional data. TEST **10** (2), 419–440 (2001)
8. Liu, R.Y.: On a notion of data depth based on random simplices. Ann. Stat. **18**, 405–414 (1990)
9. López-Pintado, S., Romo, J.: Depth-based classification for functional data. Amer. Math. Soc. DIMACS Series **72**, 103–119 (2006)
10. López-Pintado, S., Romo, J.: On the concept of depth for functional data. J. Am. Stat. Assoc. **104** (486), 718–734 (2009)
11. Mahalanobis, P. C.: On the generalized distance in statistics. Proc. Natl. Inst. Science **12**, 49–55 (1936)
12. Oja, H.: Descriptive statistics for multivariate distributions. Stat. Probab. Lett. **1**, 327–332 (1983)
13. Tukey, J.W.: Mathematics and picturing of data. Proceedings of ICM, Vancouver **2**, 523–531 (1975)
14. Zuo, Y, Serfling, R.: General notions of statistical depth function. Ann. Stat. **28** (2), 461–482 (2000)

Chapter 38
Second-Order Inference for Functional Data with Application to DNA Minicircles

Victor M. Panaretos, David Kraus, John H. Maddocks

Abstract The problem of comparison of second-order (covariance) properties of two samples of random curves is considered. The work is motivated by the study of the mechanical properties of short strands of DNA. Our test is based on the common empirical Karhunen–Loève expansion and truncated approximation of the Hilbert–Schmidt distance of the empirical covariance operators.

38.1 Introduction

The development of the statistical methods described here was motivated by a dataset consisting of reconstructed three-dimensional electron microscope images of loops (called minicircles) obtained from short strands of DNA (Amzallag, Vaillant, Jacob, Unser, Bednar, Kahn, Dubochet, Stasiak and Maddocks, 2006). There are two types (called TATA and CAP) of DNA minicircles with identical base-pair sequences, except for short susubsequence where they differ. The main question is whether this difference affects the geometry of the minicircle.

Mathematically, DNA minicircles are closed curves in $\mathbb{R}^3$. Figure 38.1 shows projections of these curves on the planes given by the axes of the coordinate system. In Figure 38.2 coordinates on the axes are plotted against the arc length of the curve. This plot suggests that the data could be analysed by means of functional data analysis.

Victor M. Panaretos
Ecole Polytechnique Fédérale de Lausanne, Lausanne, Switzerland, e-mail: victor.panaretos@epfl.ch

David Kraus
Ecole Polytechnique Fédérale de Lausanne, Lausanne, Switzerland, e-mail: david.kraus@epfl.ch

John H. Maddocks
Ecole Polytechnique Fédérale de Lausanne, Lausanne, Switzerland, e-mail: john.maddocks@epfl.ch

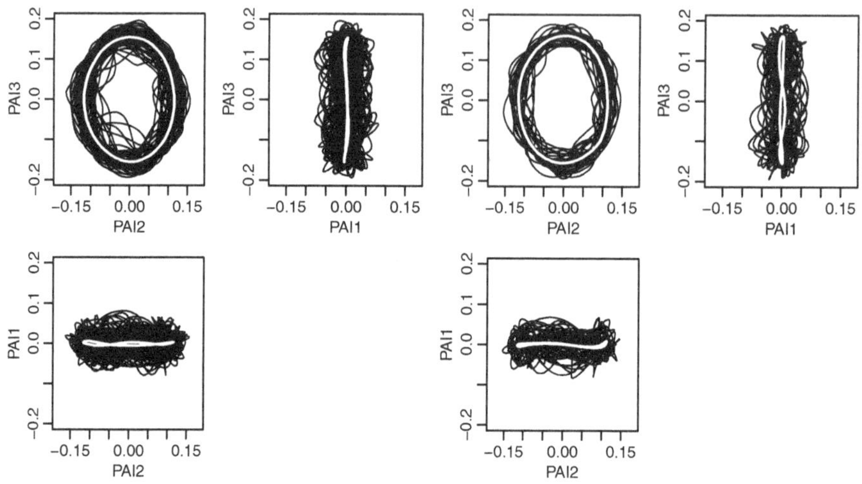

Fig. 38.1: Projections of DNA minicircles on the planes given by the principal axes of inertia (three panels on the left side: TATA curves, right: CAP curves). Mean curves are plotted in white.

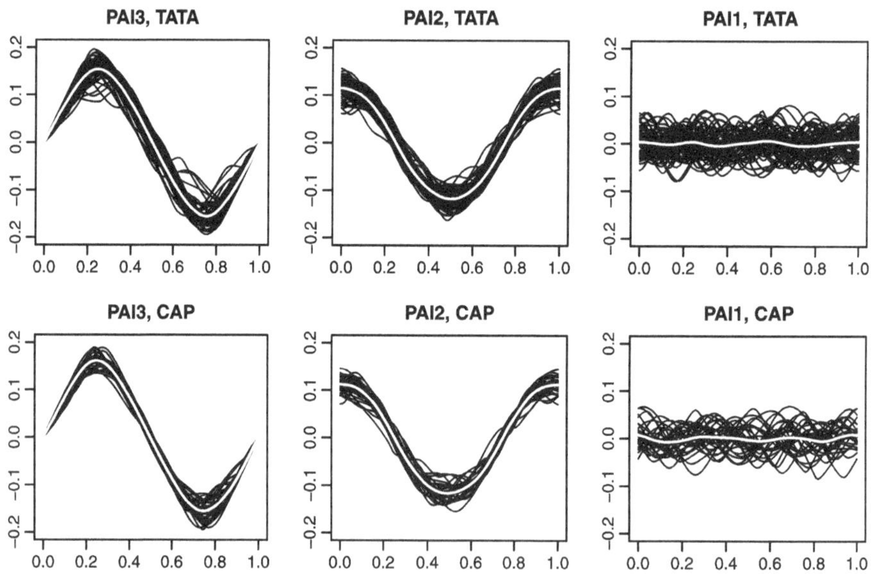

Fig. 38.2: Coordinates of DNA minicircles on the principal axes of inertia. Mean curves are plotted in white.

Plots of estimated mean functions do not suggest any difference between the two types of curves. We tested the hypothesis of equal mean functions and the results were insignificant. Therefore we focused on second-order properties and developed methods for comparing covariance operators.

In this extended abstract we sketch the main idea of the testing procedure (Section 2) and summarise results of the analysis of DNA minicircles (Section 3). Details of the statistical methods and data application mentioned here can be found in Panaretos, Kraus and Maddocks (2010).

38.2 Test

Let $X_1,\dots,X_{n_1}$ and $Y_1,\dots,Y_{n_2}$ be two independent samples of stochastic processes with paths in $L^2[0,1]$ with mean functions μ_X,μ_Y and covariance operators $\mathscr{R}_X,\mathscr{R}_Y$. The aim is to test the null hypothesis $\mathscr{R}_X=\mathscr{R}_Y$ against the general alternative $\mathscr{R}_X\neq\mathscr{R}_Y$.

The problem of comparing covariance operators of functional data has received relatively little attention in the literature. Related but different second-order problems were studied by Benko, Härdle and Kneip (2009) and Horváth, Hušková and Kokoszka (2010).

Our test is based on the comparison of the empirical covariance operators

$$\hat{\mathscr{R}}_X=\frac{n_1}{n_1+n_2}\sum_{i=1}^{n_1}(X_i-\bar{X})\otimes(X_i-\bar{X}),\qquad \hat{\mathscr{R}}_Y=\frac{n_1}{n_1+n_2}\sum_{i=1}^{n_2}(Y_i-\bar{Y})\otimes(Y_i-\bar{Y}).$$

The test will reject the null hypothesis when the operator $\mathscr{D}=\hat{\mathscr{R}}_X-\hat{\mathscr{R}}_Y$ is significantly far from the zero operator.

The distance of $\mathscr{D}$ from zero can be measured by the squared Hilbert–Schmidt norm

$$\|\mathscr{D}\|^2=\sum_{j=1}^{\infty}\sum_{k=1}^{\infty}\langle\varphi_j,\mathscr{D}\varphi_k\rangle^2$$

where $\{\varphi_j, j=1,2,\dots\}$ is any orthonormal basis of the sample Hilbert space $L^2[0,1]$. This random variable does not have a tractable asymptotic distribution. Therefore we perform dimension reduction and study the infinite-dimensional object $\mathscr{D}$ on a finite-dimensional subspace. Let Φ be the K-dimensional linear subspace generated by an orthonormal basis $\{\varphi_1,\dots,\varphi_K\}$ (where K is a finite number small than or equal to the rank of the covariance operator). Instead of measuring the difference of $\mathscr{D}$ from zero on the whole Hilbert space $L^2[0,1]$, we restrict our attention to Φ. More precisely, instead of $\mathscr{D}$ we use the operator $\pi_\Phi\mathscr{D}\pi_\Phi$ where $\pi_\Phi=\sum_{k=1}^K\varphi_k\otimes\varphi_k$ is the projection operator on Φ. The square of its Hilbert–Schmidt norm equals

$$\|\pi_\Phi \mathscr{D} \pi_\Phi\|^2 = \sum_{j=1}^{K}\sum_{k=1}^{K} \langle \varphi_j, \mathscr{D}\varphi_k\rangle^2.$$

In light of the Karhunen–Loève expansion and Mercer's theorem, it is natural to choose the functions φ_k as the first K eigenfunctions $\hat{\varphi}_k$ of the pooled sample covariance estimator $\hat{\mathscr{R}} = \frac{n_1}{n_1+n_2}\hat{\mathscr{R}}_X + \frac{n_2}{n_1+n_2}\hat{\mathscr{R}}_Y$. (Note that one cannot perform eigendecomposition of each covariance operator separately because a common basis is needed.)

The terms $S_{jk} = \langle \varphi_j, \mathscr{D}\varphi_k\rangle$ can be seen as differences of the empirical covariances of the Fourier coefficients of the observations with respect to $\varphi_1, \dots, \varphi_K$. That is, for $\beta_{ik}^X = \langle X_i, \varphi_k\rangle$, $\beta_{ik}^Y = \langle Y_i, \varphi_k\rangle$ one can see that $S_{jk} = \hat{\lambda}_{jk}^X - \hat{\lambda}_{jk}^Y$ where

$$\hat{\lambda}_{jk}^X = \frac{1}{n_1}\sum_{i=1}^{n_1}(\beta_{ij}^X - \bar{\beta}_j^X)(\beta_{ik}^X - \bar{\beta}_k^X), \qquad \hat{\lambda}_{jk}^Y = \frac{1}{n_2}\sum_{i=1}^{n_2}(\beta_{ij}^Y - \bar{\beta}_j^Y)(\beta_{ik}^Y - \bar{\beta}_k^Y).$$

The variable $\|\pi_\Phi \mathscr{D} \pi_\Phi\|^2$ thus equals the squared Frobenius norm of the difference of the empirical covariance matrices of the Fourier scores.

Instead of simply summing the squares of S_{jk}, one combines the $K(K+1)/2$ different terms S_{jk}, $1 \le j \le k \le K$ in a quadratic form reflecting their covariance structure as follows. Under certain assumptions it can be shown using the Hilbert space Central Limit Theorem that under the null hypothesis the test operator

$$\frac{n_1^{1/2} n_2^{1/2}}{(n_1+n_2)^{1/2}}\mathscr{D}$$

is asymptotically distributed as a zero-mean Gaussian random linear operator on $L^2[0,1]$. Consequently, in view of the consistency of empirical eigenfunctions the vector with components S_{jk}, $1 \le j \le k \le K$ converges to a mean zero Gaussian vector whose covariance matrix can be consistently estimated by the empirical covariance matrix, say W, of the summands in S_{jk}. Then the quadratic test statistic follows the form

$$\frac{n_1 n_2}{n_1+n_2} S^\top W S.$$

Its asymptotic distribution under the null is chi-square with $K(K+1)/2$ degrees of freedom. The test rejects H_0 when the value of the statistic is significantly large.

In the case of Gaussian data the limiting covariance structure of S simplifies. It turns out that the components S_{jk} are asymptotically independent and their limiting variances can be expressed in terms of the eigenvalues of $\mathscr{R}_1 = \mathscr{R}_2$. This leads to the statistic

$$T = \frac{n_1 n_2}{n_1+n_2}\sum_{j=1}^{K}\sum_{k=1}^{K}\frac{(\hat{\lambda}_{jk}^X - \hat{\lambda}_{jk}^Y)^2}{2(\frac{n_1}{n}\hat{\lambda}_{jj}^X + \frac{n_2}{n}\hat{\lambda}_{jj}^Y)(\frac{n_1}{n}\hat{\lambda}_{kk}^X + \frac{n_2}{n}\hat{\lambda}_{kk}^Y)}$$

with asymptotic χ^2 distribution with $K(K+1)/2$ degrees of freedom. When one a priori expects the eigenfunctions in the two samples to be equal, the test can be based only on the diagonal ($j = k$) terms in the sum above (comparing only variances, not covariances of the scores). Such a statistic is asymptotically χ^2_K-distributed. Modifications of the test statistics can be obtained by variance stabilising transformations of the summands.

The truncation level K can be selected with the help of scree plots and cumulative variance plots. We have also proposed an automatic procedure based on a penalised fit criterion.

38.3 Application to DNA minicircles

The original original (x, y, z)-coordinates of the curves were obtained from electron microscope images of a frozen liquid containing the minicircles. Therefore the original curves are randomly rotated and shifted, thus not directly comparable. So it is necessary to align them. We cannot apply landmark alignment methods because there are no landmarks (the sequence of DNA base-pairs is not observed). Warping methods are not appropriate as they could modify the second-order properties. Instead, after centering (setting the center of mass to 0) and scaling to unit length, we align each curve separately by rotating it in a way given by the moment of inertia tensor.

The moment of inertia tensor is defined as

$$J(u) = \int_{\mathbb{R}^3} \|(I - uu^\top)x\|^2 \mu(dx)$$

where u is a unit vector in $\mathbb{R}^3$ and μ is the uniform distribution of mass on the curve. By integrating the squared distance of the points on the curve from the axis, the tensor measures how difficult it is to rotate the curve around the axis given by u. The first eigenvector (corresponding to the largest eigenvalue) determines the first principal axis of inertia (PAI1) around which the curve is most difficult to rotate. The projection on the plane orthogonal to PAI1 is most spread. Then PAI2 given by the second eigenvector is the axis orthogonal to PAI1 around which the projection of the curve on the first principal plane is most difficult to rotate. Within this plane, the projection on the axis PAI3 orthogonal to PAI2 is most spread.

For each curve we computed the principal axes of inertia and rotated the curve so that its principal axes agree with the (x, y, z)-axes. This procedure is similar to the balancing of a tyre. Figure 38.1 shows the rotated minicircles. These closed curves have no starting point and no orientation. As the starting point of each curve we chose the point where the projection of the curve on the first principal plane intersects the positive horizontal semi-axis; we chose the counter-clockwise orientation. As the 'time' argument of each functional observation we use the arc length of the curve from the starting point. The resulting functional data set is plotted in Figure 38.2.

The test comparing the covariance operators suggests significant differences between the samples. For example, when applied to the projections on the first principal plane (PAI2,3) with $K = 7$ (selected by the automatic procedure), the p-value is 0.023.

References

1. Amzallag, A., Vaillant, C., Jacob, M., Unser, M., Bednar, J., Kahn, J. D., Dubochet, J., Stasiak, A., Maddocks, J. H.: 3D reconstruction and comparison of shapes of DNA minicircles observed by cryo-electron microscopy. Nucleic Acids Research **34**, e125 (2006)
2. Benko, M., Härdle W., Kneip, A.: Common functional principal components. Ann. Stat. **37**, 1–34 (2009)
3. Horváth, L., Hušková, M., Kokoszka, P.: J. Multivariate Anal. **101**, 352–367.
4. Panaretos, V. M., Kraus, D., Maddocks, J. H.: Second-order comparison of Gaussian random functions and the geometry of DNA minicircles. J. Am. Stat. Assoc. **105**, 670–682 (2010)

Chapter 39
Nonparametric Functional Time Series Prediction

Efstathios Paparoditis

Abstract We consider the problem of predicting a time series on a whole interval in terms of its own past. An approach based on a wavelet decomposition and an appropriate distance measure between time series curves is introduced. Applications of this approach to nonparametric conditional mean estimation, clustering and bootstrap of time series curves are discussed.

39.1 Wavelet-kernel based prediction

Consider the prediction problem of a time series on a whole time interval in terms of its own past. The approach that we adopt is based on functional kernel nonparametric regression estimation techniques where observations are discrete recordings of segments of an underlying stochastic process considered as curves. More specifically, we assume that segments $Z_k(t)$, $k = 1,2,\ldots,n$ of an underlying continuous time stochastic process $\{X(t), t \in \mathbf{R}\}$ are observed, where for $k = 1,2,\ldots,n$, $Z_k(t) = X(t+(k-1)\delta)$ and $t \in [0,\delta)$. δ is a parameter that controls the length of the time series curve $Z_k(t)$ and depends on the particular application at hand. Further, we assume that the the time series curves $Z_k(t)$ are observed at P discrete time points of the corresponding time interval $t+(k-1)\delta$, $k = 1,2,\ldots,n$; that is the observations consist of n time series curves $Z_k(t_j)$, $k = 1,2,\ldots,n$ and each one of them is recorded at $t_1,t_2,\ldots,t_P$ equidistant time points in the interval $t+(k-1)\delta$. One situation where this kind of data occur is that of forecasting the daily power demand of electricity. Suppose that the daily power demand is recorded every 15 minutes, then δ describes the time interval containing the load of one day and each time series curve $Z_k(t)$ consists of $P = 96$ points, the 96 quarters of a day at which the power demand is recorded.

Efstathios Paparoditis
University of Cyprus, Nicosia, Cyprus, e-mail: stathisp@ucy.ac.cy

Now, given the observed time series curves $Z_1(t), Z_2(t), \ldots, Z_n(t)$, our goal is to predict the whole feature curve $Z_{n+1}(t)$, that is the behavior of this curve at the P time points $Z_{n+1}(t_1), Z_{n+1}(t_2), \ldots, Z_{n+1}(t_P)$. Notice that because P is usually large, classical methods of time series prediction (parametric or non-parametric) are not appropriate since they lead to multi-step forecasting with a long forecasting horizon which implies an increase of the mean square error of prediction as the prediction horizon increases. In contrast to more classical methods, and in order to perform the prediction we consider $Z_n(t)$ as a time series curve and calculate the prediction by means of nonparametric kernel estimator of the conditional mean based on a wavelet decomposition of the time series curves. The use of wavelet decomposition of the segment sample paths is very useful in order to take into account the inhomogeneity and the local irregularities of the different time series curves. Based on this wavelet decomposition a notion of similarity is introduced which is used to calibrate the prediction.

To be more specific, assume that $P = 2^J$, $J \in \mathbf{N}$, and let

$$\vartheta^{(s)}_{j,k},\ s = 1, 2, \ldots, n,$$

be the discrete wavelet coefficients of the time series curve $Z_s(t) = (Z_s(t_1), Z_s(t_2), \ldots, Z_s(t_P))$ at scale (or resolution) j, $j = j_0, j_0+1, \ldots, J-1$ and location k, $k = 0, 1, \ldots, 2^j - 1$. The use of wavelet coefficients after scale j_0 implies that the scaling coefficients below this scale do not have any discriminative power. Let $\vartheta^{(s)}$ denote the set of the discrete wavelet coefficients of the time series curve $Z_s(t)$, that is,

$$\vartheta^{(s)} = (\vartheta^{(s)}_{j,k} : j = j_0, j_0+1, \ldots, J-1, \text{ and } k = 0, 1, \ldots, 2^j - 1).$$

To quantify similarity of any two time series curves $Z_r(t)$ and $Z_s(t)$ at each scale j we use the distance measure

$$d_j(\vartheta^{(r)}, \vartheta^{(s)}) = \sqrt{\sum_{k=0}^{2^j-1} (\vartheta^{(r)}_{j,k} - \vartheta^{(s)}_{j,k})^2},$$

and to combine the distances at different scales we use

$$D(\vartheta^{(r)}, \vartheta^{(s)}) = \sum_{j=j_0}^{J-1} \frac{1}{2^{j/2}} d_j(\vartheta^{(r)}, \vartheta^{(s)}). \tag{39.1}$$

Denote by $\Xi_s = (\xi^{(s)}_{J,k}, k = 0, 1, 2, \ldots, J-1)$ the scaling coefficients of the time series curve $Z_s(t)$ at the finest scale J. The first step in our procedure is to predict the set of scaling coefficients Ξ_{n+1} of the segment $Z_{n+1}(t)$ which we denote in the following by $\widehat{\Xi}_{n+1}$. This is done by means of the kernel smoothing

$$\widehat{\Xi}_{n+1} = \frac{\sum_{m=1}^{n-1} \Xi_{m+1} K(D(\vartheta^{(n)}, \vartheta^{(m)})/h)}{n^{-1} + \sum_{m=1}^{n-1} K(D(\vartheta^{n}, \vartheta^{m})/h)},$$

where $K : \mathbf{R} \rightarrow [0, +\infty)$ is a kernel function, $K((x,y)/h) = K(x/h, y/h)$ and h a bandwidth. Notice that the predicted scaling coefficients $\widehat{\Xi}_{n+1}$ are obtained as a weighted average of the scaling coefficients Ξ_{m+1} of the time series curves $Z_{m+1}(t)$, $m = 1, 2, \ldots, n-1$, where more weight is given to the scaling coefficients Ξ_{m+1} for which the set wavelet coefficients $\vartheta^{(m)}$ of the preceding time series curve $Z_m(t)$ is close (in the sense of the distance measure (39.1)) to the set of wavelet coefficients $\vartheta^{(n)}$ of the last observed time series curve $Z_n(t)$. The last step of the prediction procedure is then to transform the set of predicted scaling coefficients back to the time domain in order to obtain the predictor of the time series curve $Z_{n+1}(t)$ at the P time points. Denote the corresponding predictor by $\widehat{Z}_{n+1}(t)$, then this predictor is obtained as

$$\widehat{Z}_{n+1}(t_r) = \sum_{k=0}^{2^J-1} \widehat{\xi}_{J,k}^{(n+1)} \phi_{J,k}(t_r), \quad r = 1, 2, \ldots, P,$$

where $\widehat{\xi}_{J,k}^{(n+1)}$, $k = 0, 1, \ldots, 2^J - 1$ are the components of the predicted scaling coefficients $\widehat{\Xi}_{n+1}$. Asymptotic properties of the predictor $\widehat{Z}_{n+1}(t)$ under mild conditions and when the number of observed segments n grows to infinity has been derived by Antoniadis et al. (2006).

39.2 Bandwidth Choice

As any nonparametric procedure, the prediction procedure proposed based on a wavelet decomposition of the time series curves, heavily depends on the choice of the bandwidth parameter h which essentially controlls the number of time series curves that are effectively used for prediction. We propose a novel method to select the bandwidth which is taylor made for the problem of functional time series prediction. The idea underlying this method is to calculate the empirical risk of prediction using past segments of the observed series and to select as value of the bandwidth for performing the prediction the bandwidth which minimizes this empirical risk.

To be more specific, let $u_n = u(n)$ be the number of segments that will be used to evaluate the empirical risk, $1 < u_n \ll n$. Then the last u_n segments $Z_{n-u_n+1}, Z_{n-u_n+2}, \ldots, Z_n$ are predicted using the past $r = n - u_n$ segments. That is the segment Z_{n-u_n+s}, $s \in \{1, 2, \ldots, u_n\}$ is predicted using the segments $Z_{n-u_n+s-1}, Z_{n-u_n+s-2}, \ldots, Z_s$ for all bandwidths $h = h_n$ in the set $H_n = \{KC_n/L, 2KC_n/L, \ldots, KC_n\}$ where $C_= (\log^2(n)/n)^{1/(P+4)}$, K is a positive constant and L depends on the smallest bandwidth one wants to try.

Let $\widehat{Z}_{n-u_n+1}^{(h_n)}, \widehat{Z}_{n-u_n+2}^{(h_n)}, \ldots, \widehat{Z}_n^{(h_n)}$ be the predictions obtained using the prediction method described in the previous Section and the bandwidth h_n and define the empirical risk of prediction using the bandwidth h_n,

$$R_n(h_n) = \frac{1}{u_n P} \sum_{i=1}^{P} \sum_{j=0}^{u_n-1} \left(Z_{n-j}(t_i) - \widehat{Z}_{n-j}^{(h_n)}(t_i)\right)^2.$$

The bandwidth selected for prediction is then the one that minimizes $R_n(h_n)$, that is

$$h_n^* = \text{argmin}_{h_n \in H_n} R_n(h_n).$$

It can be shown along the same lines as in Antoniadis et al. (2009) that if

$$u_n \to \infty \quad \text{such that} \quad u_n/n \to 0, \quad \text{as } n \to \infty,$$

then the proposed bandwidth estimator imitates (asymptotically) the value of the bandwidth which minimizes the unknown theoretical risk of prediction.

39.3 Further Issues

We consider the issue of constructing pointwise prediction intervals for the trajectories predicted. For this an appropriately designed bootstrap procedure is proposed which resamples the observed curves in a nonparametric way assigning different resampling weights to these curves. We show that the prediction intervals achieves the desired pointwise coverage probability. Furthermore, the idea of a wavelet decomposition of a time series curve together with the defined distance measure $D(\vartheta^{(r)}, \vartheta^{(s)})$ between the time series curves $Z_r(t)$ and $Z_s(t)$ obtained by evaluating the distance of the corresponding set of wavelet coefficients $\vartheta^{(r)}$ and $\vartheta^{(s)}$ respectively, can be used for other purposes as well. One important application is that of functional time series clustering which will be discussed in the talk. Finally applications of the methodology proposed to real-life data will be presented.

References

1. Antoniadis, A., Paparoditis, E., Sapatinas, T.: A functional wavelet-kernel approach for time series prediction. J. Roy. Stat. Soc. B **68**, 837–857 (2006)
2. Antoniadis, A., Paparoditis, E., Sapatinas, T.: Bandwidth selection for functional time series prediction. Stat. Probab. Lett. **79**, 733–740 (2009)

Chapter 40
Wavelets Smoothing for Multidimensional Curves

Davide Pigoli, Laura M. Sangalli

Abstract We describe a wavelet-based method that provides accurate estimates of curves in more than one dimension and of their derivatives. The method is particularly attractive when the curves to be estimated have a varying smoothness and present strongly localized features. The proposed multidimensional wavelet estimation technique is thus applied to multi-lead electrocardiogram records, where strongly localized features are indeed expected.

40.1 Introduction

The estimation of smooth functions from their noisy and discrete observations is the first step in Functional Data Analysis. The choice of the basis of functions to be used in the smoothing is crucial, since its properties influence the subsequent analysis. Usual choices are Fourier bases and spline bases (see, e.g., Ramsay and Silverman, 2005). Wavelet bases have been so far mainly applied in problems where there is no interest in derivatives, because of the absence of close analytical forms for smooth wavelet bases. To overcome this limitation, in Pigoli and Sangalli (2010) we resorted to a numerical method that allows one to obtain derivatives of wavelet estimated data. We moreover extended traditional wavelet estimators to curves in general dimensions; this requires the development of a new estimation procedure which takes into account simultaneously all the space coordinates of the multidimensional curve. A stimulating application for this research has been the fitting of multi-lead electrocardiogram records, illustrated in Section 4.

Davide Pigoli
Politecnico di Milano, Italy e-mail: davide.pigoli@mail.polimi.it

Laura M. Sangalli
Politecnico di Milano, Italy e-mail: laura.sangalli@polimi.it

40.2 An overview on wavelets

In this section we briefly recall wavelet bases for $L^2(\mathbb{R})$. For a systematic introduction to wavelets, see, e.g., Nason (2008). Wavelets are defined starting from an orthogonal multiresolution:

Definition 40.1. Let $\{V_j\}_{j\in\mathbb{Z}}$ be a sequence of closed subspaces $V_j \subseteq L^2(\mathbb{R})$ and let $\varphi \in V_0$. An orthogonal multiresolution for $L^2(\mathbb{R})$ is a couple $(\{V_j\}_j, \varphi)$ such that:

1. $V_j \subset V_{j+1}$
2. $\overline{\bigcup_j V_j} = L^2(\mathbb{R})$ and $\bigcap_{j=-\infty}^{+\infty} V_j = \{0\}$
3. $\{l \mapsto f(l)\} \in V_j \;\Leftrightarrow\; \{l \mapsto f(2l)\} \in V_{j+1}$
4. $\{\varphi(l-k)\}_{k\in\mathbb{Z}}$ is an orthonormal basis for V_0 and $\int_{\mathbb{R}} \varphi \neq 0$.

The projections of $f \in L^2(\mathbb{R})$ on the sequence $\{V_j\}_j$ give a progressively better approximation of f as j increases. The function φ is called a *scaling function* or *father wavelet*. Thanks to property 3 above, $\{2^{j/2}\varphi(2^j l - k)\}_k$ is an orthonormal basis for V_j. However, it is often more useful to explore the detailed information needed to go from the space V_j to the space V_{j+1}, starting from a coarse space V_0. This is the reason for introducing the sequence of complement spaces $W_j = V_{j+1} \backslash V_j$. A *mother wavelet* is a function $\psi \in W_0$ with the property that $\{\psi(l-k)\}_k$ is a basis for W_0. As a consequence, $L^2(\mathbb{R}) = \bigoplus_{j\in\mathbb{Z}} W_j$ and $\{\psi_{j,k}(l)\}_k = \{2^{\frac{j}{2}}\psi(2^j l - k)\}_k$ is an orthonormal basis for $L^2(\mathbb{R})$. Therefore, any function $f \in L^2(\mathbb{R})$ has the following wavelet representation:

$$
\begin{aligned}
f = \sum_j \sum_k \langle f, \psi_{j,k}\rangle \psi_{j,k} &= \sum_k \langle f, \varphi_{j_0,k}\rangle \varphi_{j_0,k} + \sum_{j=j_0}^{+\infty} \sum_k \langle f, \psi_{j,k}\rangle \psi_{j,k} = \\
&= \sum_k s_{j_0,k}\varphi_{j_0,k} + \sum_{j=j_0}^{+\infty} \sum_k d_{j,k}\psi_{j,k},
\end{aligned}
$$

where $\langle \cdot,\cdot \rangle$ is the scalar product in $L^2(\mathbb{R})$, $s_{j_0,k} := \langle f, \varphi_{j_0,k}\rangle$ and $d_{j,k} := \langle f, \psi_{j,k}\rangle$. The coefficients $\{s_{j_0,k}\}_{k\in\mathbb{Z}}$, $\{d_{j,k}\}_{j\in\mathbb{Z}\cap\{j>=j_0\},k\in\mathbb{Z}}$ are called *discrete wavelet transforms* of f. It can be shown that φ and ψ satisfy the dilation/refinement equations:

$$
\varphi(l) = \sum_k h_k \sqrt{2}\varphi(2l-k) \qquad \text{and} \qquad \psi(l) = \sum_k g_k \sqrt{2}\varphi(2l-k) \tag{40.1}
$$

for sequences $\{h_k\}_k$ and $\{g_k\}_k$, named respectively *scaling filter* and *wavelet filter*. It is important to note that smooth and compactly supported wavelet bases have no analytical form, and that they are instead defined via their scaling and wavelet filters. However, thanks to a numerical method (see Strang, 1989), it is possible to compute pointwise values of scaling/wavelet functions and of their derivatives. In Pigoli and Sangalli (2010) we resort to this method to obtain pointwise values of functional data derivatives.

40.3 Wavelet estimation for p - dimensional curves

In this Section we generalize to the p-dimensional case the shrinkage estimator with universal threshold proposed by Donoho et al. (1995) for the monodimensional case.

Let $\{\mathbf{w}_k \in \mathbb{R}^p; k = 1,\dots,n = 2^J\}$ be a noisy and discrete observation of a p-dimensional parametric curve $\mathbf{f}$, with $\mathbf{f}: \mathbb{R} \ni l \mapsto \left(f_1(l),\dots,f_p(l)\right) \in \mathbb{R}^p$, on a grid of 2^J equispaced points. Assume that these data are generated by the model

$$\mathbf{w}_k = \mathbf{f}(l_k) + \boldsymbol{\varepsilon}_k \quad k = 1,\dots,n = 2^J, \tag{40.2}$$

where the error $\boldsymbol{\varepsilon}_k$ has a multivariate normal distribution with mean $\mathbf{0} \in \mathbb{R}^p$ and variance-covariance matrix $\sigma^2\mathbb{I}_p$. Our goal is to accurately estimate the p-dimensional curve $\mathbf{f}$ and its derivatives. Analogously to the 1-dimensional case, we thus consider the corresponding model on the space of wavelet coefficients. Thanks to the orthogonality of the wavelet transform, this is given by

$$\mathbf{d}_{j,k} = \mathbf{d}^0_{j,k} + \boldsymbol{\rho}_{j,k}, \tag{40.3}$$

with $\mathbf{d}_{j,k}, \mathbf{d}^0_{j,k}, \boldsymbol{\rho}_{j,k} \in \mathbb{R}^p$, where $\mathbf{d}_{j,k}$ are the vectors of the empirical wavelet coefficients corresponding to the data, $\mathbf{d}^0_{j,k}$ are the vectors of the true wavelet coefficients of $\mathbf{f}$, and $\boldsymbol{\rho}_{j,k}$ are the wavelet transforms of the noise and have a multivariate normal distribution with mean $\mathbf{0}$ and variance-covariance matrix $\sigma_d^2\mathbb{I}_p$. As commonly done in the 1D case, we thus aim at obtaining an estimator of the curve $\mathbf{f}$, via estimation of its true wavelet coefficients $\mathbf{d}^0_{j,k}$, starting from the empirical ones $\mathbf{d}_{j,k}$. This could be achieved by considering separately the p-entries of the wavelet coefficients, and thus applying to each of the p-corresponding models on wavelets coefficients a 1D soft-thresholding technique, such as the popular one described by Donoho et al. (1995); this would hence lead to a separate estimation of each of the coordinate functions $f_1,\dots,f_p$. However, we notice that if the curve $\mathbf{f}$ has a significant feature at some point of the physical space, we expect that this will be reflected on all p coordinates concurrently. For this reason, in Pigoli and Sangalli (2010) we develop an estimation technique that works jointly on the p coordinate functions, taking into account the vectorial structure of the function to be estimated. Specifically, the proposed estimation technique is such that the same wavelet basis functions are used for estimation of all coordinate functions $f_1,\dots,f_p$ of $\mathbf{f}$; a specific wavelet basis function, with a specific frequency and location, is either used for each of the coordinate functions, in order to capture a feature of the p-dimensional function $\mathbf{f}$, or is not used for any of the coordinate functions, if unnecessary to capture relevant features of $\mathbf{f}$. In particular, extending the rationale behind the 1D soft-thresholding technique to the p-dimensional case, we set a threshold on the square norm of wavelet vector coefficient $||\mathbf{d}_{j,k}||_2^2$ and proceed as follows. If the empirical wavelet coefficients $\mathbf{d}_{j,k}$ have square norm smaller than the threshold, then they are considered as coming only from noise, and the corresponding true coefficients $\mathbf{d}^0_{j,k}$ are thus estimated by the null vector $\mathbf{0}$; otherwise, the estimator of the true coefficients $\mathbf{d}^0_{j,k}$ is obtained from the empirical coefficients $\mathbf{d}_{j,k}$ applying a shrinkage operation, with the aim of removing the part due to noise; the proposed shrinkage takes accurately into account all p-coordinates concurrently.

To fix a threshold on $||\mathbf{d}_{j,k}||_2^2$, we follow and extend the argument used by Donoho et al. (1995). Since $||\boldsymbol{\rho}_{j,k}/\sigma_d||_2^2 \sim \chi^2(p)$, we look for a threshold which contains with high probability n observations from a random variable having the $\chi^2(p)$ distribution. In Pigoli and Sangalli (2010) we prove the following result.

Proposition 40.1. *Let $\{Y_n\}_n$ be a sequence of i.i.d. $\chi^2(p)$ random variables and $A_n = \{\max_{i=1,\dots,n} Y_i \leq c_p \log n\}$, where $c_p = 2$ if $p = 1$ and $c_p = 3$ if $p \geq 2$. Then*

$$\mathbb{P}(A_n) \rightarrow 1 \ \text{ for } \ n \rightarrow +\infty.$$

This proposition leads to the universal threshold proposed by Donoho et al. (1995) in the case $p = 1$. In the case $p \geq 2$, the same proposition support instead to use the following threshold on $||\mathbf{d}_{j,k}||_2^2$ (see Pigoli and Sangalli, 2010):

$$t_p = \hat{\sigma}_d^2(3\log n),$$

where the standard deviation is estimated using the median of the absolute deviation from the median on the wavelet coefficients of level $J-1$, which are assumed to be pure noise (see e.g. Donoho et al., 1995). The proposed soft-thresholding estimator is

$$\hat{\mathbf{d}}_{j,k} = \left(1 - \frac{\sqrt{t_p}}{||\mathbf{d}_{j,k}||_2}\right)_+ \mathbf{d}_{j,k}. \tag{40.4}$$

Geometrically, this soft-thresholding procedure works as follows. Consider a p-dimensional sphere with radius $\sqrt{t_p}$ and centered in the origin; if the p-dimensional vector $\mathbf{d}_{j,k}$ lies completely inside the sphere, then the estimated wavelet coefficient $\hat{\mathbf{d}}_{j,k}$ is set to $\mathbf{0}$; otherwise, $\hat{\mathbf{d}}_{j,k}$ is obtained from $\mathbf{d}_{j,k}$ by removing the part of $\mathbf{d}_{j,k}$ that lies inside the sphere. Figure 40.1, left panel, gives a visual representation of this procedure for $p = 3$. Notice that the shrinkage estimator (40.4) gives an estimate of $\mathbf{d}_{j,k}^0$ that has the same direction of the empirical coefficient $\mathbf{d}_{j,k}$. This is justified by the hypothesis that the variance of the error on the coefficients is the same in all p directions, so that the direction of the vector $\mathbf{d}_{j,k}$ can be consider to be mainly determined by that of the true coefficient $\mathbf{d}_{j,k}^0$.

In Pigoli and Sangalli (2010), by means of simulation studies, we compare the proposed method with another locally adaptive regression technique for multidimensional curves, based on free-knots regression splines. The simulations show that wavelet based methods are particularly attractive when the data are characterized by strongly localized features. In the absence of these characteristics, wavelet method provide estimates that have a level of accuracy comparable to that of free-knots regression splines.

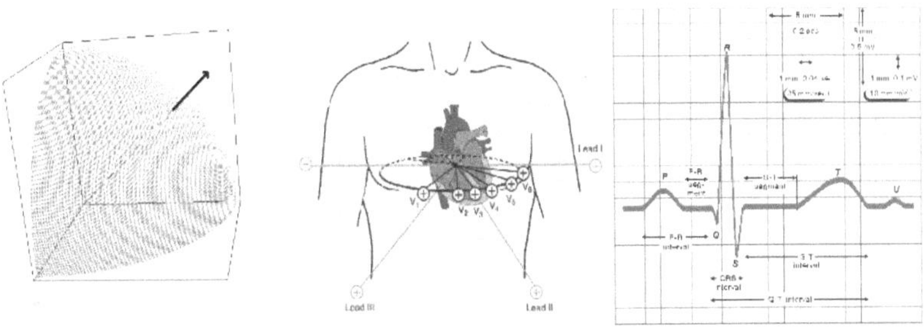

Fig. 40.1: Left: Visual representation in three dimensions of the soft-thresholding procedure: only the part of the vector $\mathbf{d}_{j,k}$ that lies outside the sphere with radius $\sqrt{t_p}$ is retained as significant. Center: Scheme of the directions along which the potential difference is measured for every lead. Right: Template of a physiological ECG record on Lead I.

40.4 Application to ECG data

In this section we briefly illustrate an application of the described multidimensional wavelet fitting technique to the estimation of Electro Cardio Gram (ECG) records. The data come from the 118 Dispatch Center, the medical operating emergency unit, operating in Milano, Italy. These records were collected as part of the PROMETEO (PROgetto Milano Ecg Teletrasmessi ExtraOspedaliero) Project, the aim of which is to anticipate diagnostic time in heart attacks, in order to improve the prognosis of reperfusive treatments and reduce infarction complications. The processing of ECG records as functional data is becoming increasingly important with the advent of statistical techniques that exploit curve shapes in the analysis of these records (see, e.g, Boudaoud et al., 2007). These data have a multidimensional nature, because ECG records provide potential differences, named leads, between multiple electrodes. In particular, ten electrodes are used for a standard "12-leads" ECG. Figure 40.1, central panel, shows the positions of electrodes and leads. Eight of these leads are jointly needed to describe the complex heart dynamics, both on the sagittal plane (Leads I and II) and on the horizontal plane (Leads V1, V2, V3, V4, V5 and V6). When smoothing these data it is thus appropriate to use a technique which takes into account all eight significant leads simultaneously; moreover, this helps in detecting significant features, which reflect on more than one lead. Furthermore, wavelet bases are particularly suited to capturing ECG shapes, because these are characterized by localized strong oscillations. Figure 40.1, right panel, gives a scheme of the typical structure of Lead I. Figure 40.2 shows one of the ECG records stored in the PROMETEO database. The eight figure panels display the raw data of the eight significant ECG leads for a patient affected by ST Elevation Myocardial Infarction. Superimposed are the estimates of these eight-dimensional functional data, obtained by the proposed technique using Daubechies wavelet basis with 10 vanishing moments (see, e.g., Nason, 2008) and the generalized soft-thresholding estimator (40.4). Figure 40.3 shows the estimated first and second derivatives of

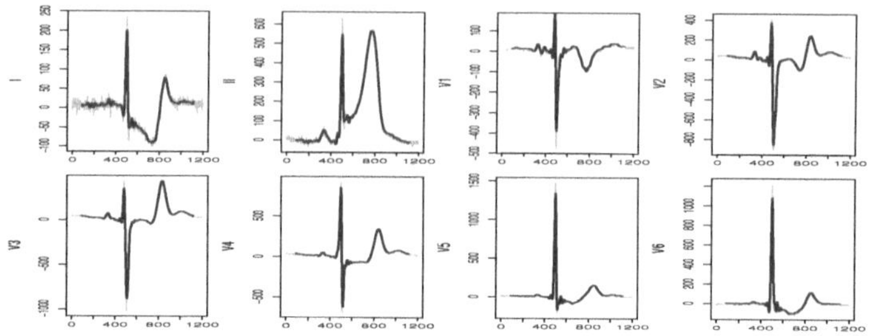

Fig. 40.2: Eight significant leads in a 12-leads ECG for a patient affected by ST Elevation Myocardial Infarction; raw data (grey) and multidimensional wavelet estimate (blue).

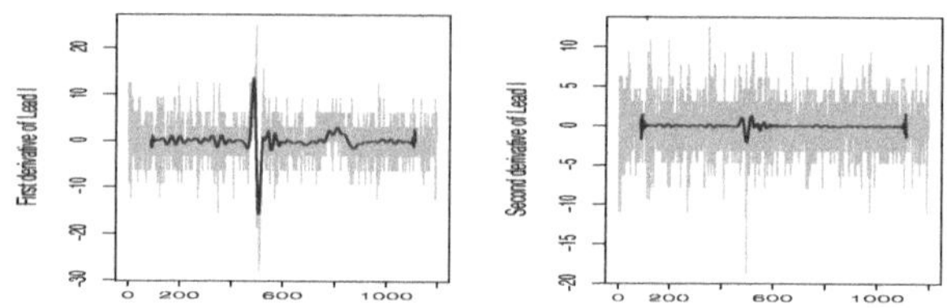

Fig. 40.3: Left: Estimate of first derivative of Lead I (blue), superimposed to first central differences of raw data (grey). Right: Estimate of second derivative of Lead I (blue), superimposed to second central differences of raw data (grey).

Lead I for this patient. The obtained estimates of the eight-lead traces and of their derivatives, for the records in the PROMETEO database, are the starting point of extensive analyses that aim at identifying existing pathologies via ECG shapes, as well as exploring epidemiologic correlations among different cardiovascular diseases. A first promising result in this respect is for instance the identification of patients affected by Bundle Branch Blocks (see Ieva et al., 2010).

Acknowledgements Laura Sangalli acknowledges funding by MIUR Ministero dell'Istruzione dell'Università e della Ricerca, *FIRB* research project "Advanced statistical and numerical methods for the analysis of high dimensional functional data in life sciences and engineering".

References

1. Boudaoud, S., Rix, H., Meste, O., Heneghan, C., O'Brien, C.: Corrected Integral Shape Averaging Applied to Obstructive Sleep Apnea Detection from the Electrocardiogram. EURASIP Journal on Advances in Signal Processing **41**, 909–996 (2007)
2. Donoho, D.L., Johnstone, I.M., Kerkyacharian, G. and Picard, D.: Wavelet Shrinkage: Asymptopia. J. Roy. Stat. Soc. B **57**, 301–369 (1995)

3. Ieva, F., Paganoni, A.M., Pigoli, D., Vitelli, V.: Statistics on ECGs: wavelet smoothing, registration and classification. Tech. Rep. No. 04/2011 MOX, Dipartimento di Matematica, Politecnico di Milano. http://mox.polimi.it/it/progetti/pubblicazioni (2010)
4. Nason, G.P.: Wavelet Methods in Statistics with R. Springer, New York (2008)
5. Pigoli, D., Sangalli, L.M.: Wavelets in Functional Data Analysis: estimation of multidimensional curves and their derivatives. Tech. Rep. No. 09/2011 MOX, Dipartimento di Matematica, Politecnico di Milano. http://mox.polimi.it/it/progetti/pubblicazioni. *Submitted* (2010)
6. Ramsay, J.O., Silverman, B.W.: Functional Data Analysis (Second Edition), Springer, New York (2005)
7. Strang, G.: Wavelets and dilation equations: a brief introduction. SIAM Review **31**, 614–627 (1989)

Chapter 41
Nonparametric Conditional Density Estimation for Functional Data. Econometric Applications

Alejandro Quintela-del-Río, Frédéric Ferraty, Philippe Vieu

Abstract We present some recent results about nonparametric conditional density estimation, when we consider a functional explanatory variable, and some applications of these techniques to econometrics. In a first part, we construct a test to check the parametric form of the conditional density function. In a second part, we estimate two well-known risk measures, the conditional Value-at-Risk and the conditional expected shortfall. Some simulations are shown.

41.1 Introduction

In a functional data setting, the conditioning variable is allowed to take its values in some abstract semi-metric space. In Ferraty and Vieu (2006) kernel nonparametric estimators of the conditional density and the conditional distribution are defined, and they give the rates of convergence (in an almost complete sense) to the corresponding functions. In this book, the practical interest of these kind of techniques are shown in different scientific fields, such as econometrics. Several papers in the last decades use functional data ideas to deal with econometric data; see e.g. Bugni et al. (2009) for some examples or references.

Alejandro Quintela-del-Río
Universidade da Coruña, Spain, e-mail: aquintela@udc.es

Frédéric Ferraty
Institut de Mathématiques de Toulouse, France, e-mail: ferraty@math.univ-toulouse.fr

Philippe Vieu
Institut de Mathématiques de Toulouse, France, e-mail: vieu@math.univ-toulouse.fr

41.2 The conditional density estimator

Let $\{(\chi_i, Y_i), i = 1, \ldots n\}$ a sample of n independent random pairs, each one being distributed like $(\mathscr{X}, Y)$. We consider that $\mathscr{X}$ is a random variable valued in some-metric space (E, d) which is (possibly) infinitely dimensioned. The response variable Y is scalar (i.e. Y takes its values in $\mathbb{R}$). The conditional cumulative distribution of Y given $\mathscr{X}$ is defined for any $y \in \mathbb{R}$ and any $\chi \in E$ by:

$$F(y/\chi) = P[Y \leq y/\mathscr{X} = \chi],$$

while the conditional density, denoted by $f(y/\chi)$, is defined to be the density of this distribution with respect to Lebesgue measure on $\mathbb{R}$. The conditional density $f(y/\chi)$ can be estimated using kernel functions. Such an estimate can be defined as follows:

$$\hat{f}_n(y/\chi) = \frac{\frac{1}{g}\sum_{i=1}^n K\left(\frac{d(\chi,\chi_i)}{h}\right) K_0\left(\frac{y-Y_i}{g}\right)}{\sum_{i=1}^n K\left(\frac{d(\chi,\chi_i)}{h}\right)}, \tag{41.1}$$

where K and K_0 are kernel functions, and where g and h are sequences of smoothing parameters. The conditional distribution $F(\cdot/\chi)$ can be estimated by

$$\hat{F}_n(y/\chi) = \frac{\sum_{i=1}^n K\left(\frac{d(\chi,\chi_i)}{h}\right) H\left(\frac{y-Y_i}{g}\right)}{\sum_{i=1}^n K\left(\frac{d(\chi,\chi_i)}{h}\right)}, \tag{41.2}$$

with the function $H(\cdot)$ defined by $H(x) = \int_{-\infty}^{x} K_0(u)du$.

41.3 Testing a parametric form for the conditional density

Any information on the conditional density function is always of great practical interest. We propose a statistic test for the null hypothesis

$$H_0 : \exists\theta_0, f(y/\chi) = f_{\theta_0}(y/\chi),$$

where $\{f_\theta(./\chi),\ \theta \in \Theta\}$ is a specific parametric family of densities. We propose a local alternative of the form

$$H_1 : \inf_{\theta\in\Theta} |f(y/\chi) - f_\theta(y/\chi)| = \varepsilon_n s(y),$$

The test statistic proposed has the form

$$I_n = \int \left[\frac{1}{n\varphi_\chi(h)}\sum_{i=1}^n K\left(\frac{d(\chi,\mathscr{X}_i)}{h}\right)\left(\frac{1}{g}K_0\left(\frac{y-Y_i}{g}\right) - f_{\hat{\theta}_0}(y/\chi)\right)\right]^2 w(y)dy,$$

and our testing procedure is based on the following normalized version of this statistic:

$$J_n = \frac{\sqrt{\frac{n\varphi_\chi(h)}{g^4}}\left(I_n - g^4 B\right)}{\sqrt{V}}, \tag{41.3}$$

with B and V depending on K_0, K and the derivatives of the function f. Under certain general assumptions, we can prove the two following theorems.

Theorem 41.1. *Under H_0 we have: $J_n \longrightarrow N(0,1)$, as $n \to \infty$.*

Theorem 41.2. *Under H_1 we have: $|J_n| \longrightarrow \infty$, in probability.*

41.4 Value-at-risk and expected shortfall estimation

A major concern for regulators and owners of financial institutions is the risk analysis. The Value-at-Risk (VaR) (Embrechts et al., 1997) is one of the most common risk measures used in finances. It measures down-side risk and is determined for a given probability level α. In a typical situation, measuring losses, the VaR is the lowest value which exceeds this level (that is, the quantile of the loss distributions). The expected shortfall (ES) (Acerbi, 2002) is the average of the 100(1-α)% worst losses. It takes into account all possible losses that exceed the severity level corresponding to the VaR. Since the first Basel Accord (1996), the VaR (and recently the ES) forms the essential basis of the determination of market risk capital.

In this work, we consider the question of how to estimate the VaR and the ES when auxiliary information about returns Y_t is available, through a set of predictor variables. The conditional information can contain economic and market (exogenous) variables and past observed returns. In a general way, if Y_t is the return of a portfolio at time t, the VaR (conditional) is defined, for a fixed level α, as the value v_α such that $P(Y_t < v_\alpha / F_{t-1}) = \alpha$, with $F_{t-1}$the information available at time $t-1$. The (conditional) ES is defined as $\mu_\alpha = E[Y_t / Y_t < v_\alpha, F_{t-1}]$. Most studies estimate VaR through quantile estimation (Gaglianone et al., 2009). When we summarize the conditional information through a functional data χ_0, the VaR estimate can be computed as

$$\hat{v}_\alpha(\chi_0) = \hat{F}_n^{-1}(\alpha/\chi_0), \tag{41.4}$$

that is, the conditional quantile (41.2), and we can also compute the CES by

$$\hat{\mu}_\alpha(\chi_0) = \alpha^{-1} \int_{v_\alpha(\chi_0)}^{+\infty} y \hat{f}_n(y/\chi_0) dy. \tag{41.5}$$

Classical nonparametric kernel estimates has been used in VaR and CES estimation (Scaillet (2004), Cai and Wang (2008)). But the functional data analysis allows us to treat the conditional information in a more successful form by taking into account the functional feature of the data.

Theorem 41.3. *Under general conditions, we have that*

$$\hat{v}_\alpha(\chi_0) \longrightarrow v_\alpha(\chi_0) \;\; and \;\; \hat{\mu}_\alpha(\chi_0) \longrightarrow \mu_\alpha(\chi_0) \;\; a.co.$$

41.5 Simulations

Let us consider an autoregressive model and let Z_t be the process such that

$$Z_t = 0.9Z_{t-1} + a_t, \;\; a_t \sim \mathcal{N}(0, 0.1), \tag{41.6}$$

and next consider the average

$$Z_t^* = \frac{Z_t + Z_{t+1} + Z_{t+19}}{20}, \;\; t = 1, ..., N. \tag{41.7}$$

We take a total sample size of $N = 100 \times n$ consecutive times. These data can be splitted into n paths of size 100, leading to a set of n trajectories $\chi_i = \{\chi_i(t) = Z^*(t + 10 \times (i-1)),\, t = 1, \ldots, 100\}$, $i = 1, \ldots, n$. These n paths χ_i are the functional explanatory variables in our experiment. The real valued responses Y are generated by

$$Y_i = r(\mathscr{X}_i) + \varepsilon_i, \;\; i = 1, \ldots n, \quad r(\mathscr{X}) = \int \mathscr{X}(t)\cos(t)dt. \tag{41.8}$$

The various parameters of the methods were chosen in the following ways:
- The kernel function K was taken to be the uniform density on $(0,1)$ and K_0 was taken to be the usual Epanechnikov kernel.
- The semi-metric d was taken to be the usual L_2 distances between curves.
- The bandwidth parameters were selected in the intervals $[d_1, d_2]$, being $d_1 = \min d_{ij}$ and $d_2 = \max d_{ij}$ ($d_{ij} = \{d(\chi_i, \chi_j)\}_{i,j=1}^n$) for h and g in $[g_1, g_2]$ with $g_1 = (max(Y_i) - min(Y_i))/20$ and $g_1 = (max(Y_i) - min(Y_i))/2$.

41.5.1 Results for the hypothesis testing.

The null hypothesis consists in simulating the innovation errors as $H_0 : \varepsilon \sim \mathcal{N}(0, \sigma_0)$. Under H_0 this conditional distribution is just $\mathcal{N}(\mu_\chi, \sigma_0)$. The alternative H_1 is constructed by simulating the errors as a mixture of $\mathcal{N}(-d, \sigma_1)$ and $\mathcal{N}(d, \sigma_1)$ distributions.

We tried various null hypotheses by changing the value of σ_0. Precisely, denoting the signal-to-noise ratio by snr we used values of σ_0 of the form $\sigma_0 = c \times 0.1 \times snr$, and we express the results as a function of c. Tables 41.1 and 41.2 show that, even for moderate sample sizes ($n = 100$) and for high signal-to-noise ratio, the testing procedure is good since the percentage of acceptation is always close to the true theoretical level.

Now, we check the results under the alternative hypothesis H_1. We look at how the power of the test procedure changes. We choose various null hypotheses H_1,

c	1	2	3	4	5
% acceptance of H_0	0.91	0.90	0.95	0.98	1

Table 41.1: Results under H_0 when σ_0 varies ($n = 100$).

Sample size	% acceptance of H_0	% acceptance of H_1
100	0.91	0.13
200	0.93	0.03
500	0.98	0

Table 41.2: Results under H_0 when the sample size n varies ($c = 1$).

constructed from different values of d and σ_1. As before, the values of σ_1 are chosen of the form $\sigma_1 = c \times 0.1 \times snr$. Tables 41.3 and 41.4 show the evolution of the percentage of rejection as a function of d and σ_1.

d	0.1	0.25	0.3	0.5
% Rejection of H_0	1	0.96	0.94	0.38

Table 41.3: Results under H_1 when d varies ($n = 100$ and $c = 1$).

c	1	2	4	5	10	20
% Rejection of H_0	1.0	0.93	0.91	0.86	0.91	0.86

Table 41.4: Results under H_1 when σ_1 varies ($n = 100$ and $d = 0.1$).

For moderate sample size and high signal-to-noise ratio, the power of the test is rather high. The only counterexample is when the distance between both modes of the density is high (see Table 3 and large value of d) because the nonparametric estimate tends to oversmooth the distribution. Except for this situation, which is rather unusual in practice (and easy to detect empirically), the procedure has good power.

41.5.2 Results for the CVaR and CES estimates

For χ_0 fixed, the conditional density $f(y/\chi_0)$ is a Gaussian density with mean $r(\chi_0)$ and standard deviation σ_0. Several econometric models for volatility dynamics assume the conditional normality, as for instance J.P. Morgan's Riskmetrics (Riskmetrics, 1995). Thus, the CVaR can be exactly computed as

$$v_\alpha(\chi_0) = F^{-1}(\alpha/\chi_0), \quad F \sim \mathcal{N},$$

that is, the conditional quantile of a normal distribution, and we can also compute exactly the conditional expected shortfall by calculating (or approximating) the integral

$$\mu_\alpha(\chi_0) = \alpha^{-1} \int_{\nu_\alpha(\chi_0)}^{+\infty} y \frac{1}{\sigma_0\sqrt{2\pi}} e^{(y-r(\chi_0))^2/2\sigma_0^2} dy.$$

Now, each time we generate the sample, $\{(\mathscr{X}_i, Y_i), i = 1, \ldots n\}$, we also generate a grid of new functional points $\chi_0^1, \ldots, \chi_0^m$, and we can calculate the performance of the estimators in terms of the mean absolute deviation error, defined as $m^{-1}\sum_{k=1}^{m} \left|\nu_\alpha(\chi_0^k) - \hat{\nu}_\alpha(\chi_0^k)\right|$ (the same definition for the conditional expected shortfall). We checked several values for n from 100 to 250, for $m = 10$ to 50, and replicating the experiment $r = 500$ times (the level α was 0.05 and 0.01). We observed that the errors of the estimates decreased when sample size increased, making estimations become stable as the sample size increases. We show a specific example, when the sample size was $n = 250$ and for a value of the level $\alpha = 0.05$. The grid consisted of $m = 50$ new points χ_0^k. Table 41.5 shows the results.

	Min.	1st Quartile	Median	Mean	3rd Quartile	Max.
CVaR	0.02066	0.02961	0.03385	0.03412	0.03776	0.05277
CES	0.05809	0.09274	0.10360	0.10420	0.11530	0.16570

Table 41.5: Summary statistics for CVar and CES estimates.

References

1. Acerbi, C.: Spectral measures of risk: a coherent representation of subjective risk aversion. J. Bank. Financ. **26**, 1505–1518 (2002)
2. Basel Committee on Banking Supervision: Amendment to the capital accord to incorporate market risks. Bank for International Settlements (1996)
3. Bugni, F.A., Hall, P., Horowitz, J.L., Neumann, G.R.: Goodness-of-fit tests for functional data. Econometrics J. **12**, 1–18 (2009)
4. Cai, Z., Wang, X.: Nonparametric estimation of conditional VaR and expected shortfall. J. Econometrics **147**, 120-130 (2008)
5. Embrechts, P., Kluppelberg, C., Mikosch, T.: Modeling extremal events for finance and insurance. Springer Verlag, New York (1997)
6. Ferraty, F. and Vieu, P.: Nonparametric functional data analysis. Series in Statistics, Springer, New York (2006)
7. Ferraty, F., Quintela del Río, A., Vieu, P.: Specification test for conditional distribution with functional variables. Economet. Theor., in press.
8. Gaglianone, W.P., Lima, L.R., Linton, O., Smith, D.: Evaluating value-at-risk models via quantile regression. Working paper 09-46, Economic Series (25), Universidad Carlos III de Madrid, Spain (2009)
9. Riskmetrics: RiskMetrics Technical Document, J.P. Morgan (1995)
10. Scaillet, O. (2004), Nonparametric estimation and sensitivity analysis of expected shortfall. Math. Financ. **14**, 115–129 (2004)

Chapter 42
Spatial Functional Data Analysis

James O. Ramsay, Tim Ramsay, Laura M. Sangalli

Abstract We describe a spatial spline regression model, that efficiently deals with data distributed over irregularly shaped regions featuring complex boundaries. The model also accounts for covariate information. Efficient spline bivariate smoothing is achieved by resorting to the finite element method.

42.1 Introduction

Our work aims at extending techniques typical of functional data analysis (FDA), such as penalized smoothing, to the analysis of spatial and spatio-temporal arguments. There exists in fact a natural link between FDA methodology and the modeling of spatial (spatio-temporal) phenomena that still remains largely unexplored.

In Ramsay et al. (2011) we describe a spatial spline regression model for data distributed over complex bidimensional domains. One of the main limitations of classical methods for spatial data analysis is in fact that they cannot efficiently deal with data distributed over irregularly shaped regions, featuring complex boundaries, concavities and interior holes. Consider for instance the data over the C-shaped domain displayed in the top right panel of Figure 1; the lower values of the response variable in the lower arm of the C are faced by higher values in the higher arm (size of the point marker proportional to data value). Most spatial methods would in this case smooth across the boundary between the two arms of the C, overestimating the response in the lower arm and underestimating it in the higher.

James O. Ramsay
McGill University, Montreal, Canada, e-mail: ramsay@psych.mcgill.ca

Tim Ramsay
Ottawa Health Research Institute, Canada, e-mail: tramsay@ohri.ca

Laura M. Sangalli
Politecnico di Milano, Italy, e-mail: laura.sangalli@polimi.it

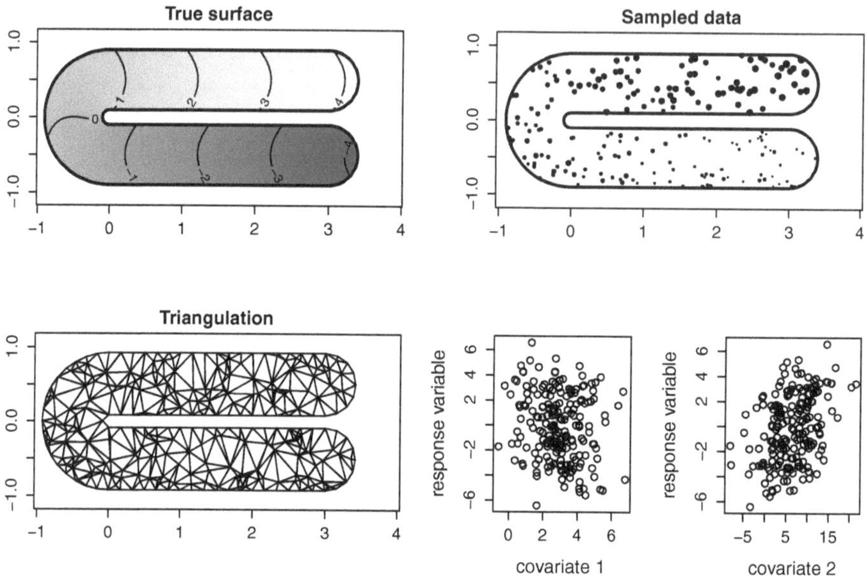

Fig. 42.1: Top left: true surface function. Top right: sampled data (size of the point marker proportional to sampled data value), replicate 1. Bottom left: domain triangulation, replicate 1. Bottom center and right: scatter plots of response variable vs covariates, replicate 1.

The described spatial spline regression model is able to accurately handle data distributed over complex domains, and also accounts for covariate information. The model incorporates and ameliorates the penalized bivariate spline smoother introduced by Ramsay (2002); in this smoother, the roughness penalty consists in a Laplace operator that is integrated only over the region of interest thanks to a *finite element* formulation.

42.2 Data, model and estimation problem

Consider a set of n points $\{\mathbf{p}_i = (x_i, y_i); i = 1\ldots, n\}$, on a polygonal domain $\Omega \subset \mathbb{R}^2$. The domain Ω can be quite complex; for instance, it can be non-convex and have holes and islands. Let z_i be the value of a real valued variable of interest, observed at point $\mathbf{p}_i$, and let $\mathbf{w}_i = (w_{i1}, \ldots, w_{iq})^t$ be a q-vector of covariates associated to observation z_i at $\mathbf{p}_i$. Summarizing, our data are

$$\{(\mathbf{p}_1, z_1, \mathbf{w}_1), \ldots, (\mathbf{p}_n, z_n, \mathbf{w}_n)\} \subset \Omega \times \mathbb{R} \times \mathbb{R}^q.$$

Assume the model

$$z_i = \mathbf{w}_i^t \boldsymbol{\beta} + f(\mathbf{p}_i) + \varepsilon_i \qquad i = 1, \ldots, n$$

where ε_i, $i = 1,\ldots,n$, are i.i.d. errors with mean 0 and variance σ_ε^2, and where $\boldsymbol{\beta} \in \mathbb{R}^q$ is a vector of coefficients and f is a twice differentiable real-valued function on Ω. We are thus considering an additive model, with a parametric part that is a regression on the covariates, and a non-parametric part that is defined as a surface over the region of interest Ω, and deals with the spatial structure of the phenomenon.

Building on the work of Ramsay (2002), in Ramsay et al. (2011) we propose to estimate $\boldsymbol{\beta}$ and f by minimizing the following penalized sum-of-square-error functional

$$J_\lambda(\boldsymbol{\beta}, f) = \sum_{i=1}^{n} \left(z_i - \mathbf{w}_i^t \boldsymbol{\beta} - f(\mathbf{p}_i)\right)^2 + \lambda \int_\Omega (\Delta f)^2 d\Omega \tag{42.1}$$

where Δf is the Laplacian of f, i.e.,

$$\Delta f = \frac{\partial^2 f}{\partial x^2} + \frac{\partial^2 f}{\partial y^2}.$$

Recall that Δf is a measure of the local curvature of f. Moreover, the Laplace operator is invariant to rotations and translations of the coordinate system. This property, which is thus inherited by the estimator, is of paramount importance, being the coordinate system often arbitrary. Notice that the penalty in (1), $\int_\Omega \{(f_{xx})^2 + 2 f_{xx} f_{yy} + (f_{yy})^2\} dx dy$, is similar to the one used by thin-plate splines, i.e., $\int_{\mathbb{R}^2} \{(f_{xx})^2 + 2(f_{xy})^2 + (f_{yy})^2\} dx dy$, the latter being though integrated over the entire plane $\mathbb{R}^2$ (here, $f_{xy} := \partial^2 f / \partial x \partial y$).

Let $H^m(\Omega)$ be the set of all continuous functions on Ω whose mth-order partial derivatives are all square integrable and whose partial derivatives of order less than m are all continuous; moreover, let $H^m_{n0}(\Omega)$ be the subset of $H^m(\Omega)$ consisting of those functions whose normal derivatives are 0 on the boundary of Ω. The functional (1) is guaranteed to have a unique minimizer over $\boldsymbol{\beta} \in \mathbb{R}^q$ and $f \in H^2_{n0}(\Omega)$ (see Ramsay et al., 2011).

Denote by W the $n \times q$ matrix whose ith row is given by $\mathbf{w}_i^t$ and assume that W has full rank. Moreover, denote by H the orthogonal projection matrix $H = W(W^t W)^{-1} W^t$. Furthermore, set $\mathbf{z} := (z_1, \ldots, z_n)^t$ and, for a given function g on Ω denote by $\mathbf{g}_n$ the vector of evaluations of g at the n data locations, $\mathbf{g}_n := (g(\boldsymbol{p}_1), \ldots, g(\boldsymbol{p}_n))^t$. In Ramsay et al. (2011) we show that the estimators $\hat{\boldsymbol{\beta}}$ and $\hat{f}$ that jointly minimize (1), over $\boldsymbol{\beta} \in \mathbb{R}^q$ and $f \in H^2_{n0}(\Omega)$, are given by

- ▷ $\hat{\boldsymbol{\beta}} = (W^t W)^{-1} W^t (\mathbf{z} - \hat{\mathbf{f}}_n)$
- ▷ $\hat{f}$ is solution of the variational problem: find $(f, g) \in H^2_{n0}(\Omega) \times H^2(\Omega)$ that satisfy

$$\begin{cases} \mathbf{u}_n^t (I - H) \mathbf{f}_n - \lambda \int_\Omega (\nabla u \cdot \nabla g) = \mathbf{u}_n^t (I - H) \mathbf{z} \\ \int_\Omega v g - \int_\Omega (\nabla v \cdot \nabla f) = 0 \end{cases} \tag{42.2}$$

for all $(u, v) \in H^2_{n0}(\Omega) \times H^2(\Omega)$, where ∇ is the gradient operator.

42.3 Finite element solution of the estimation problem

In this section we briefly describe how an approximate solution to the estimation problem can be obtain using the finite element method.

Finite elements analysis has been mainly developed and used in engineering applications (for an introduction to finite element analysis see, e.g., Braess, 2007). Its strategy is very similar in spirit to that of univariate splines. In fact, it consists of partitioning the problem domain in small disjoint sub-domains and then constructing a separate polynomial function on each of these sub-domains, in such a way that the union of these pieces closely approximates the solution. This simplified problem is made computationally tractable by a clever choice of the basis functions for the space of piecewise polynomials on the domain partition. Each piece of the partition, equipped with the basis functions defined over it, is named *finite element*.

In our context, a convenient partition of the domain Ω is given by a triangulation $\mathcal{T}$ where each data point $\mathbf{p}_i$ is a triangle vertex. Figure 1, bottom left panel, shows the triangulation of the C-domain corresponding the the data sample shown in the top right panel of the same figure. We thus consider a base of piecewise polynomials over the domain triangulation. In particular, each triangle vertex is associated a simple basis function, defined in such a way that the set of all basis spans the space $H^1_{\mathcal{T}}(\Omega)$ of all continuous functions on Ω which are linear when restricted to any triangle of $\mathcal{T}$. Ramsay et al. (2011) review the construction of this finite element space.

Now, notice that the variational problem (2) involve no second-order partial derivatives and is therefore well defined in $H^1(\Omega)$; moreover, $H^1_{\mathcal{T}}(\Omega) \subset H^1(\Omega)$. An approximated solution of the variational problem can thus be obtained looking for $(f,g) \in H^1_{\mathcal{T}}(\Omega) \times H^1_{\mathcal{T}}(\Omega)$ that satisfy (2) for all $(u,v) \in H^1_{\mathcal{T}}(\Omega) \times H^1_{\mathcal{T}}(\Omega)$. Thanks to the clever choice for domain partition and function basis, this reduces to solving a linear system. The approximate solution $\hat{f}$ thus obtained is named Finite Element L-spline (FELspline). The corresponding estimator $\hat{\mathbf{f}}_n$, and thus also $\hat{\boldsymbol{\beta}}$, turn out to be linear functions of the data values $\mathbf{z}$, so that their properties can be easily derived. An estimator of the error variance σ^2_ε is also easily obtained, and the selection of the smoothing parameter λ is straightforward to accomplish via generalized cross-validation. See Ramsay et al. (2011).

42.4 Simulations

In this Section we briefly illustrate the advantages of the proposed spatial spline regression model with respect to a more classical model based instead on thin-plate splines. In Ramsay et al. (2011) we describe a larger comparison study that also considers filtered kriging and the soap film smoother introduced by Wood et al. (2008). We shall here consider the same surface test function used by Wood et al. (2008), but also include covariates. The surface test function f, defined on a C-shaped domain, is shown in Figure 1, top left. For $N = 50$ replicates, we simulate data as follows. We sample n locations, $\mathbf{p}_1, \dots, \mathbf{p}_N$, uniformly on this domain. Independently for each $\mathbf{p}_i$, we sample two independent covariates w_{i1} and w_{i2}; these are generated respectively from a $N(\mu_1, \sigma_1^2)$ and a $N(\mu_2, \sigma_2^2)$ distribution. The values of the response variable, $z_1, \dots, z_n$, at the sampled data points and with the sampled covariates, are thus obtained as follows:

$$z_i = \beta_1 w_{i1} + \beta_2 w_{i2} + f(\mathbf{p}_i) + \varepsilon_i \qquad i = 1, \dots, n$$

where ε_i, $i = 1, \dots, n$, are independent errors with $N(0, \sigma_\varepsilon^2)$ distribution.

The parameter values used in the simulation are: $n = 200$, $\beta_1 = -0.5$, $\beta_2 = 0.2$, $\sigma_\varepsilon = 0.5$, $\mu_1 = 3$, $\sigma_1 = 1.5$, $\mu_2 = 7$, $\sigma_2 = 5$. Figure 1, top right, shows the data sampled in the first replicate, with the size of the point marker proportional to sampled data value. The center and right bottom panels of the same figure display the corresponding scatter plots of the response variable versus the two covariates.

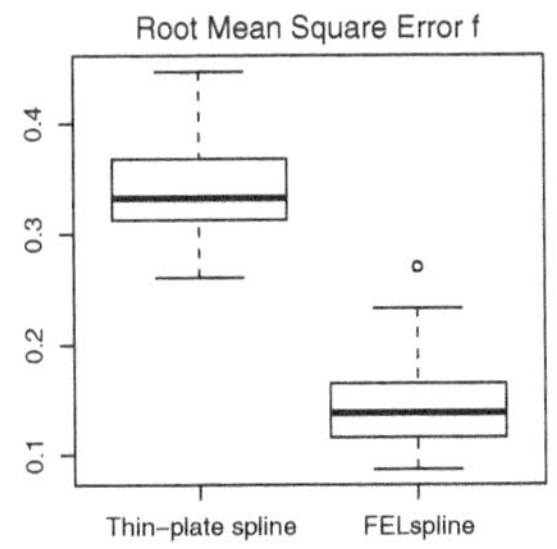

Fig. 42.2: Boxplots of RMSE for the estimators of f.

The values of the smoothing parameters for the two methods are selected, for each replicate, by generalized cross-validation. The Root Mean Square Errors (RMSE) of the estimators of the parameters β_1, β_2 and σ_ε are smaller for the model based on FELsplines than on thin-plate splines (0.023 vs 0.025 for β_1; 0.008 vs 0.009 for β_2; 0.029 vs 0.061 for σ_ε). Figure 2 compares the boxplots of the RMSE for $\hat{f}$ evaluated on a fine grid of points over the C-domain, showing that the RMSE associated to FELspline model is stochastically lower. With the proposed model, approximate confidence intervals for the coefficients $\boldsymbol{\beta}$ and approximate confidence volumes for the surface f can also be obtained (see Ramsay et al., 2011).

Figure 3, top panels, compares the estimated surfaces, for the first replicate. The bottom panels of the same figure display the absolute residuals of the estimated surfaces with respect to the true surface f. Note the high absolute errors near the inner borders of the C arms, for the surface estimate provided by the thin-plate spline model: this is due to the fact that this method smooth across the boundaries. The proposed spatial spline regression model instead efficiently deals with this complex domain.

42.5 Discussion

We briefly illustrated as penalized smoothing, a technique that is typical of FDA, can be profitably extended to deal with spatial phenomena. Simulation studies detailed in Ramsay et al. (2011) show that the described model outperforms thin-plate

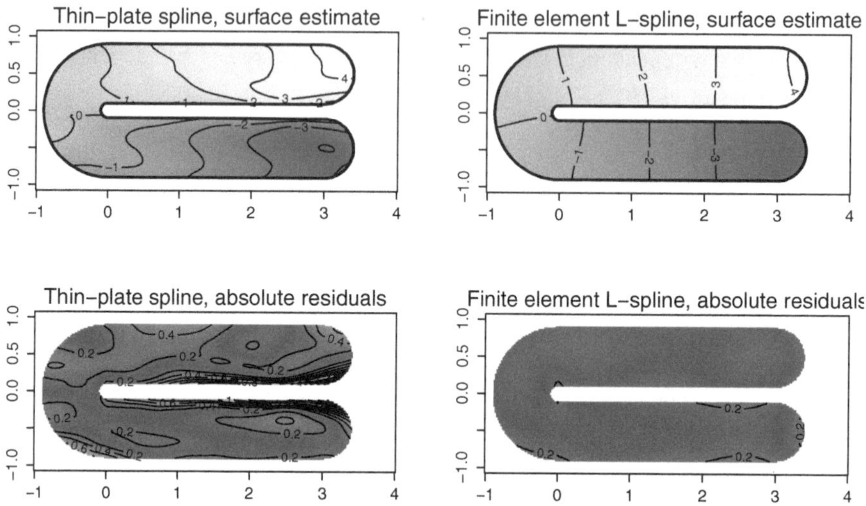

Fig. 42.3: Top: surface estimates provided by thin-plate spline model (left) and FELspline model (right), replicate 1. Bottom: corresponding absolute residuals.

splines, filtered kriging and other state-of-the-art methods for spatially distributed data.

The proposed model can comply with different boundary conditions, such as fixed surface values on the domain boundary or along part of it (see Ramsay et al., 2011). This possibility is per se of great interest, also in more standard surface estimation problems, not involving covariates. Moreover, the described technique can be generalized in several directions, for instance

- to the case of repeated observations at each data location,
- to general link functions, such as the logit,
- to loss-functions other than the classical sum-of-squared-error,
- to roughness penalties based on more complex differential operators,

thus allowing for very large potential application. The covariates themselves, when having a spatial structure, can be modeled as surfaces. Moreover, this technique can be extended to three-variate functional data, i.e. volumes; this would have a great impact on forefront applications such as brain-imaging, where methods able to efficiently comply with the organ external and internal boundaries would be highly desirable. Furthermore, this approach also constitutes a very promising line of research for the modeling of spatio-temporal phenomena, including dynamical functional data, such us curves and surfaces deforming over time.

Acknowledgements L.Sangalli acknowledges funding by Ministero dell'Istruzione dell'Università e della Ricerca, *FIRB* research project "Advanced statistical and numerical methods for the analysis of high dimensional functional data in life sciences and engineering".

References

1. Braess, D.: Finite elements (Third Edition). Cambridge University Press, Cambridge (2007)
2. Ramsay, J. O., Ramsay, T., Sangalli, L. M.: Spatial spline regression models. Tech. Rep. MOX, Dipartimento di Matematica, Politecnico di Milano (2011)
3. Ramsay, T. Spline Smoothing over Difficult Regions. J. Roy. Stat. Soc. B **64**, 307–319 (2002)
4. Wood, S. N., Bravington, M. V., Hedley, S. L.: Soap film smoothing. J. Roy. Stat. Soc. B **70**, 931–955 (2008)

Chapter 43
Clustering Spatially Correlated Functional Data

Elvira Romano, Ramon Giraldo, Jorge Mateu

Abstract In this paper we discuss and compare two clustering strategies: a hierarchical clustering and a dynamic clustering method for spatially correlated functional data. Both the approaches aim to obtain clusters which are internally homogeneous in terms of their spatial correlation structure. With this scope they incorporate the spatial information into the clustering process by considering, in a different manner, a measure of spatial association ables to emphasize the average spatial dependence among curves: the trace-variogram function.

43.1 Introduction

The study of techniques for spatial functional data has recently attracted the interest of the functional data analysis research community due to the fact that many real applications manage data observed in the space that change continuously with respect to time. It is the case when samples of functions are observed in different sites of a region (the so called spatially correlated functional data) or when these functions are observed over a discrete set of time points (temporally correlated functional data). Recent contributions in this framework deal with different topics (see Delicado et al., 2010).

In this paper we focus on clustering. The purpose of clustering methods in the spatial functional framework is to find subgroups of spatial homogeneous curves. To the best of our knowledge, very few clustering methods incorporating spatial

Elvira Romano
Seconda Universitá degli Studi di Napoli, Italy, e-mail: elvira.romano@unina2.it

Ramon Giraldo
Universidad Nacional de Colombia, Bogota, Colombia, e-mail: rgiraldoh@unal.edu.co

Jorge Mateu
Universitat Politècnica de Catalunya, Barcelona, Spain, e-mail: mateu@mat.uji.es

dependence information between curves exist (see Giraldo et al., 2009, Romano et al., 2010a, 2010b, Jiang and Serban, 2010).

We describe and compare two of these clustering strategies that are based on a common variability structure: the trace-variogram function. Both the strategies allow to find spatially homogeneous groups of sites when the observations at each sampling location consist of samples of random functions by considering the spatial association among functions. However the approaches are different.

The first (Giraldo et al., 2009) is a hierarchical approach based on a spatial weighted dissimilarity measure between curves. The weight is the trace-variogram function or the multivariable variogram calculated on the coefficients of the basis function.

The second (Romano et al., 2010b) is a Dynamic approach which looks for the best partition optimizing a criterion of spatial association among functional data, moreover it is such that a summary of the variability structure of each cluster is discovered.

The following sections describe briefly the spatial functional data structure, the clustering procedures and their main characteristics.

43.2 Spatially correlated functional data

Let $\left\{\chi_s(t) : s \in D \subset \mathbf{R^d}, t \in T \subset R\right\}$ be a stationary isotropic functional random process. We assume to observe a realization of this random process observed at n locations, $\chi_{s_1}(t), \ldots, \chi_{s_i}(t), \ldots, \chi_{s_n}(t)$ for $s_i \in D$.

The observed data for a fixed site s_i, follows the model:

$$\chi_{\mathbf{s_i}}(t) = \mu_{s_i}(t) + \varepsilon_{s_i}(t), \quad i = 1, \ldots, n \tag{43.1}$$

where $\varepsilon_{s_i}(t)$ are residuals with independent zero mean and $\mu_{s_i}(\cdot)$ is the mean function which summarizes the main structure of χ_{s_i}.

The assumption of isotropy and stationarity (that is, the mean and variance functions are constant and the covariance depends only on the distance between sampling sites) imply that exist a function $\gamma(h,t)$ called semivariogram of $\chi_{\mathbf{s}}(\mathbf{t})$ such that:

$$\gamma(h,t) = \gamma_{s_i s_j}(t) = \frac{1}{2}\mathbb{V}(\chi_{s_i}(t) - \chi_{s_j}(t)) = \frac{1}{2}\mathbb{E}\left[\chi_{s_i}(t) - \chi_{s_j}(t)\right]^2 \tag{43.2}$$

where $h = \left\|s_i - s_j\right\|$ and all $s_i, s_j \in D$. which by using Fubini's theorem, becomes $\gamma(h) = \int_T \gamma_{s_i s_j}(t)dt$ for $\left\|s_i - s_j\right\| = h$.

This function, called trace-variogram function can be estimated by the classical methods of the moments by means of (Giraldo et al. 2010):

$$\hat{\gamma}(h) = \frac{1}{|2N(h)|} \sum_{i,j \in N(h)} \int_T \left(\chi_{s_i}(t) - \chi_{s_j}(t)\right)^2 dt \tag{43.3}$$

where $N(h) = \{(s_i;s_j) : \|s_i - s_j\| = h\}$ for regular spaced data and $|N(h)|$ is the number of distinct elements in $N(h)$. When data are irregularly spaced the $N(h)$ becomes $N(h) = \{(s_i;s_j) : \|s_i - s_j\| \in (h-\varepsilon, h+\varepsilon)\}$ with $\varepsilon \geq 0$ being a small value.

We consider that the functions are expanded in terms of some basis functions by:

$$\chi_{s_i}(t) = \sum_{l=1}^{Z} a_{il} B_l(t) = \mathbf{a}_i{}^T \mathbf{B}(t),\ i = 1,\ldots,n \tag{43.4}$$

The coefficients of the curves can be consequently organized in a matrix as follows:

$$A = \begin{pmatrix} a_{1,1} & a_{1,2} & \cdots & a_{1,Z} \\ a_{2,1} & a_{2,2} & \cdots & a_{2,Z} \\ \vdots & \ddots & \cdots & \cdots \\ a_{n,1} & a_{n,2} & \cdots & a_{2,Z} \end{pmatrix}_{n\times Z}$$

thus, the empirical trace-variogram function can be expressed by

$$\gamma(h) = \frac{1}{2|N(h)|} \sum_{i,j\in N(h)} \left[(\boldsymbol{a}_i - \boldsymbol{a}_j)^T \mathbf{W} (\boldsymbol{a}_i - \boldsymbol{a}_j) \right] \forall i,j \mid \|s_i - s_j\| = h$$

where $\boldsymbol{a}_i, \boldsymbol{a}_j$ are vectors of the basis coefficients for the χ_{s_i} and χ_{s_j} curves, and $W = \int_T \boldsymbol{B}(t)\boldsymbol{B}(t)^T dt$ is the Gram matrix that is the identity matrix for any orthonormal basis while for other basis as B-Spline basis function, W is computed by numerical integration.

43.3 Hierarchical clustering of spatially correlated functional data

Hierarchical clustering is one of the most well known approaches for clustering. There are two main types of hierarchical clustering methods: agglomerative and divisive. The agglomerative techniques proceed by series of fusions of the n objects into groups, while the divisive methods separate the n objects successively into finer groupings. The results may be represented by a two dimensional diagram known as dendrogram which illustrates the fusions or divisions made at each successive stage of analysis. In this framework a central role is played by the measure of similarity/dissimilarity among the object.

The hierarchical method we refer (Giraldo et al., 2009) for spatial functional data is a natural extension to the functional framework of the approaches proposed for geostatistical data, where the the L_2 norm between curves χ_{s_i}, χ_{s_j} is replaced by a weighted norm among the georeferenced functions. Especially two alternatives are proposed, respectively for univariate and multivariate context. The first one is based on a weighted dissimilarity measure among the georeferenced curves expressed by

$$d_w(\chi_{s_i},\chi_{s_j}) = \sum_{j=1}^{n} d(\chi_{s_i},\chi_{s_j})\gamma_{ij}(h) \tag{43.5}$$

where $d\left(\chi_{s_i}(t),\chi_{s_j}(t)\right) = \sqrt{\int_T (\chi_{s_i}(t)-\chi_{s_j}(t))^2 dt}$ is the distance between the curves without considering the spatial component, and $\gamma_{ij}(h)$ corresponds to the trace-variogram function calculated for the distance between sites s_i and s_j. The main characteristic of this approach is that several distances among curves can be considered. Once the trace-variogram function is estimated for a sequence of values of h, a parametric model is fitted by any of the classical and widely used models. The parametric trace-variogram is always valid because its properties are those of a parametric variogram fitted from a univariate gestotastical data set.

The second for the multivariate spatial process consists in estimating variogram and cross-variograms of coefficients basis functions used for smoothing the observed data and applying a hierarchical approach to a dissimilarities measure constructed by using the variogram of the first principal component or a sum of variograms of the first principal components.

43.4 Dynamic clustering of spatially correlated functional data

Dynamic clustering algorithm (DCA) or *Nueés Dynamiques* (Diday, 1971) is an unsupervised batch training algorithm. Like in the classical clustering techniques the aim is to find groups that are internally dense and sparsely connected with the others.

Let E be a set of n objects, the Dynamic Clustering Algorithm finds a partition $P^* = (C_1,\ldots,C_k,\ldots,C_K)$ of E in K non empty clusters and a set of representative prototypes $L^* = (G_1,\ldots,G_k,\ldots,G_K)$ for each C_k cluster of P so that both P^* and L^* optimize a criterion Δ:

$$\Delta(P^*,L^*) = Min\{\Delta(P,L)\ /\ P \in P_k, L \in \Lambda^k\} \tag{43.6}$$

with P_k the set of all the K-clusters partitions of E and Λ^k the representation space of the prototypes. $\Delta(P,L)$ is a function which measures how well the prototype G_k represents the characteristics of objects of the cluster and it can be usually interpreted as an heterogeneity or a dissimilarity measure of goodness of fit between G_k and C_k.

The aim is to partition a sample of curves $\{\chi_{s_1}(t),\ldots,\chi_{s_i}(t),\ldots,\chi_{s_n}(t)\}$ for $t \in T$ into a set of K clusters such that the obtained clusters can be described by a trace-variogram model (Romano et al., 2010b).

In this context it is assumed as criterion function:

$$\Delta(P,L) = \sum_{k=1}^{K} \sum_{h \in N_k(h)} (\gamma_k(h) - \gamma_k^*(h))^2 \tag{43.7}$$

where $N_k(h) \subseteq N(h)$ is the set of ordered $h = \|s_i - s_j\|$ for all $\chi_{s_j}(t), \chi_{s_j}(t) \in C_k$. This criterion is based on the best fit function $\phi(\gamma_k(h), \gamma_k^*(h))$ between the empirical trace-variogram $\gamma_k(h)$ and the theoretical trace-variogram $\gamma_k^*(h)$, for each cluster C_k, according to a chosen model $\gamma^*(h)$ of variogram.

Thus it is assumed that the estimated trace-variogram functions $\gamma_k^*(h)$ (for $k = 1, \ldots, K$) as prototypes of the clusters.

Starting from a random initialization partition P of the of n functions in K clusters, this algorithm alternatively performs a *representation* and an *allocation* step.

The *representation* step estimates the theoretical trace-variogram $\gamma_k^*(h)$, for each cluster C_k, by Ordinary Least Square method. In order to allocate a curve to a cluster, a natural way consistent with the objective function is to calculate again the emphirical trace-variogram and to evaluate the fitting with the prototype. It is note worthy that the emphirical trace-variogram can be expressed by:

$$\gamma_k(h) = \frac{1}{\sum_{m=1}^{|C_k|} |N_m(h)|} \sum_{m}^{|C_k|} \delta_{m,k}(h) * |N_{m,k}(h)| \tag{43.8}$$

where for each $\chi_{s_m}(t)$ the function $\delta_{m,k}(h)$ is expressed by:

$$\delta_{m,k}(h) = \int_T \delta_{m,k}(h,t)dt = \int_T V(\chi_{s_m}(t) - \chi_{s_i}(t))dt \tag{43.9}$$

with: $\chi_{s_i}(t) \in C_k$ and $\| s_m - s_i \| = h_{m,i} \in N_{m,k}(h)$, where $N_{m,k}(h)$ is the set of ordered $h_{m,i} = \| s_m - s_i \|$ for all $\chi_{s_i}(t) \in C_k$. The function $\delta_{m,k}(h)$ is a spatial variability function obtained by comparing all the curves of the cluster with the candidate curve to allocate.

Then, $\chi_{s_m}(t)$ is assigned to a cluster C_k according to the minimum mean squared distances between $\delta_{m,k}(h)$ and $\gamma_k^*(h)$ (for $k = 1, \ldots, K$):

$$\sum_{h \in N_{m,k}(h)} (\delta_{m,k}(h) - \gamma_k^*(h))^2 \leq \sum_{h \in N_{m,k'}(h)} (\delta_{m,k'}(h) - \gamma_{k'}^*(h))^2 \quad k' \neq k. \tag{43.10}$$

43.5 Discussion

The two clustering methods we have presented are to our knowledge among the first proposals to solve the problem of clustering spatially correlated functional data.

The hierarchical clustering approach uses a weighted distance measure for the function comparison. This approach as the advantage that the number of clusters are not priorly established and the cluster structure can be observed at different level of similarity.

On the contrary the Dynamic approach depends from the chosen number of clusters but is able to discover both the spatial partition of the data and the spatial variability structures representative of each cluster.

The performance of the methods will be analyzed by means of several simulation experiments and will be illustrated by means of real data examples.

References

1. Delicado, P., Giraldo, R., Comas, C., Mateu, J.:Statistics for spatial functional data: some recent contributions. Environmetrics **21**, 224-239 (2010)
2. Diday, E.: La méthode des Nuées dynamiques. Revue de Statistique Appliquée **19** (2), 19–34 (1971)
3. Giraldo, R., Delicado, P., Comas, C., Mateu, J.: Hierarchical clustering of spatially correlated functional data. Technical Report. Available at www.ciencias.unal.edu.co/unciencias/data-file/estadistica/RepInv12.pdf (2009)
4. Giraldo, R., Delicado, P., Mateu, J.: Ordinary kriging for function-valued spatial data. J. Environment. Ecol. Stat. To appear (2010)
5. Jiang, H., Serban, N.: Clustering Random Curves Under Spatial Interdependence: Classification of Service Accessibility. Technometrics, to appear (2010)
6. Ramsay, J.E., Silverman, B.W.: Functional Data Analysis (Second Edition). Springer (2005)
7. Romano E., Balzanella A., Verde R.: Clustering Spatio-functional data: a model based approach. Studies in Classification, Data Analysis, and Knowledge Organization. Springer, Berlin-Heidelberg, New York (2010a)
8. Romano E., Balzanella A., Verde R.: A new regionalization method for spatially dependent functional data based on local variogram models: an application on environmental data. In: Atti delle XLV Riunione Scientifica della Societá Italiana di Statistica Universitá degli Studi di Padova Padova. Padova, 16 -18 giugno 2010. CLEUP, ISBN/ISSN: 978 88 6129 566 7 (2010b)

Chapter 44
Spatial Clustering of Functional Data

Piercesare Secchi, Simone Vantini, Valeria Vitelli

Abstract We propose a new algorithm for clustering spatially dependent functional data that accounts for spatial dependence by repeatedly clustering functional local representatives of a random system of neighborhoods. The algorithm output is the frequency distribution of cluster assignment for each site of a given map. We illustrate different implementations of the algorithm by analyzing synthetic and real data.

44.1 Introduction

We consider the problem of unsupervised classification of spatially dependent functional data, where each curve of a finite, albeit possibly very large, data set is indexed by the sites of a spatial finite lattice S, defining the region of interest for the analysis. The problem consists in associating to each site $\mathbf{x} \in S$ a label $l \in \{1,...,K\}$, such that sites in subregions of S homogeneous with respect to the distribution generating the functional data are labeled the same. The cardinality K of the label set is given; the aim of the analysis is the reconstruction of the latent field of labels.

We propose a new clustering procedure that exploits spatial dependence by repeatedly generating random connectivity maps and by clustering, at each iteration, local representatives of neighboring functional data. More precisely, given the number K of clusters, the algorithm iteratively proceeds along four basic steps: generation of a spatial Voronoi tessellation, identification of a representative for each

Piercesare Secchi
Politecnico di Milano, Italy, e-mail: piercesare.secchi@polimi.it

Simone Vantini
Politecnico di Milano, Italy, e-mail: simone.vantini@polimi.it

Valeria Vitelli
Politecnico di Milano, Italy, e-mail: valeria.vitelli@mail.polimi.it

of the n elements of the tessellation, p-dimensional reduction and clustering of the representatives, cluster matching. For each site of the lattice S, the final output is a frequency distribution of assignment to each of the K clusters; this can be summarized in a classification map via a majority vote on the cluster assignment. Moreover, we propose an evaluation of the quality of the final classification based on entropy.

The fact that our data is functional is not irrelevant to the computational cost of standard procedures for the analysis of lattice data; hence the motivation for our method, which implicitly performs a reduction both in the dimension of the sample (by clustering a small number n of representatives) and in the infinite dimension of functional data (through the p–dimensional reduction of the representatives). The performances of the algorithm have been tested in various situations; a selected number of them are illustrated in Section 3. Finally, Section 4 describes an application of the algorithm to irradiance data analysis.

44.2 A clustering procedure for spatially dependent functional data

We propose an iterative algorithm whose four basic steps are hereafter summarized:

1. capture potential spatial dependence through the generation of a random Voronoi tessellation of the spatial lattice of sites indexing the functional data. Different choices for the distance between sites are allowed (e.g., Euclidean, geodesic, ...);
2. identify a local representative for each element of the random tessellation (e.g., a weighted mean value, a medoid, a loess curve, ...); the representative sums up local information, since neighboring functional data are most likely drawn from the same functional distribution;
3. perform functional dimensional reduction on the sample of local representatives (e.g., functional principal components, wavelets, ...), to select relevant functional features in the data and cluster the projections of local representatives on the space spanned by the previously obtained basis, via a suitable classification technique (e.g., K-means clustering, PAM, hierarchical clustering, ...).
4. match the cluster labels of the current iteration with those of the previous iteration (this step is missing in the first iteration).

The procedure repeats these four basic steps M times, obtaining at each iteration a different tessellation, and thus different local representatives of functional data: the greater is M, the higher is the algorithm accuracy.

More precisely, the algorithm is initialized by fixing the number M of iterations of the four basic steps, the number n of elements of the Voronoi tessellation, the dimension p of the basis for dimensional reduction of the local representatives and the number K of clusters considered in the clustering procedure.

Hence, for $m = 1, \ldots, M$, steps 1-4 are repeated. A set of nuclei $\Phi_n^m = \{\mathbf{Z}_1^m, \ldots, \mathbf{Z}_n^m\}$ is sampled at random among the sites in S, and a random Voronoi tessellation of the lattice S, $\{V(\mathbf{Z}_i^m|\Phi_n^m)\}_{i=1}^n$, is obtained by assigning each site $\mathbf{x} \in S$ to the nearest nucleus $\mathbf{Z}_i^m$, according to a specified distance (e.g., Euclidean, geodesic, ...). For $i = 1,\ldots,n$, the local representative $g_i^m(t)$, corresponding to the nucleus $\mathbf{Z}_i^m$ of the i-th element $V(\mathbf{Z}_i^m|\Phi_n^m)$ of the tessellation, is obtained; for instance, through a weighted mean of the functional data associated to the sites belonging to the Voronoi element (a Gaussian density with diagonal covariance matrix and centered in the nucleus $\mathbf{Z}_i^m$ is a standard option for the weighting function). Other possible implementations of the algorithm identify the representative as the medoid, or estimate it through loess, of the curves corresponding to sites belonging to the tessellation element. Then, dimensional reduction of the n representatives $g_1^m(t),\ldots g_n^m(t)$ is performed by projecting them on the space spanned by a proper p–dimensional functional basis, thus generating the p-dimensional scores vectors $\{\mathbf{g}_1^m,\ldots,\mathbf{g}_n^m\}$, which are then clustered in K groups according to a suitable unsupervised method, depending on the application. Let $\Gamma_1^m,\ldots,\Gamma_n^m$ denote the labels of the local representatives; for $i = 1,\ldots,n$, all sites $\mathbf{x}$ in $V(\mathbf{Z}_i^m|\Phi_n^m)$ get the label Γ_i^m. For $l = 1,\ldots,K$, indicate with C_l^m the set of $\mathbf{x} \in S$ whose label is equal to l. Finally, if $m \geq 2$, the labels identifying the clusters $C_1^m,\ldots,C_K^m$ are renamed by matching them with the cluster assignments $C_1^{m-1},\ldots,C_K^{m-1}$ obtained at the previous iteration; indeed the algorithm looks for the label permutation $\{l_1,\ldots,l_K\}$ in the set $\{1,\ldots,K\}$ that minimizes the total sum of the off-diagonal frequencies in the contingency table describing the joint distribution of sites along the two classifications $C_1^{m-1},\ldots,C_K^{m-1}$ and $C_{l_1}^m,\ldots,C_{l_K}^m$. Other different procedures for cluster matching are conceivable.

The final classification map of the lattice S is obtained by considering the frequency distribution of assignment of each site to each of the K clusters along the M iterations. In fact, for each site $\mathbf{x} \in S$, we compute $f_{\mathbf{x}}^l = \#\{m \in \{1,\ldots,M\} : \mathbf{x} \in C_l^m\}/M$, for $l = 1,\ldots,K$. A final assignment of site $\mathbf{x}$ to *one* of the K clusters can be obtained by selecting that corresponding to a mode of the distribution $f_{\mathbf{x}} = (f_{\mathbf{x}}^1,\ldots,f_{\mathbf{x}}^K)$. Moreover, the final classification can be evaluated via a quality index based on entropy, $\eta_{\mathbf{x}}^K = -\sum_{k=1}^K f_{\mathbf{x}}^k \cdot \log(f_{\mathbf{x}}^k)$, which assumes minimum value 0 when the classification is neat, and maximum value $\log(K)$ when it is highly uncertain. For comparisons, the quantity $\eta_{\mathbf{x}}^K$ is commonly normalized by its maximum value; indeed, if K is not known, a comparison of the normalized entropy for different values of K is performed in order to choose the optimum value of K.

44.3 A simulation study on synthetic data

In this Section we report the results of a simulation study performed to test the proposed spatial clustering algorithm under stronger/weaker spatial dependence in the latent field of labels, while also varying the choice for the parameter n controlling the density of the nuclei sets Φ_n^m, i.e., the coarseness of the Voronoi tessellation; the functional signals associated to the sites of the field belong to a finite dimensional functional space.

The space S is a two-dimensional square lattice of 50×50 sites and the latent field of labels is generated by a Ising Markov Random field $L : S \to \{-1,1\}$, where

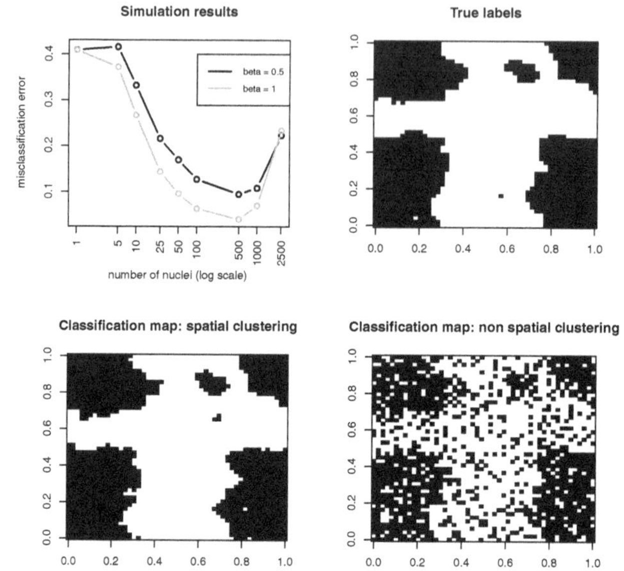

Fig. 44.1: Results of simulation study: misclassification error of spatial clustering over different choices for n and β – mean over 30 repetitions of the procedure (top, left); true labels, Ising field with $\beta = 1$ (top, right); final classification map obtained via spatial clustering with $n = 500$ (bottom, left); final classification map obtained via non-spatial clustering (bottom, right).

the strength of spatial dependence is controlled by the parameter β, higher values of β implying a stronger spatial dependence (see Cressie, 1993 for details). For each site $\mathbf{x} \in S$, denote by $L_{\mathbf{x}}$ its true label drawn from a Ising field. For our simulation studies, we let β range in the interval $(0.5, 1)$.

Conditionally on the realization of the latent field, in each site of S we generate independently a random function belonging to a p-dimensional space, spanned by a Fourier basis; the distribution of the random function depends exclusively on the site label. For $\mathbf{x}$ ranging in S, we thus obtain a sample of dependent functions $f_{\mathbf{x}}(t) = \sum_{h=1}^{p} c_{\mathbf{x}h}\varphi_h(t)$, with $\mathbf{c}_{\mathbf{x}}|L_{\mathbf{x}} = l \sim N_p(\boldsymbol{\mu}_l, \Sigma)$; $\varphi_h(t)$ is the h-th element of the Fourier basis, and $\mathbf{c}_{\mathbf{x}} \in \mathbb{R}^p$ is the vector of basis coefficients, drawn from a multivariate Gaussian distribution; $\Sigma = \sigma \mathbf{I}_p$, with $\mathbf{I}_p$ being the identity matrix in p-dimensions, $\sigma = 2$, and $\boldsymbol{\mu}_{-1} = \mathbf{0}$ while $\boldsymbol{\mu}_1 = (-2, -1, 0, 1, 2)$.

The parameters controlling the algorithm are fixed as follows: $M = 50$, $K = 2$ and $n \in \{1, 5, 10, 25, 50, 100, 500, 1000, 2500\}$. The Voronoi tessellations are generated by means of the Euclidean distance, the n representatives are identified as weighted means with Gaussian isotropic weights while no dimensional reduction of the representatives is performed since they already belong to a finite dimensional functional space; finally, clusterization of the representatives is obtained through K-means on the basis coefficients. The final classification map is obtained through a majority vote on cluster assignment, and the result is evaluated by computing a misclassification error rate with respect to the true realization of the field. The final evaluation of the algorithm is obtained by repeating the simulation 30 times, and by calculating a mean misclassification rate.

Results are illustrated in Figure 44.1. Consider the top/left panel of the picture, showing the mean misclassification error rate for different values of β and n: we appreciate the existence of a value of n which minimizes the misclassification error. Moreover, misclassification error is uniformly smaller (with respect to n) for higher values of β: hence the improvement introduced by the algorithm is stronger in the presence of a stronger spatial dependence in the latent field of labels. Note that the limiting case $n = 50 \times 50 = 2500$ corresponds to a non-spatial clustering procedure, which gives worse results than spatial clustering for nearly all values of n.

Finally, in the top/right panel, the true latent field of labels for one of the simulated data sets is shown, while in the bottom panels appear the results obtained via spatial clustering with $n = 500$ (left), and via non-spatial clustering (right). It is evident from the picture that, although the final map obtained via non-spatial clustering is suggestive of the true label pattern, the final result obtained via spatial clustering is far more precise, and the classification map nearly coincides with the true label field (misclassification errors are 3.28% and 23.08% for spatial and non-spatial clustering, respectively).

44.4 A case study: clustering irradiance data

We now summarily report on an application of our classification algorithm to irradiance data, carried out to investigate exploitation of solar energy in different areas of the planet. In particular, power production via collectors that are able to track the sun diurnal course is strongly influenced by solar irradiance and atmospheric conditions. In fact, solar thermal power employs only direct sunlight and it is therefore best positioned in areas, such as deserts, steppe or savannas, where large amounts of humidity, fumes or dust, that may deviate the sunbeams, do not occur (Richter *et al.*, 2009).

We try to identify these optimal areas by analyzing a quantity related both to the average solar irradiance and to the maximal number of consecutive no-sun days along the year, observed over a period of 22 years from 1983 to 2005, in 47880 worldwide non-polar districts (see NASA, *Surface meteorology and Solar Energy database*). In particular, we examine the *annual pattern of the maximum solar radiation deficit below expected value incident on a horizontal surface over a consecutive–day period* (kWh/m^2), which is increasing in the monthly average irradiance (we call this quantity *buffer capacity*, since it is directly related to the amount of energy that needs to be stored in a solar power plant in order to successfully cover for gaps in energy supply due to unfavorable environmental and weather conditions).

For this application, we set $M = 100$ and $n = 300$; the set of nuclei is repeatedly drawn from a uniform distribution on the sphere and Voronoi tessellation is based

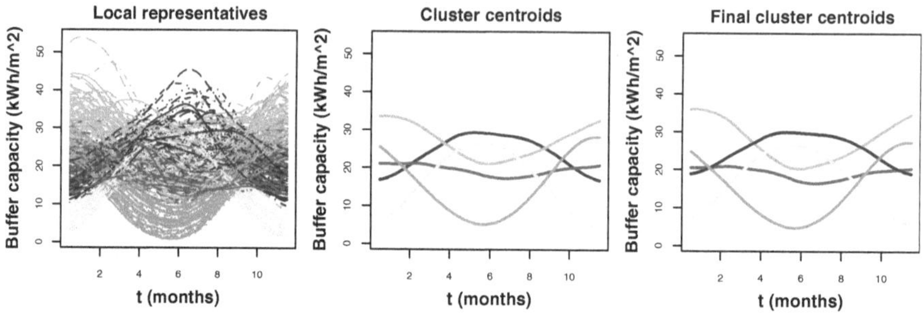

Fig. 44.2: Results of spatial clustering of irradiance data: sample of local representatives obtained via spatial clustering procedure with $n = 300$ (left panel), and corresponding cluster centroids (centre panel). In the right panel are shown cluster centroids of the final classification map shown in Figure 44.3: different colors correspond to different cluster labels.

on geodesic distance. We choose the first $p = 3$ functional principal components to project data; we then perform hierarchical clustering using the L^2 semi-metric induced by the principal components and using a Ward linkage. Finally, we choose the optimal K through the evaluation of the classification map by means of entropy.

The spatial clustering algorithm identifies different homogeneous macro-areas which – prima facie – seem interpretable in terms of the observed phenomenon, even though a climatological analysis, which is beyond the scopes of this paper, could deepen their explanation; indeed, the same macro-areas are not captured by customary unsupervised classification procedures, that do not take into proper account the spatial dependence among data. The best classification, according to spatial entropy evaluation (Figure 44.3, bottom), is obtained for $K = 5$: final results for this choice of K are shown in Figures 44.2 and 44.3. In Figure 44.2 a sample of local representatives is shown (left panel), together with cluster centroids corresponding to cluster assignments of the same local representatives (centre); in the right panel of the same picture cluster centroids corresponding to the final classification map (Figure 44.3) are depicted: each centroid color correspond to the color of the macro-area in the map it belongs to. The similarity between the last two pictures confirms the fact that – due to the spatial dependence among close data – no important information is lost in the replacement of the entire dataset with a suitable number of local representatives.

The red cluster is characterized by a non-seasonal pattern, and by high average buffer capacity along the year; it covers Africa, Middle-East and equatorial America and its presence is not explained only in terms of latitude. From North to South we can then identify four clusters with seasonal patterns depending on the hemisphere and on the average buffer capacity along the year: north-low (yellow), north-high (blue), south-high (violet), south-low (green). It is interesting to note the absence of the north-high and south-high clusters in Europe and Africa, the absence of the non-seasonal cluster in Asia and Australia, and the presence of all clusters in the Americas.

Acknowledgements We thank Alessia Pini for her invaluable help in the analysis of the *Surface Meteorology and Solar Energy* data set.

References

1. Cressie, N.: Statistics for Spatial Data. Wiley Series in Probability and Mathematical Statistics (1993)
2. NASA, Surface meteorology and Solar Energy: A renewable energy resource web site (release 6.0), `http://eosweb.larc.nasa.gov/cgi-bin/sse/sse.cgi?#s01` [accessed on the 25th of November, 2010]
3. Richter, C., Teske S., Nebrera, J. A.: Concentrating Solar Power Global Outlook 09. Greenpeace International / European Solar Thermal Electricity Association (ESTELA) / IEA SolarPACES, Report (2009)

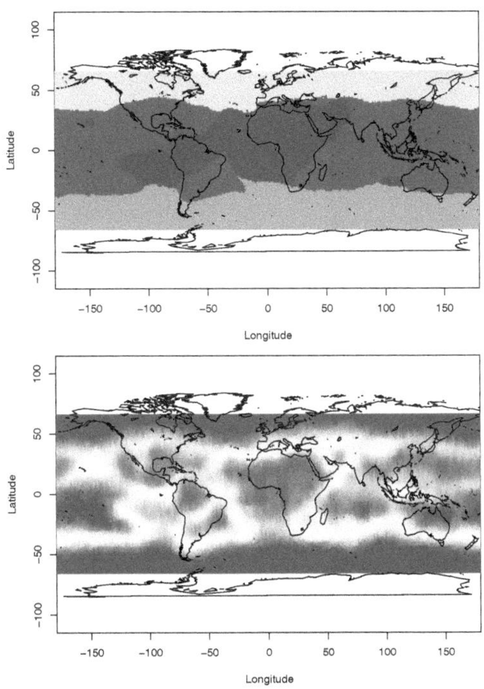

Fig. 44.3: Results of spatial clustering on buffer capacity data from the *Surface meteorology and Solar Energy database*: in the top panel, final classification map obtained by setting $K = 5$ via a majority vote on frequencies of assignment; in the bottom panel, normalized spatial entropy associated to the classification with $K = 5$. In the bottom panel, colors from red to white correspond to values from 0 to 1; higher values identify areas where classification is more uncertain.

Chapter 45
Population-Wide Model-Free Quantification of Blood-Brain-Barrier Dynamics in Multiple Sclerosis

Russell Shinohara, Ciprian Crainiceanu

Abstract The processes by which new white matter lesions in multiple sclerosis (MS) develop are only partially understood. Recently developed lesions tend to enhance on magnetic resonance imaging (MRI) scans following the intravenous administration of a contrast agent. In this paper, we develop a model-free framework for the analysis of these data that provides biologically meaningful quantification of this blood-brain barrier opening.

45.1 Introduction

Multiple sclerosis (MS) is an inflammatory disease that causes demyelinating lesions in the central nervous system. Although gray-matter lesions are common (Calabrese et al., 2010), white-matter lesions are easiest to identify, both pathologically and radiologically, due to their loss of normal myelin and often high degree of inflammation. In the first clinical stage of MS, these lesions appear relatively frequently and can occur in unpredictable locations at unpredictable times. The disease-modifying drugs that are currently used to treat MS can reduce the incidence of these lesions (Calabresi et al., 1997).

The processes by which new lesions develop are only partially understood. Much of this understanding has come through magnetic resonance imagine (MRI) of the brain. These lesions have long been known to form around veins (Dawson, 1916) where inflammatory cells, especially T lymphocytes, form perivenular cuffs. One of the hallmarks of newly forming lesions is enhancement on MRI following the intravenous administration of gadolinium-based contrast agents that shorten the longitudinal (T_1) relaxation time of the tissue (Grossman et al., 1988). This visible

Russell Shinohara
Johns Hopkins University, Baltimore, USA, e-mail: rshinoha@jhsph.edu

Ciprian Crainiceanu
Johns Hopkins University, Baltimore, USA, e-mail: ccrainic@jhsph.edu

enhancement in the MRI results from opening of the blood-brain barrier (BBB) and reveals areas of active inflammation. Lesion enhancement typically lasts 4 to 8 weeks and may be accompanied by neurological signs and symptoms, but new lesions are often asymptomatic (Capra et al., 1992). The incidence and number of existing enhancing lesions are common outcome measures used in MS treatment clinical trials.

The exact nature of BBB opening in new MS lesions and the selectivity of the resulting permeability remain unclear. The analysis of contrast-agent uptake can provide only limited insight into these issues. Dynamic-contrast-enhanced MRI (DCE-MRI) has been used for the past two decades to quantify the rate at which contrast agents pass from the plasma to MS lesions as a measure of BBB permeability (Kermode et al., 1990).

DCE-MRI data are typically analyzed using deterministic pharmacokinetic modeling techniques based on multi-compartment tissue models with exchange (Davidian and Giltinan, 1995). These techniques are limited for four major reasons. The first is that the tissue composition, specifically the number of compartments in the pharmacokinetic model, is unknown, posing technical and interpretive difficulties. Secondly, the number of compartments may vary within and between tissue types, which makes the a-priori choice of a number of compartments for every single voxel in the brain a difficult proposition. Thirdly, saturation of these models leads to interpolation, which in itself does not help with the quantification and dimension reduction. Finally, when fitting these models to the DCE-MRI data from our study, the standard deterministic algorithms fail to converge in over 80% of the voxels.

In this paper, we consider a subject with relapsing-remitting multiple sclerosis selected because of active disease, as evidenced by the development of contrast-enhancing lesions on a monthly scan. We observed one DCE-MRI scan recorded during a single clinical visit. The DCE-MRI consisted of short T_1-weighted scans recorded as the contrast agent flows through the brain; details concerning the acquisition of these data can be found in our complete paper (Shinohara et al., 2010). Our goal was to provide a statistically principled platform for the quantification of observed lesion enhancement. To achieve these goals, in Section 3 we use functional principal components analysis (FPCA) (Ramsay and Silverman, 2002, 2005) to study directions of variation in the voxel-level time series of intensities. We finish the paper with a brief description of further work that we have detailed in our complete work (Shinohara et al., 2010).

45.2 Methods and Results

We start by introducing some prominent characteristics of the data. As the contrast agent propagates through the areas under observation via MRI, the signal intensity on T_1-weighted images increases because the gadolinium shortens the T_1 relaxation time of the tissue. This increase in the signal is related to the concentration of the contrast agent in the tissue. However, exact calibration is not possible without careful T_1 mapping, which we explicitly avoided in order to decrease scan time, reduce variability, and limit the number of assumptions of our analysis. Without such mapping as well as knowledge of the relaxivity properties of gadolinium, units of MRI signal cannot be taken as indicative of gadolinium concentration (Tofts, 1997). The interpretation of the recorded intensity varies with respect to the location and baseline magnetic properties of the various voxels in the brain. Quantifying the temporal and spatial behavior of the signal intensity in white matter is the primary goal of this paper.

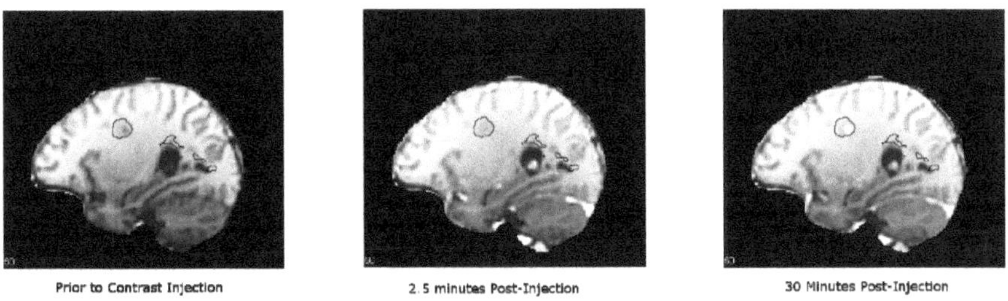

Fig. 45.1: DCE-MRI scans three time points both before and after contrast injection. Black contour lines indicate the spatial extent of the lesions as seen on T_2-weighted FLAIR scans obtained during the same session.

For illustration, the intensity maps in a sagittal slice are displayed in Figure 1 at three time points: before the injection and 2.5 and 30 minutes afterward. Although we only show three time points, many more volumes are typically observed for each subject. The subject analyzed in this paper was scanned over 155 minutes, and 67 volumes were acquired during the single scan. The solid black contours in Figure 1 are the reconstructed in-slice boundaries of the lesions obtained using the Lesion-TOADS automatic segmentation algorithm (Shiee et al., 2010). Most of the delineated lesions had been present on previous scans of the same subject and did not enhance with contrast.

Several characteristics of the data are immediately apparent. First, in the time point measured 2.5 minutes after contrast injection, the blood vessels are bright, indicating a high concentration of the contrast agent. The rest of the brain remains essentially unchanged at this time. Second, as time progresses some of the voxels in regions of interest (ROI) within the lesions enhance.

Another way of looking at the data is to plot the time series for each voxel. More specifically, the data from a single subject can be written as a T x V matrix, where T is the number of time points and V is the number of voxels. For the first subject, T=67 and V=7.2 million (corresponding to the volume of dimension 182 x 218 x 182, where each voxel is interpolated to 1mm x 1mm x 1mm cuts from an acquired resolution of 2mm^3). The skull-stripping procedure [Carass et al., 2007] reduces V from 7.2 million to 1.6 million. The time series for 0.1% of these 1.6 million voxels are displayed in Figure 2. Unfortunately, the sheer number of voxels masks important features in the data, such as the large spikes in some of the voxels immediately following injection (time 0).

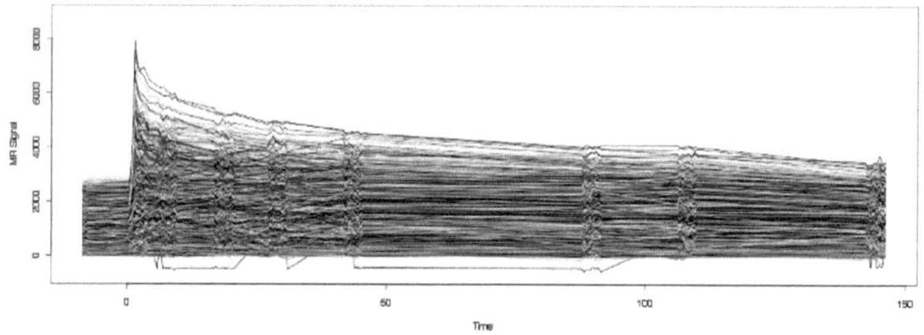

Fig. 45.2: Raw MR signal time series plotted over time. Intensity is measured in arbitrary units.

Given the complexity and size of the data, a natural next step in the exploratory data analysis is to find the number and shape of patterns at the subject level. Our primary goal is to quantify these patterns. We start by applying FPCA to the collection of time series. The first three principal components (PCs) from this analysis are depicted in Figure 3. The first PC (orange) is roughly a vertical shift; this corresponds to baseline discrepancies between voxels. For example, the intensity in gray matter voxels and NAWM voxels changes little over time; however, the gray matter voxel intensities tend to be shifted downwards compared to the white matter due to their longer intrinsic T_1. Similarly, there is variance in the baseline intensity within each of these sections in the brain; some parts of the gray matter are darker than other parts. We conclude that the first PC captures natural differences in the magnetic properties of voxels that are independent of the contrast agent's presence. The second PC (red) depicts a sudden increase in intensity after injection followed by an exponential decline. This behavior is identical to that seen in blood vessels in Figure 1. In terms of physiology, this is consistent with the delivery of the contrast agent in high concentrations immediately following injection, followed by its efficient clearance. The third PC (blue) is a gradual increase in intensity followed by a plateau, which is strikingly similar to the shape of the time series in the enhancing ROI. This indicates that blood is not rapidly introduced to these regions; rather, it slowly seeps in over time.

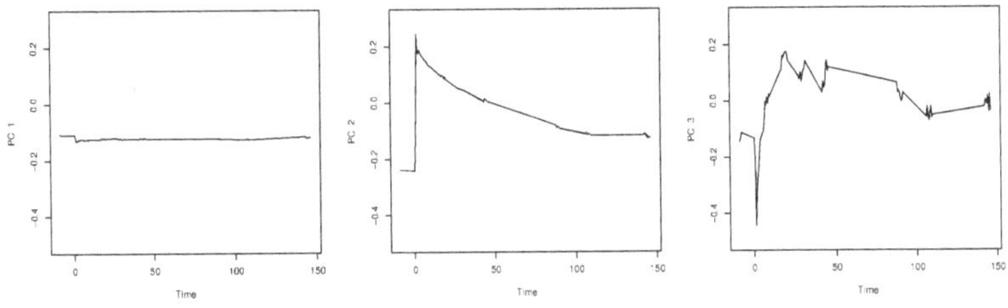

Fig. 45.3: First three PCs from the FPCA. The color indicates the index of the PC. The noticeable jumps in intensity are noise, likely related to scanner drift, onset of scanning, subject movement, and possibly other factors.

The first three PCs, which explain 99% of the variation in the data, are interpretable and apparently correspond to real features in the observed time series. To further investigate our empirical findings, we analyze the spatial patterns associated with the loadings of the voxel time series. To accomplish this, we calculate the PC loadings on each of these components for each voxel. Specifically, for a voxel v with corresponding observed time series $Y^O(t,v)$ and a principal component $\phi_j(t)$, we find the loading $\xi(v) = < Y^O(t,v), \phi_j(t) > = \sum_t Y^O(t,v)\phi_j(t)$. We then map these scores, $\xi(v)$, back to the three-dimensional brain volume. Figure 4 is a map of the spatial patterns of these loadings for the second and third PCs in the same sagittal slice from Figure 1. The first PC as it only shows baseline differences and is not of general interest. The second PC loads heavily only in the blood vessels (yellow spots), as expected. The third PC loads in the enhancing ROI and in residual highly vascularized extracranial tissues (such as the scalp).

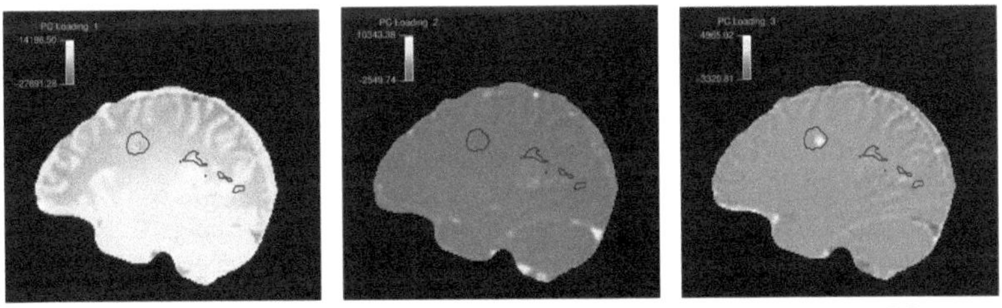

(a) First PC (b) Second PC (c) Third PC

Fig. 45.4: Maps indicating the first through third PC loadings in a sagittal slice.

45.3 Conclusions

The above subject-by-subject analysis is enlightening, but the analysis is subject-specific and the measures defined therein are only valid within the particular subject.

In our work (Shinohara et al., 2010), our primary goal is to quantify these subject-specific patterns using measures that are meaningful across subjects. Thus, we: 1) normalize and interpolate the data to a common grid; 2) obtain population-level PCs; 3) ensure that the features identified by the above subject-level analyses are also identified by the population-level method; and 4) generate hypotheses concerning the nature of enhancement patterns and outline appropriate statistical methods. We also consider spatiotemporal modeling that quantifies centripetal and centrifugal enhancement properties described in Gaitan et al. [2010]. This work opens several directions for future studies, including extension of the analysis to large populations of subjects.

References

1. Calabrese, M., Filippi, M., Gallo, P.: Cortical lesions in multiple sclerosis. Nat. Rev. Neurol. **6**, 438–444 (2010)
2. Calabresi, P.A., Stone, L.A., Bash, C.N., Frank, J.A., McFarland, H.F.: Interferon beta results in immediate reduction of contrast-enhanced MRI lesions in multiple sclerosis patients followed by weekly MRI. Neurology **48** (5), 1446 (1997)
3. Capra, R., Marciano, N., Vignolo, L.A., Chiesa, A. Gasparotti, R.: Gadolinium-Pentetic Acid Magnetic Resonance Imaging in Patients With Relapsing Remitting Multiple Sclerosis. Archives of Neurology **49** (7), 687 (1992)
4. A., Wheeler, M.B., Cuzzocreo, J., Bazin, P.L., Bassett, S.S., Prince, J.L.: A joint registration and segmentation approach to skull stripping. In: Biomedical Imaging: From Nano to Macro (2007)
5. Davidian, M., Giltinan, D.M.: Nonlinear models for repeated measurement data. Chapman & Hall/CRC (1995)
6. Dawson, J.W.: The histology of disseminated sclerosis. Trans R Soc Edinb **50**, 517 (1916)
7. C.Z. Di, C.M. Crainiceanu, B.S. Caffo, and N.M. Punjabi. Multilevel functional principal component analysis. Ann. Appl. Stat. **3** (1), 458–488 (2009)
8. Greven, S., Crainiceanu, C., Caffo, B., Reich, D.: Longitudinal functional principal component analysis. Electron. J. Stat. **4**, 1022–1054 (2010)
9. Grossman, R.I., Braffman, B.H., Brorson, J.R., Goldberg, H.I., Silberberg, D.H., Gonzalez-Scarano, F.: Multiple sclerosis: serial study of gadolinium-enhanced MR imaging. Radiology **169** (1), 117 (1988)
10. Kermode, A.G., Tofts, P.S., Thompson, A.J., MacManus, D.G., Rudge, P., Kendall, B.E., Kingsley, D.P.E., Moseley, I.F., Boulay, E., McDonald, W.I.: Heterogeneity of blood-brain barrier changes in multiple sclerosis: an MRI study with gadolinium-DTPA enhancement. Neurology **40** (2), 229 (1990)
11. Kurtzke, J.F.: Rating neurologic impairment in multiple sclerosis: an expanded disability status scale (EDSS). Neurology **33** (11), 1444 (1983)
12. Ramsay, J.O., Silverman, B.W.: Functional data analysis (Second Edition). Springer Verlag (2005)
13. Ramsay, J.O., Silverman, B.W.: Applied functional data analysis: methods and case studies. Springer Verlag (2002)
14. Shiee, N., Bazin, P.L., Ozturk, A., Reich, D.S., Calabresi, P.A., Pham, D.L.: A topology-preserving approach to the segmentation of brain images with multiple sclerosis lesions. NeuroImage **49** (2), 1524–1535 (2010)
15. Tofts, P.S.: Modeling tracer kinetics in dynamic Gd-DTPA MR imaging. J. Magnetic Resonance Imaging **7** (1), 91–101 (1997)

Chapter 46
Flexible Modelling of Functional Data using Continuous Wavelet Dictionaries

Leen Slaets, Gerda Claeskens, Maarten Jansen

Abstract A random effects model for functional data based on continuous wavelet expensions is proposed. It incorporates phase variation without the use of warping functions. Both coarse-scale features and fine-scale information are modelled parsimoniously, yet flexible. The regularity of the estimated function can be controlled, creating a joint framework for Bayesian estimation of smooth as well as spiky and possibly sparse functional data.

46.1 Introduction

Functional data have been around for centuries, but the availability of methodology recognizing their functional nature and corresponding features has blossomed more recently. Overviews and great contributions in the field, both for academic researchers and practitioners, are the works by Ramsay and Silverman (2006) and Ferraty and Vieu (2006).

A great deal of attention has been devoted to the study of variation in samples of curves with the purpose of gaining insight in the mechanisms that drive the data. Such samples $y_{nj} = y_n(t_j)$ are often encountered when observing a process over a certain time interval (at discrete time points t_j, $j = 1, \ldots, T_n$) for several subjects or instances $n = 1, \ldots, N$. A key element of the functional data framework is the recognition of phase variation (variation in timing of features) as a source of variability in the data, in addition to amplitude variation (variation in amplitude of features).

Leen Slaets
Katholieke Universiteit Leuven, Belgium, e-mail: leen.slaets@econ.kuleuven.be

Gerda Claeskens
Katholieke Universiteit Leuven, Belgium, e-mail: gerda.claeskens@econ.kuleuven.be

Maarten Jansen
Université Libre de Bruxelles, Belgium, e-mail: maarten.jansen@ulb.ac.be

A monotone increasing function transforming the time-axis, called a warping function, is typically used to take phase variation into account, before or joint with the analysis of amplitudes. These warping functions behave differently than the actual curves in the sample, complicating a combined analysis and proper understanding of the total variation as a mixture of the two. With clustering in mind, Liu and Yang (2009) circumvented the warping function by representing the curves as B-splines with randomly shifted basis functions.

Along that line we introduce a model which incorporates phase variation in a natural and intuitive way, by avoiding the use of warping functions, while still offering a good and controllable degree of complexity and flexibility. By building a model around wavelet transformations, we can use the location and scale notion of wavelet functions to model phase variation. The coefficients corresponding to the wavelet functions represent amplitude. Wavelets have already greatly shown their efficiency for the representation of single functions, and those strengths are exactly what we aim to generalize towards samples of curves. Morris and Carroll (2006) recently used wavelet transformations in a functional context to generalize a classic mixed effects model. They use the wavelet transformation as an estimation tool, while our goal is to use wavelet functions for direct modelling of the data, not to fit general functional mixed effects models. An additional advantage of using wavelets is that by choosing an appropriate wavelet many types of data can be analyzed, ranging from smooth processes to spiky spectra. The proposed model serves as a basis for a variety of applications, such as (graphical) exploration and representation, clustering and regression with functional responses.

46.2 Modelling Functional Data by means of Continuous Wavelet dictionaries

The proposed model is built around a scaling function ϕ and wavelet function ψ, the latter often used to form an orthonormal basis ψ_{jk}, $j,k \in \mathbb{Z}$, by shifting and rescaling the mother wavelet ψ, subject to the dyadic constraints: $\psi_{jk}(t) = 2^{j/2}\psi\left(2^j t - k\right)$. A downside of obtaining orthonormality, is the fact that functions need to be observed on an equidistant grid of time points. Therefore continuous wavelet transformations, using an overcomplete set of wavelet functions with arbitrary locations and scales, continue to gain popularity. In a functional setting, an overcomplete wavelet dictionary can represent the sample of curves in the following way:

$$y_n(t_j) = \sum_{m=1}^{M} c_{n,m}\sqrt{a_{n,m}}\phi\left(a_{n,m}(t_{n,j} - b_{n,m})\right) + \sum_{k=M+1}^{M+K} c_{n,k}\sqrt{a_{n,k}}\psi\left(a_{n,k}(t_{n,j} - b_{n,k})\right) + e_{n,j}, \tag{46.1}$$

with random scales $a_{n,m}$,$a_{n,k}$, random shifts $b_{n,m}$, $b_{n,k}$, random amplitudes $c_{n,m}$,$c_{n,k}$ and independent random errors $e_{n,j}$. Take $a_{n,k} \geq a_{n,m}$, $\forall m = 1,\ldots,M, k = M+1,\ldots,K$ and denote:

$$\mathbf{a}_{n,M} = (a_{n,1}, a_{n,2}, \ldots, a_{n,M}),\ \mathbf{a}_{n,K} = (a_{n,M+1}, a_{n,M+2}, \ldots, a_{n,M+K}),$$
$$\mathbf{b}_{n,M} = (b_{n,1}, b_{n,2}, \ldots, b_{n,M}),\ \mathbf{b}_{n,K} = (b_{n,M+1}, b_{n,M+2}, \ldots, b_{n,M+K}),$$
$$\mathbf{c}_{n,M} = (c_{n,1}, c_{n,2}, \ldots, c_{n,M}),\ \mathbf{c}_{n,K} = (c_{n,M+1}, c_{n,M+2}, \ldots, c_{n,M+K}),$$

which have the following random effects distributions:

$$(\mathbf{a}_{n,M}, \mathbf{b}_{n,M}, \mathbf{c}_{n,M}) \sim \mathcal{N}_K(\boldsymbol{\mu}_M, \Sigma_M), \quad \text{for } n = 1, \ldots, N$$
$$(\mathbf{a}_{n,K}, \mathbf{b}_{n,K}, \mathbf{c}_{n,K}) \sim \mathcal{N}_K(\boldsymbol{\mu}_K, \Sigma_K), \quad \text{for } n = 1, \ldots, N$$
$$e_{n,j} \sim \mathcal{N}(0, \sigma^2), \quad \text{for } n = 1, \ldots, N \text{ and } j = 1, \ldots, T_n,$$

with $\boldsymbol{\mu}_K = (\boldsymbol{\alpha}_K, \boldsymbol{\beta}_K, \boldsymbol{\gamma}_K) = (\alpha_1, \alpha_2, \ldots, \alpha_K, \beta_1, \beta_2, \ldots, \beta_K, \gamma_1, \gamma_2, \ldots, \gamma_K)$ and likewise for M. The index K (and M) refers to the dimensionality of the vector which depends on the number of wavelet functions, K (or scale functions M), in expansion (46.1). While M is a fixed constant, K is also a parameter in the model. We will assume $\Sigma_K = \sigma_\psi^2 I_K$.

The scale functions can be interpreted as representing the main features in a homogeneous functional data sample. Plugging in the estimated mean of the random effects, the functions $\gamma_m \sqrt{\alpha_m} \phi\,(\alpha_m(t - \beta_m)$ give an idea of the underlying pattern in the sample. The random effects $a_{n,m}$, $b_{n,m}$, $c_{n,m}$ allow for curve-specific deviations in respectively scale, location and amplitude from these average features, while maintaining parsimony. Phase and amplitude variation are thus being modelled in an intuitive way, by means of a random scale, location (both representing phase) and amplitude of scale functions.

The covariance matrix Σ_M explains how the random effects corresponding to a certain feature relate to others. Σ_M can thus uncover complicated patterns in a functional data sample, which are often impossible to detect by eye or by more simple methods. Widths of initial peaks could be related to increased amplitudes of peaks at later time points. For the special case $\Sigma_M = \sigma_\phi^2 I_M$ all data features are independent of each other.

In this model there is no need for a fixed or equispaced grid of time points, as continuous wavelets are being used and information is borrowed within and across curves by means of the random wavelet functions. This makes the method suitable for the analysis of sparse data as well.

For a single curve y ($N = 1$), model (46.1) fits the framework introduced in Abramovich et al. (1999). They established conditions on the model parameters under which the smoothness of the expansion can be controlled. In the Bayesian framework, Chu et al. (2009) do so by an appropriate choice of priors on the model parameters. For the estimation they use a reversible jump Markov Chain Monte Carlo algorithm to improve computational efficiency. The ideas in both papers are used to generalize these results to those of a random effects models for samples of curves.

The idea behind model (46.1) is that the data follow one main pattern with curve-specific deviations in location, scale and amplitude. In case the data are heterogeneous, the model can be used for a clustering procedure following a k-centers type algorithm. The model can also be extended by incorporating additional covariates, giving rise to a regression model with functional responses. Extensions of (46.1) with a continuous regressor x include:

$$y_n(t_j) = \zeta x_n + \sum_{m=1}^{M} c_{n,m}\sqrt{a_{n,m}}\phi\left(a_{n,m}(t_{n,j} - b_{n,m})\right) + \sum_{k=M+1}^{M+K} c_{n,k}\sqrt{a_{n,k}}\psi\left(a_{n,k}(t_{n,j} - b_{n,k})\right) + e_{n,j},$$

$$y_n(t_j) = \sum_{m=1}^{M} (c_{n,m} + \zeta_m x_n)\sqrt{a_{n,m}}\phi\left(a_{n,m}(t_{n,j} - b_{n,m})\right) + \sum_{k=M+1}^{M+K} c_{n,k}\sqrt{a_{n,k}}\psi\left(a_{n,k}(t_{n,j} - b_{n,k})\right) + e_{n,j},$$

$$y_n(t_j) = c_{n,1}\sqrt{\zeta_2 x_n \cdot a_{n,1}}\phi\left(\zeta_2 x_n \cdot a_{n,1}(t_{n,j} - (b_{n,1} + \zeta_1 x_n^2))\right) + \sum_{m=2}^{M} c_{n,m}\sqrt{a_{n,m}}\phi\left(a_{n,m}(t_{n,j} - b_{n,m})\right) + \sum_{k=M+1}^{M+K} c_{n,k}\sqrt{a_{n,k}}\psi\left(a_{n,k}(t_{n,j} - b_{n,k})\right) + e_{n,j},$$

In summary, we create a framework to analyze many different types of functional data (smooth, spiky, sparse), while still being flexible and easy to understand, estimate and use.

References

1. Abramovich, F., Sapatinas, T., Silverman, B.W.: Stochastic expansions in an overcomplete wavelet dictionary. Probab. Theor. Rel. **117**, 133–144 (2000)
2. Chu, J.-H., Clyde, M.A., Liang, F. Bayesian function estimation using continuous wavelet dictionaries. Stat. Sinica **19**, 1419–1438 (2009)
3. Ferraty, F., Vieu, P.: Nonparametric functional data analysis: theory and practice. Springer, New York (2006)
4. Liu, X., Yang, M.C.K.: Simultaneous curve registration and clustering for functional data. Comput. Stat. Data An. **53**, 1361–1376 (2009)
5. Morris, J.S., Carroll, R.J.: Wavelet-based functional mixed models. J. Roy. Stat. Soc. B **68**, 179–199 (2006)
6. Ramsay, J.O., Silverman, B.W.: Functional data analysis (Second Edition). Springer, New York (2005)

Chapter 47
Periodically Correlated Autoregressive Hilbertian Processes of Order p

Ahmad R. Soltani, Majid Hashemi

Abstract We consider periodically correlated autoregressive processes of order p in Hilbert spaces. Our studies on these processes involve existence, strong law of large numbers, central limit theorem and parameter estimation.

47.1 Introduction

The Hilbertian autoregressive model of order 1 (ARH(1)) generalizes the classical AR(1) model to random elements with values in Hilbert spaces. This model was introduced by Bosq (1991), then studied by several authors, as Mourid (1993), Besse and Cardot (1996), Pumo (1999), Mas (2002, 2007), Horvath, Huskova and Kokoszka (2010). Periodically correlated processes in general and PC autoregressive models in particular have been widely used as underlying stochastic processes for certain phenomena.

PC Hilbertian processes, of weak type, were introduced and studied by Soltani and Shishehbor (1998, 1999). These processes assume interesting time domain and spectral structures. The periodically correlated autoregressive Hilbertian processes of order one were introduced by Soltani and Hashemi (2010). The existence, covariance structure, strong law of large numbers and central limit theorem, are the topics that are covered by them.

In this work, we consider PC autoregressive Hilbertian processes of order $p \geq 1$. We defined periodically correlated autoregressive Hilbertian processes of order p (PCARH(p)) as follows:

A centered discrete time second order Hilbertian process $\mathscr{X} = \{X_n, n \in \mathbb{Z}\}$ is called PCARH(p) with period T, associated with $(\varepsilon, \rho_1, \rho_2, \cdots, \rho_p)$ if it is periodically

---

Ahmad R. Soltani
Kuwait University, Safat, Kuwait, e-mail: soltani@kuc01.kuniv.edu.kw

Majid Hashemi
Shiraz University, Shiraz, Iran, e-mail: maryam3752@yahoo.com

correlated and satisfies

$$X_n = \rho_{1,n}(X_{n-1}) + \rho_{2,n}(X_{n-2}) + \cdots + \rho_{p,n}(X_{n-p}) + \varepsilon_n, \tag{47.1}$$

where $\varepsilon_n = \{(\varepsilon_{nT}, \cdots, \varepsilon_{nT+T-1})', n \in \mathbb{Z}\}$ is a zero mean, strongly second order process with orthogonal values and orthogonal components, $\rho_i = (\rho_{i,0}, \cdots, \rho_{i,T-1}), i = 1, \cdots, p$ and for each $i = 1, \cdots, p$, $\{\rho_{i,n}, n \in \mathbb{Z}\}$ is a T-periodic sequence in $\mathscr{L}(\mathbf{H})$ with respect to n, with $\rho_{p,n} \neq 0$. Condition $\rho_{p,n} \neq 0$ is of course necessary for identifiability of p.
We let Hilbert space $\mathbf{H}^p$ to be the Cartesian product of p copies of $\mathbf{H}$, equipped with the inner product

$$< (x_1, \cdots, x_p), (y_1, \cdots, y_p) >_p := \sum_{j=1}^{p} < x_j, y_j >, \tag{47.2}$$

where $x_1, \cdots, x_p, y_1, \cdots, y_p \in \mathbf{H}$. We denote the norm in $\mathbf{H}^p$ by $\| \cdot \|_p$, the Hilbert space of bounded linear operators over $\mathbf{H}^p$ by $\mathscr{L}(\mathbf{H}^p)$.
Now let us set $\mathscr{Y} = \{\mathbf{Y}_n,\ n \in \mathbb{Z}\}$, where

$$\mathbf{Y}_n = (X_n, X_{n-1}, \cdots, X_{n-p+1})',\ n \in \mathbb{Z}, \tag{47.3}$$

and

$$\xi = \{\xi_n,\ n \in \mathbb{Z}\}, \tag{47.4}$$

with $\xi_n = (\varepsilon_n, 0, \cdots, 0)'$, $n \in \mathbb{Z}$, where 0 appears p-1 times. We define the following operator on $\mathbf{H}^p$

$$\boldsymbol{\Pi}_n = \begin{pmatrix} \rho_{1,n} & \rho_{2,n} & \cdots & 0 & \rho_{p,n} \\ I & 0 & \cdots & 0 & 0 \\ 0 & I & \cdots & 0 & 0 \\ \vdots & \vdots & \ddots & \vdots & \vdots \\ 0 & 0 & \cdots & I & 0 \end{pmatrix}, \tag{47.5}$$

where I denotes the identity operator.
We have the following simple but crucial lemma.

Lemma 47.1. *Let $\mathscr{X}$ be a PCARH(p) with period T, associated with $(\varepsilon, \rho_1, \rho_2, \cdots, \rho_p)$. Then $\mathscr{Y}$ is an $PCARH^p(1)$ process associated with $(\gamma, \boldsymbol{\Pi})$ where $\boldsymbol{\Pi}$ is given in (5) and $\gamma_n = (\xi_{nT}, \xi_{nT+1}, \cdots, \xi_{nT+T-1})'$.*

For expressing limiting theorems we need some extra notations. Let $\mathscr{W} = \{W_n; n \in \mathbb{Z}\}$, where

$$\boldsymbol{W}_n = (\boldsymbol{Y}_{nT}, \boldsymbol{Y}_{nT+1}, \cdots, \boldsymbol{Y}_{nT+T-1})', \tag{47.6}$$

and

$$\delta_n = (\delta_{n,0}, \delta_{n,1}, \cdots, \delta_{n,T-1})', \;\; n \in \mathbb{Z}, \tag{47.7}$$

where $\delta_{n,i} = \sum_{k=0}^{i} A_{k,i}\xi_{nT-k+i}$ for $i = 0, \cdots, T-1$, and $A_{k,i} = \boldsymbol{\Pi}_i \cdots \boldsymbol{\Pi}_{i-k+1}$, $k = 1, 2, \cdots$ and $A_{0,i} = I_p$. We note that $\delta_n = \boldsymbol{V}\gamma_n$, and

$$\boldsymbol{C}_{\gamma_n} = \begin{pmatrix} C_{\xi_{nT}} & 0 & 0 \cdots & 0 \\ 0 & C_{\xi_{nT+1}} & 0 \cdots & 0 \\ \vdots & \vdots & \vdots \ddots & \vdots \\ 0 & 0 & 0 \cdots & C_{\xi_{nT+T-1}} \end{pmatrix}, \tag{47.8}$$

and

$$\boldsymbol{V} = \begin{pmatrix} I_p & 0 & 0 & \cdots 0 \\ A_{1,1} & I_p & 0 & \cdots 0 \\ A_{2,2} & A_{1,2} & I_p & \cdots 0 \\ \vdots & \vdots & \vdots & \ddots 0 \\ A_{T-1,T-1} & A_{T-2,T-1} & A_{T-3,T-1} & \cdots I_p \end{pmatrix}. \tag{47.9}$$

Let us set $\boldsymbol{U} = \boldsymbol{V}^{-1}$. The we easily see that

$$\boldsymbol{U} = \begin{pmatrix} I_p & 0 & 0 & 0 & \cdots & 0 \\ -\boldsymbol{\Pi}_1 & I_p & 0 & 0 & \cdots & 0 \\ 0 & -\boldsymbol{\Pi}_2 & I_p & 0 & \cdots & 0 \\ \vdots & \vdots & \vdots & \vdots & \ddots & 0 \\ 0 & 0 & 0 & \cdots & -\boldsymbol{\Pi}_{T-1} & I_p \end{pmatrix}. \tag{47.10}$$

Also for given $\boldsymbol{\Pi}_0, \cdots, \boldsymbol{\Pi}_{T-1}$, we define the following operators on $\mathbf{H}^{Tp}$

$$\boldsymbol{\Delta} = \begin{pmatrix} 0 & 0 & \cdots & 0 & \boldsymbol{\Pi}_0 \\ 0 & 0 & \cdots & 0 & \boldsymbol{\Pi}_1\boldsymbol{\Pi}_0 \\ \vdots & \vdots & \ddots & \vdots & \vdots \\ 0 & 0 & \cdots & 0 & \boldsymbol{\Pi}_{T-1}\cdots\boldsymbol{\Pi}_0 \end{pmatrix}. \tag{47.11}$$

and $\alpha = \boldsymbol{U}\boldsymbol{\Delta}\boldsymbol{U}^{-1}$,

$$\alpha = \begin{pmatrix} \boldsymbol{\Pi}_0\boldsymbol{\Pi}_{T-1}\cdots\boldsymbol{\Pi}_1 & \boldsymbol{\Pi}_0\boldsymbol{\Pi}_{T-1}\cdots\boldsymbol{\Pi}_2 & \cdots & 0 & \boldsymbol{\Pi}_0 \\ 0 & 0 & \cdots & 0 & 0 \\ \vdots & \vdots & \ddots & \vdots & \vdots \\ 0 & 0 & \cdots & 0 & 0 \end{pmatrix}. \tag{47.12}$$

We will use the natural "projector" of $\mathbf{H}^p$ onto $\mathbf{H}$ defined as

$$\pi(x_1,\cdots,x_p) = x_1, \quad (x_1,\cdots,x_p) \in \mathbf{H}^p \tag{47.13}$$

Assumption A_1: There are integers $k_0,\cdots,k_{T-1} \in [1,\infty)$ such that $\sum_{i=0}^{T-1} \| \boldsymbol{\Pi}_i \|^{k_i} < 1$.

Theorem 47.1. *Under the assumption $\mathbf{A}_1$, the equation $X_n = \rho_{1,n}(X_{n-1}) + \rho_{2,n}(X_{n-2}) + \cdots + \rho_{p,n}(X_{n-p}) + \varepsilon_n$ has a unique solution given by*

$$X_{nT+i} = \sum_{j=0}^{\infty} (\pi A_{j,nT+i}) \xi_{nT+i-j}, \quad n \in \mathbb{Z}, \tag{47.14}$$

where $A_{j,t} = \boldsymbol{\Pi}_t \boldsymbol{\Pi}_{t-1} \cdots \boldsymbol{\Pi}_{t-j+1}$ and the series (14) converges in $L^2_{\mathbf{H}}(\Omega,\mathscr{F},P)$ and with probability one as well.

Theorem 47.2. *Let X_n be a PCARH(p) process with period T. Suppose that there exist $\nu \in \mathbf{H}$ and $\{\alpha_{i,j}\}$, $i = 1,\cdots,p$, $j = 0,\cdots,T-1$ such that*

$$\rho^*_{i,j}(\nu) = \alpha_{i,j}\nu, \ i = 1,\cdots,p, \ j = 0,\cdots,T-1,$$

and $\min_n E < \varepsilon_n, \nu >^2 > 0$. Then $\{< X_n, \nu >, \ n \in \mathbb{Z}\}$ is a PCAR(p) process that satisfies

$$< X_n, \nu > = \alpha_{1,n} < X_{n-1}, \nu > + \alpha_{2,n} < X_{n-2}, \nu > + \cdots + \alpha_{p,n} < X_{n-p}, \nu > + < \varepsilon_n, \nu > .$$

$\mathscr{X}$ is said to be a standard PCARH(p) if assumption $\mathbf{A}_1$ is satisfied.

47.2 Large Sample Theorems

Theorem 47.3. *(SLLN). Let $\mathscr{X}$ be a standard PCARH(p) and $X_0,\cdots,X_{n-1}$ be a finite segment from this model, and $S_n(X) = \sum_{i=0}^{n-1} X_i$. Then as $n \longrightarrow \infty$,*

$$\frac{n^{\frac{1}{4}}}{(\log n)^{\beta}} \frac{S_n(X)}{n} \xrightarrow{a.s} 0, \qquad \beta > \frac{1}{2}. \tag{47.15}$$

By defining $\boldsymbol{I}_{Tp}$ as follows, we have Lemma 2 given below.

$$\boldsymbol{I}_{Tp} = \begin{pmatrix} \boldsymbol{I}_p & 0 & \cdots & 0 \\ 0 & \boldsymbol{I}_p & \cdots & 0 \\ \vdots & \vdots & \ddots & \vdots \\ 0 & 0 & \cdots & \boldsymbol{I}_p \end{pmatrix}. \tag{47.16}$$

Lemma 47.2. $\mathbf{I}_{Tp} - \alpha$ *is invertible if and only if $I_p - A_{T,T}$ is invertiable in $\mathbf{H}^p$.*

We now give a Central Limit Theorem.

Theorem 47.4. *Let $\mathscr{X}$ be a standard PCARH(p) associated $(\varepsilon, \rho_1, \cdots, \rho_p)$, and ε_n are independent and identical distributed and such that $I_p - A_{T,T}$ is invertible. Then*

$$\frac{S_n(X)}{\sqrt{n}} \xrightarrow{\mathscr{D}} \mathscr{N}(0, \Gamma), \tag{47.17}$$

where

$$\Gamma = \frac{1}{T}\pi \boldsymbol{A}' \boldsymbol{U}^{-1}(\mathbf{I}_T - \Delta)^{-1} \boldsymbol{C}_{\gamma_n}(\mathbf{I}_T - \Delta^*)^{-1} \boldsymbol{U}^{*-1} \boldsymbol{A}\pi, \tag{47.18}$$

and $\boldsymbol{A} = (I_p, I_p, \cdots, I_p)'$; and π is defined in (13).

47.3 Parameter estimation

In applications it is indeed crucial to estimate the PCARH(p) coefficients $\rho_1, \rho_2, \cdots, \rho_p$. For estimating the parameters, we use from $\mathscr{Y}$ processes and define $R(n,m) = C_{\boldsymbol{Y}_n, \boldsymbol{Y}_m}$ in (n,m), we obtain the following equations

$$R(\ell - 1, \ell) = \boldsymbol{\Pi}_\ell R(\ell - 1, \ell - 1), \quad \ell = 0, ..., T-1, \ k \geq 1; \tag{47.19}$$

Now set $D^{\boldsymbol{Y}}_{\ell-1} = E\boldsymbol{Y}_{\ell-1} \otimes \boldsymbol{Y}_\ell$, then

$$D^{\boldsymbol{Y}}_{\ell-1} = \boldsymbol{\Pi}_\ell C^{\boldsymbol{Y}}_{\ell-1}, \quad \ell = 0, ..., T-1, \ k \geq 1; \tag{47.20}$$

when the inference on $\boldsymbol{\Pi}_l$ is based on the moment equation (20), identifiability holds if and only if $ker(C^{\boldsymbol{Y}}_{\ell-1}) = 0$.
$C^{\boldsymbol{Y}^{-1}}_{\ell-1}$ is extremely irregular, we should propose a way to regularize it i.e. find out $C^{\boldsymbol{Y}^{\perp}}_{\ell-1}$ say, a linear operator close to $C^{\boldsymbol{Y}^{-1}}_{\ell-1}$ and having additional continuity properties. We set

$$C^{\boldsymbol{Y}^{\perp}}_{\ell-1} = \sum_{j<k_n} \frac{1}{\lambda_{j,\ell-1}} e_{j,\ell-1} \otimes e_{j,\ell-1}, \tag{47.21}$$

where $(k_n)_{n\in\mathbb{N}}$ is an increasing sequence tending to infinity. If (20) is the starting point in our estimation procedure, replacing the unknown operators by their empirical counterparts gives:

$$\hat{\boldsymbol{\Pi}}_\ell = \hat{D}^{\boldsymbol{Y}}_{\ell-1} \hat{C}^{\boldsymbol{Y}^{\perp}}_{\ell-1}, \quad \ell = 0, ..., T-1, \ k \geq 1; \tag{47.22}$$

where

$$\hat{C}^{\boldsymbol{Y}}_{\ell-1}(x) = \frac{1}{N} \sum_{k=0}^{N-1} < \boldsymbol{Y}_{\ell-1+kT}, x > \boldsymbol{Y}_{\ell-1+kT}, \quad x \in \mathbf{H}^p. \tag{47.23}$$

$$\hat{D}^{\boldsymbol{Y}}_{\ell-1}(x) = \frac{1}{N}\sum_{k=0}^{N-1} < \boldsymbol{Y}_{\ell-1+kT}, x > \boldsymbol{Y}_{\ell+kT}, \quad x \in \mathbf{H}^p. \tag{47.24}$$

$$\hat{C}^{\boldsymbol{Y}\perp}_{\ell-1} = \sum_{j<k_n} \frac{1}{\hat{\lambda}_{j,\ell-1}} \hat{e}_{j,\ell-1} \otimes \hat{e}_{j,\ell-1}, \tag{47.25}$$

and $\hat{\lambda}_{j,\ell-1}$, $\hat{e}_{j,\ell-1}$ are eigenvalue and eigenvector of $\hat{C}^{\boldsymbol{Y}}_{\ell-1}$.
By estimating Π_ℓ, in fact we estimate $\rho_{1,\ell}, \rho_{2,\ell}, \cdots, \rho_{p,\ell}$.

References

1. Besse, P., Cardot, H.: Approximation spline de la prévision d'un processus fonctionnel autorégressif d'ordre. Canad. J. Stat. **24**, 467–487 (1996)
2. Bosq, D.: Linear Processes in Function Spaces. Theory and Applications. Lecture Notes in Statistics, 149, Springer Verlag, Berlin (2000)
3. Horvath, L., Huskova, M., Kokoszka, P.: Testing the stability of the functional autoregressive process. J. Multivariate Anal. **101**, 352–367 (2010)
4. Mas, A.: Weak convergence for the covariance operators of a Hilbertian linear process. Stoch. Proc. Appl. **99**, 117–135 (2002)
5. Mas, A.: Weak convergence in the functional autoregressive model. J. Multivariate Anal. **98**, 1231–1261 (2007)
6. Mourid, T.: Processus autoregressifs d'ordre superieur. Acad. Sci. 167–172 (1993)
7. Pumo, B.: Prediction of continuous time processes by C[0,1]-valued autoregressive process. Stat. Infer. Stoch. Proc. **3**, 1–13 (1999)
8. Soltani, A.R., Shishehbor, Z.: A spectral representation for weakly periodic sequences of bounded linear transformations. Acta. Math. Hungar **80**, 265–270 (1998)
9. Soltani, A.R., Shishehbor, Z.: Weakly periodic sequences of bounded linear transformations: A spectral characterization. J. Georgian Math. **6**, 91–98 (1999)
10. Soltani, A.R., Hashemi, M.: Periodically Correlated Autoregressive Hilbertian Processes. Stat. Infer. Stoch. Proc., to appear (2010)

Chapter 48
Bases Giving Distances. A New Semimetric and its Use for Nonparametric Functional Data Analysis

Catherine Timmermans, Laurent Delsol, Rainer von Sachs

Abstract The BAGIDIS semimetric is a highly adaptive wavelet-based semimetric. It is particularly suited for dealing with curves presenting horizontally- and vertically-varying sharp local patterns. One can advantageously make use of this semimetric in the framework of nonparametric functional data analysis.

48.1 Introduction

This communication aims firstly at highlighting a new semimetric for measuring dissimilarities between regularly discretized curves, typically time series or spectra. Its main originality is that it is based upon the expansion of each series of a dataset into a *different* wavelet basis, one that is particularly suited for its description. Measuring dissimilarities in such a way implies comparing not only the projections of the series onto the bases, as usual, but also the bases themselves. As a consequence of this feature, our semimetric has the ability to capture the variations of patterns occurring in series along both the vertical and the horizontal axis. This property makes the semimetric particularly powerful when dealing with curves with sharp local features that might be affected simultaneously by horizontal shifts and vertical amplification.

Secondly, this communication aims at illustrating how we can advantageously make use of our semimetric in the framework of nonparametric functional data analysis (as developped in *Ferraty and Vieu* (2006)) when the curves we are dealing with

Catherine Timmermans
Université Catholique de Louvain, Belgium, e-mail: catherine.timmermans@uclouvain.be

Laurent Delsol
Université d'Orléans, France, e-mail: laurent.delsol@univ-orleans.fr

Rainer von Sachs
Université Catholique de Louvain, Belgium, e-mail: rainer.vonsachs@uclouvain.be

are characterized by some horizontally- and vertically-varying sharp patterns. Simulated examples are shown as well as a real data example.

48.2 Definition of the semimetric

Our semimetric has been introduced in *Timmermans and von Sachs* (2010), under the acronym BAGIDIS, that stands for *BAses GIving DIStances*. Key ideas are as follows.

Preliminary observation. When we evaluate dissimilarities between series visually, we intuitively investigate first the global shapes of the series for estimating their ressemblance, before refining the analysis by comparing the smaller features of the series. In other words our comparison is based upon a hierarchical comprehension of the curves. This visual approach inspired us to define our semimetric: we expand each series in a (different, series-adapted) basis that describes its features hierarchically, in the sense that the first basis vectors carry the main features of the series while subsequent basis vectors support less significant patterns; afterwards, we compare both the bases and the expansions of the series onto those bases, rank by rank, according to the hierarchy.

Expanding each series of the dataset in the *Unbalanced Haar Wavelet Basis* that is best suited for the hierarchical description of its shape. The family of *Unbalanced Haar Wavelet Bases* has been introduced by *Girardi and Sweldens* (1997). It consists in orthonormal bases that are made of one constant vector and a set of Haar-like (i.e. *up-and-down shaped*) orthonormal wavelets whose discontinuity point (hereafter the breakpoint) between the positive and negative parts is not necessarily located at the middle of its support. The *Bottom Up Unbalanced Haar Wavelet Transform* (BUUHWT), an algorithm that was developped in *Fryzlewicz* (2007), allows to select amongst this family of bases the best basis for describing a given series hierarchically. Besides this hierarchical organisation, the selected basis inherits the good capacity of Haar wavelets to efficiently capture sharp patterns.

We denote the expansion of a series $\boldsymbol{x}^{(i)}$ in that basis as $\boldsymbol{x}^{(i)} = \sum_{k=0}^{N-1} d_k^{(i)} \boldsymbol{\psi}_k^{(i)}$, where the coefficients $d_k^{(i)}$ (hereafter the *detail* coefficients) are the projections of the series $\boldsymbol{x}^{(i)}$ on the corresponding basis vectors $\boldsymbol{\psi}_k^{(i)}$ and where the set of vectors $\{\boldsymbol{\psi}_k^{(i)}\}_{k=0\ldots N-1}$ is the Unbalanced Haar wavelet basis that is best suited to the series $\boldsymbol{x}^{(i)}$, as obtained using the BUUHWT algorithm. Besides, we denote $b_k^{(i)}$, the breakpoint of the wavelet $\boldsymbol{\psi}_k^{(i)}$, at every rank $k \neq 0$.

Defining a semimetric by taking advantage of the hierarchy of those expansions. As shown in *Fryzlewicz* (2007), the ordered set of breakpoints $\{b_k^{(i)}\}_{k=1\ldots N-1}$ determines the basis $\{\boldsymbol{\psi}_k^{(i)}\}_{k=0\ldots N-1}$ uniquely. As a consequence, the set of points $\{y_k^{(i)}\}_{k=1\ldots N-1} = \{(b_k^{(i)}, d_k^{(i)})\}_{k=1\ldots N-1}$ determines the shape of the series $\boldsymbol{x}^{(i)}$ uniquely - i.e. it determines the series, except for a change of the mean level of the series, that is encoded by the additional coefficient $d_0^{(i)}$. Given that, we define the BAGIDIS

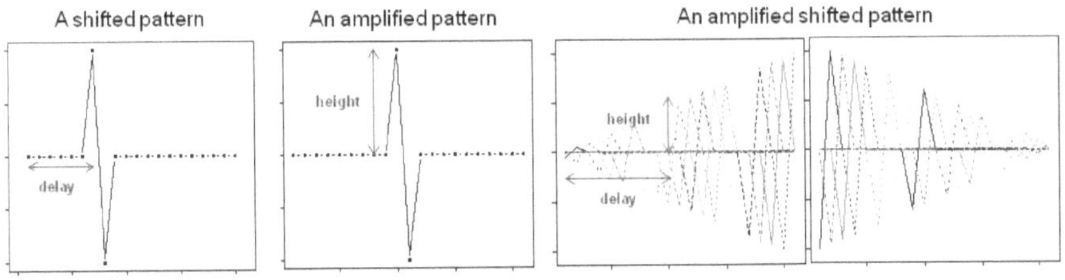

Fig. 48.1: Illustration of the simulated examples described in Section 3.

semimetric as a p-norm (weighted) distance in the *breakpoints-details* plane:

$$d_p^{\text{BAGIDIS}}(\boldsymbol{x}^{(1)},\boldsymbol{x}^{(2)}) = \sum_{k=1}^{N-1} w_k \left\| \boldsymbol{y}_k^{(1)} - \boldsymbol{y}_k^{(2)} \right\|_p = \sum_{k=1}^{N-1} w_k \left(\left| b_k^{(1)} - b_k^{(2)} \right|^p + \left| d_k^{(1)} - d_k^{(2)} \right|^p \right)^{1/p}$$

with $p = 1,2,\ldots,\infty$, and where w_k is a well suited weight function. As such, this semimetric takes advantage of the hierarchy of the well adapted unbalanced Haar wavelet bases: breakpoints and details of similar rank k in the hierarchical description of each series are compared to each other, and the resulting differences can be weighted according to that rank. As the breakpoints point to level changes in the series, the term $\left| b_k^{(1)} - b_k^{(2)} \right|$ can be interpreted as a measure of the difference of location of the features, along the horizontal axis. Being a difference of the projections of the series onto wavelets that encode level changes, the term $\left| d_k^{(1)} - d_k^{(2)} \right|$ can be interpreted as a measure of the differences of the amplitudes of the features, along the vertical axis.

Investigating the balance between breakpoints and details differences. We introduce an extension of the BAGIDIS semimetric as follows:

$$d_{p\lambda}^{\text{BAGIDIS}}(\boldsymbol{x}^{(1)},\boldsymbol{x}^{(2)}) = \sum_{k=1}^{N-1} w_k \left(\lambda \left| b_k^{(1)} - b_k^{(2)} \right|^p + (1-\lambda) \left| d_k^{(1)} - d_k^{(2)} \right|^p \right)^{1/p}$$

with $\lambda \in [0;1]$. This parameter λ actually defines a scaling in the *breakpoints-details* plane, and hence in the original units of the problem. Setting λ at its extreme values 0 or 1 allows to investigate the contributions of the breakpoints differences and details differences separately. In a prediction setting, λ can easily be optimized using cross-validation. Besides, the presence of this parameter allows the semimetric to be robust with respect to scaling effects: if λ is optimized according to a given criteria (such as the mean square error of a prediction model), the relative dissimilarities between the series of a dataset will remain the same, whatever the scales of measurements along the horizontal and vertical axes, so that the predictive qualities of the model will not be affected by such a change in the units of measurements. **Choosing the weights.** In a prediction setting, weights should ideally be 1 at rank k if that rank carries information for discriminating the series, and 0 otherwise. This is easily obtained using a cross-validation procedure. When no prediction criterium is at hand, or in order to get a first idea of how the dissimilarities do behave, we suggest in *Timmermans and von Sachs* (2010) to *a priori* use the weight function $w_k = \frac{\log(N+1-k)}{\log(N+1)}$. This allows to associate a large weight to the comparison of features encoded at the first rank of the hierarchy, and a decreasing weight to the smaller features at the end of the hierarchy, which is empirically what we expect.

48.3 Nonparametric functional data analysis

We want to make use of our semimetric together with the extended Nadaraya-Watson estimators for regression and classification on functional data that have been proposed in *Ferraty and Vieu* (2006). Under quite general hypotheses, we have shown that a curve $\boldsymbol{x}$ is of fractal order $N^* - 1$ with respect to the BAGIDIS semimetric, where N^* is the number of non-zero weights w_k. This ensures we can reach a quite good rate of convergence for predictions based upon those estimators used together with the BAGIDIS semimetric, provided N^* is small enough - i.e. provided that the number of significantly discriminative features in the curves of the dataset is not too large - and provided that the general theoretical properties stated in *Ferraty and Vieu* (2006) are satisfied. Three simulated examples of nonparametric functional regressions are presented here. They are illustrated in Figure 48.1.

First, we investigate how BAGIDIS handles horizontal shift and vertical amplification of patterns, separately:
A shifted pattern is related to its delay. The first example involves series of length 21, being zero-valued except for the presence of an *up-and-down* pattern $(10, -10)$ that is horizontally shifted from one series to the next one. Each series is related to the delay at which the *up-and-down* pattern occurs.
An amplified pattern is related to its height. The second example involves series of length 21 being zero-valued except for an *up-and-down* pattern at abscissa 10 and 11, that is more or less amplified from one series to the next one, from amplitude 1 to amplitude 20. Each series is associated with the height of the *up-and-down* pattern.

The series are affected by a gaussian noise with standard deviation $\sigma = 0.1$ and the responses are affected by a gaussian noise with standard deviation $\sigma = 1$. The following test is performed 100 times: we generate 60 time series following each the above described models, with 3 noisy replications for each possible value of the delay and/or height. Then, we select randomly 45 series out of those 60 series, and use them as a training set to calibrate the regression model. Using the model for predicting the responses associated with the 15 remaining series and comparing it with their 'true' associated responses, we calculate the associated mean square error (MSE). The performances of the BAGIDIS semimetric with various values of λ, and with the '*a priori*' weights are compared to the ones of classical semimetrics: the functional pca-based semimetric with with various number of principal components,

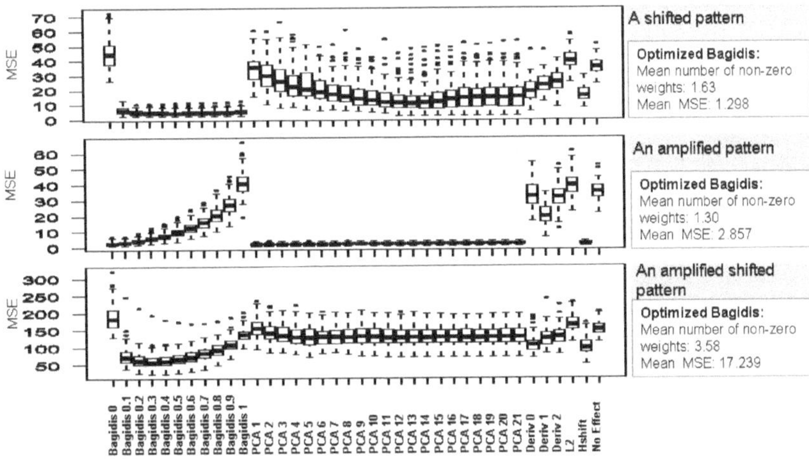

Fig. 48.2: Results of the simulated examples described in Section 3.

the derivative-based semimetric with various order of derivation, the *hshift* semimetric (all these semimetrics are described in *Ferraty and Vieu* (2006)). Moreover, performances of a simple L_2 distance are provided, as well as a *no effect* prediction (i.e. a prediction by the mean of the response values of the training set), that acts as a benchmark for the performances. Resulting empirical distributions of the MSE for each semimetric are presented in Figure 48.2. Afterwards, the parameter λ and the weigths w_k are simultaneously optimized using a cross-validation procedure for each sampled training set. The resulting mean number of selected weights and the mean MSE achieved in such a way are also indicated in Figure 48.2.

As expected, BAGIDIS has excellent performances, compared to all competitors, for dealing with the *shifted* pattern, as soon as $\lambda > 0$ - i.e. as soon as the differences between the breakpoints are taken into account. The performance is further improved when λ is optimized and the discriminative ranks are selected. On average only $N^* = 1.63$ non-zero weights are needed. Not surprisingly, the *not-shifted* amplified pattern is best tackled by the PCA semimetrics. Nevertheless, it is interesting to note that our semimetric BAGIDIS performs quite well too, with λ close to 0 - i.e where only amplitude differences are taken into account.

Our third example involves both horizontal shifts and vertical amplification: **an amplified and shifted pattern in time series is related to a value depending on its height and delay.** We consider series of length 21, being zero-valued except for the presence of an *up-and-down* pattern. That pattern appears after a certain delay and at a certain amplitude, this amplitude increasing with the delay for one half of the dataset and decreasing with it for the other half of the dataset. The responses associated with those curves are defined so as to not depend only on the height nor on the delay. For the family of curves with an amplitude of the pattern increasing with the delay, the response is the *delay*. For the family of curves with an amplitude of the patterns decreasing with the delay, the response is $delay - 20$. The series are

	Occurrence of 0 error out of 43 predictions.	Occurrence of 1 error out of 43 predictions.
PCA-based semimetric, q ≥ 6.	40 times out of 80 **50%**	40 times out of 80 **50%**
Non optimized Bagidis: weights 'a priori', λ =0.5.	48 times out of 80 **60%**	32 times out of 80 **40%**
Optimized Bagidis: →1 non-zero weight. →λ = 0.	12 times out of 12 **100%**	0 times out of 12 **0%**

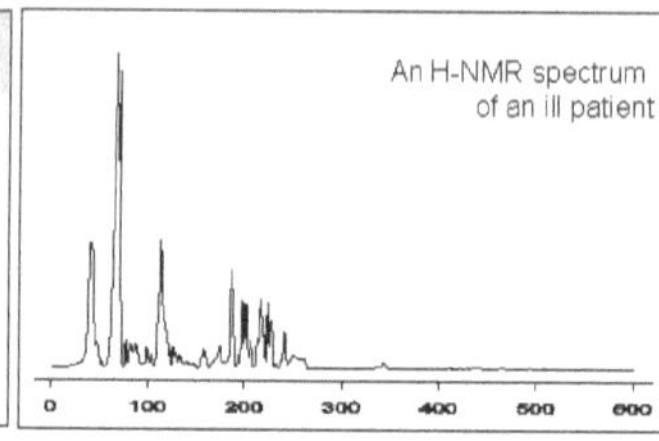

Fig. 48.3: Results and illustration of the H-NMR spectra analysis.

affected by a gaussian noise with standard deviation $\sigma = 0.1$ and the responses are affected by a gaussian noise with standard deviation $\sigma = 1$.

A regression model is estimated for 100 randomly generated training set of 60 series out of 80, and the corresponding validation sets of 20 series are used to estimate the MSE, for various semimetrics. Results are presented in Figure 48.2. Once again, BAGIDIS performs very well, and significantly better than competitors. As expected, an intermediate value of λ seems to be the best choice as both differences in the localizations and in the amplitudes are informative for the prediction. The performance is significantly improved when λ and w_k are optimized. On average only $N^* = 3.48$ non-zero weights are selected.

A real data example. We consider 193 H-NMR serum spectra of length 600, as illustrated in Figure 48.3, 94 of which corresponding to patients suffering from a given illness, the other ones corresponding to healthy patients. We aim at predicting from the spectrum if a patient is healthy or not. A training set of 150 spectra is randomly selected and a functional nonparametric discrimination model is adjusted, with various semimetrics. In order to avoid a confusion of the features in such long series, we make use of the BAGIDIS semimetric together with a sliding window of length 30. This test is repeated 80 times. In each case, the number of misclassification observed on the remaining 43 spectra is recorded. First, the BAGIDIS semimetric is used with its *'a priori'* weight function and with $\lambda = 0.5$. Results are summarized in Figure 48.3. We observe that the non-optimized BAGIDIS obtains *no error* 10% more often than its best competitor, being a pca-based semimetric with at least 6 components. Afterwards, we optimize the weights and the λ parameter of the BAGIDIS semimetric using a cross-validation procedure within the training set, and the resulting model is tested on the remaining 43 series. This test is repeated 12 times on different randomly selected training set, and no prediction error occurs. At each repetition, only 1 non-zero weight is selected, and λ is chosen to be zero. This indicates that horizontal shifts do affect the series but only the amplitudes of the patterns are discriminative. It illustrates well the ability of BAGIDIS to take into account both horizontal and vertical variations of the patterns, as well as its flexibility in the use of those informations.

Acknowledgements This database was collected and preprocessed for a study lead by *P. de Tullio, M. Frédérich* and *V. Lambert* (Université de Liège). Their agreement to use the database is gratefully acknowledged. The name of the concerned illness remains temporarily confidential.

References

1. Ferraty, F., Vieu, P.: Nonparametric Functional Data Analysis: Theory and Practice. Series in Statistics, Springer (2006)
2. Fryzlewicz, P.: Unbalanced Haar Technique for Non Parametric Function Estimation. J. Am. Stat. Assoc. **102**, 1318-1327 (2007)
3. Girardi, M., Sweldens, W.: A new class of Unbalanced Haar Wavelets that form an Unconditional Basis for Lp on General Measure Spaces. J. Fourier Anal. Appl. **3**, 457-474 (1997)
4. Timmermans, C., von Sachs, R.: Bagidis, a new method for statistical analysis of differences between curves with sharp discontinuities. Submitted, URL: http://www.stat.ucl.ac.be/ISpub/dp/2010/DP1030.pdf (2010)

List of Contributors

Ana **Aguilera** (chapter 1, p1)
Universidade de Granada, Spain, e-mail: aaguiler@ugr.es
Carmen **Aguilera-Morillo** (chapter 1, p1)
Universidade de Granada, Spain, e-mail: caguiler@ugr.es
Germán **Aneiros** (chapters 2, p9; 3, p17)
Universidade da Coruña, Spain, e-mail: ganeiros@udc.es
Ramón **Artiaga** (chapter 36, p231)
University of A Coruña, Spain, e-mail: rartiaga@udc.es
John A. D. **Aston** (chapter 4, p23)
University of Warwick, UK, e-mail: j.a.d.aston@warwick.ac.uk
Mohammed K. **Attouch** (chapter 5, p27)
Université de Sidi Bel Abbès, Algeria, e-mail: attou_kadi@yahoo.fr
Alexander **Aue** (chapter 6, p33)
University of California, Davis, USA, e-mail: alex.aue@gmail.com
Neil **Bathia** (chapter 31, p203)
London School of Economics, London, UK, e-mail: neilbathia@googlemail.com
Mareike **Bereswill** (chapter 7, p41)
University of Heidelberg, Germany, e-mail: mareike.bereswill@web.de
Graciela **Boente** (chapter 8, p49)
Universidad de Buenos Aires and CONICET, Argentina, e-mail: gboente@dm.uba.ar
Paula **Bouzas** (chapter 9, p55)
University of Granada, Spain, e-mail: paula@ugr.es
Brian **Caffo** (chapter 23, 149)
Johns Hopkins University, Baltimore, USA, e-mail: bcaffo@jhsph.edu
Stéphane **Canu** (chapter 29, p189)
INSA de Rouen, St Etienne du Rouvray, France, e-mail: scanu@insa-rouen.fr
Ricardo **Cao** (chapter 2, p9)
Universidade da Coruña, Spain, e-mail: rcao@udc.es

Gerda **Claeskens** (chapter 46, p297)
Katholieke Universiteit Leuven, Belgium, e-mail: gerda.claeskens@econ.kuleuven.be
Ciprian **Crainiceanu** (chapter 23, 149; 45, p291)
Johns Hopkins University, Baltimore, USA, e-mail: ccrainic@jhsph.edu
Rosa **Crujeiras** (chapter 10, p63)
University of Santiago de Compostela, Spain, e-mail: rosa.crujeiras@usc.es
Pedro **Delicado** (chapter 11, p71)
Universitat Politècnica de Catalunya, Spain, e-mail: pedro.delicado@upc.edu
Laurent **Delsol** (chapters 12, p77; 48, p307)
Université d'Orléans, France, e-mail: laurent.delsol@univ-orleans.fr
Jacques **Demongeot** (chapter 13, 85)
UniversitJ́. Fourier, Grenoble, France, e-mail: Jacques.Demongeot@imag.fr
Emmanuel **Duflos** (chapter 29, p189)
INRIA Lille - Nord Europe/Ecole Centrale de Lille, Villeneuve d'Ascq, France, e-mail: emmanuel.duflos@inria.fr
Manuel **Escabias** (chapter 1, p1)
Universidade de Granada, Spain, e-mail: escabias@ugr.es
Manuel **Febrero-Bande** (chapter 14, 91)
University of Santiago de Compostela, Spain, e-mail: manuel.febrero@usc.es
Frédéric **Ferraty** (chapters 3, p17; 12, p77; 15, p97; 16, p103; 17, p111; 41, p263)
Université de Toulouse, France, e-mail: ferraty@math.univ-toulouse.fr
Liliana **Forzani** (chapter 18, p117)
Instituto de Matemática Aplicada del Litoral - CONICET, Argentina, e-mail: liliana.forzani@gmail.com
Ricardo **Fraiman** (chapter 18, p117; 19, p123)
Universidad de San Andrés, Argentina, and Universida de la República, Uruguay, e-mail: rfraiman@udesa.edu.ar
Mario **Francisco-Fernández** (chapter 36, p231)
University of A Coruña, Spain, e-mail: mariofr@udc.es
Alba M. **Franco-Pereira** (chapter 20, p131)
Universidad de Vigo, Spain, e-mail: alba.franco@uvigo.es
Laurent **Gardes** (chapter 21, p135)
INRIA Rhône-Alpes and LJK, Saint-Imier, France, e-mail: Laurent.Gardes@inrialpes.fr
Gery **Geenens** (chapter 22, p141)
University of New South Wales, sydney, Australia, e-mail: ggeenens@unsw.edu.au
Abdelkader **Gheriballah** (chapter 5, p27)
Université de Sidi Bel Abbès, Algeria, e-mail: gheribaek@yahoo.fr
Ramon **Giraldo** (chapter 43, p277)
Universidad Nacional de Colombia, Bogota, Colombia, e-mail: rgiraldoh@unal.edu.co
Stéphane **Girard** (chapter 21, p135)
INRIA Rhône-Alpes and LJK, Saint-Imier, France, e-mail: Stephane.Girard@inrialpes.fr
Aldo **Goia** (chapters 15, p97)
Università del Piemonte Orientale, Novara, e-mail: aldo.goia@eco.unipmn.it

Wenceslao **González-Manteiga** (chapter 14, 91)
University of Santiago de Compostela, Spain, e-mail: wenceslao.gonzalez@usc.es
Sonja **Greven** (chapter 23, 149)
Ludwig-Maximilians-Universität München, Munich, Germany, e-mail: sonja.greven@stat.uni-muenchen.de
Oleksandr **Gromenko** (chapter 24, 155)
Utah State University, Logan, USA, e-mail: agromenko@gmail.com
Jaroslaw **Harezlak** (chapter 25, 161)
Indiana University School of Medicine, Indianapolis, USA, e-mail: harezlak@iupui.edu
Majid **Hashemi** (chapter 47, 301)
Shiraz University, Shiraz, Iran, e-mail: maryam3752@yahoo.com
Siegfried **Hörmann** (chapters 6, p33; 26, p169)
Université Libre de Bruxelles, e-mail: shormann@ulb.ac.be
Ivana **Horová** (chapter 27, 177)
Masaryk University, Brno, Czech Republic, e-mail: horova@math.muni.cz
Lajos **Horváth** (chapter 6, p33)
University of Utah, Salt Lake City, USA, e-mail: horvath@math.utah.edu
Marie **Huškova** (chapter 6, p33)
Charles University of Prague, Czech Republic, e-mail: huskova@karlin.mff.cuni.cz
Maarten **Jansen** (chapter 46, p297)
Université Libre de Bruxelles, Belgium, e-mail: maarten.jansen@ulb.ac.be
Jan **Johannes** (chapter 7, p41)
Université Catholique de Louvain, Belgium, e-mail: jan.johannes@uclouvain.be
Sungkyu **Jung** (chapter 28, p183)
University of Chapel Hill, Chapel Hill, USA, e-mail: sungkyu@email.unc.edu
Hachem **Kadri** (chapter 29, p189)
INRIA Lille - Nord Europe/Ecole Centrale de Lille, Villeneuve d'Ascq, France, e-mail: hachem.kadri@inria.fr
Claudia **Kirch** (chapter 4, p23)
Karlsruhe Institute of Technology, Germany, e-mail: claudia.kirch@kit.edu
Alois **Kneip** (chapter 30, p197)
Universität Bonn, Bonn, Germany, e-mail: akneip@uni-bonn.de
Piotr **Kokoszka** (chapters 24, 155; 26, p169)
Utah State University, USA, e-mail: piotr.kokoszka@usu.edu
David **Kraus** (chapter 38, p245)
Ecole Polytechnique Fédérale de Lausanne, Lausanne, Switzerland, e-mail: david.kraus@epfl.ch
Ali **Laksaci** (chapters 5, p27; 13, p85)
Université de Sidi Bel Abbès, Algeria, e-mail: alilak@yahoo.fr
Clifford **Lam** (chapter 31, p203)
London School of Economics, London, UK, e-mail: C.Lam2@lse.ac.uk
Rosa E. **Lillo** (chapter 20, p131)
Universidad Carlos III de Madrid, Spain, e-mail: rosaelvira.lillo@uc3m.es

Pamela **Llop** (chapter 18, p117)
Instituto de Matemática Aplicada del Litoral - CONICET, Argentina, e-mail: lloppamela@gmail.com
Jorge **López-Beceiro** (chapter 36, p231)
University of A Coruña, Spain, e-mail: jlopezb@udc.es
Sara **López-Pintado** (chapter 32, p209)
Columbia University, New York, USA, e-mail: sl2929@columbia.edu
Fethi **Madani** (chapter 13, p85)
Université P. Mendès France, Grenoble, France, e-mail: Fethi.Madani@imag.fr
John H. **Maddocks** (chapter 38, p245)
Ecole Polytechnique Fédérale de Lausanne, Lausanne, Switzerland, e-mail: john.maddocks@epfl.ch
Adela **Martínez-Calvo** (chapter 16, p103)
Universidade de Santiago de Compostela, Spain, e-mail: adela.martinez@usc.es
Jorge **Mateu** (chapter 43, p277)
Universitat Politècnica de Catalunya, Barcelona, Spain, e-mail: mateu@mat.uji.es
Ian W. **McKeague** (chapter 33, p213)
Columbia University, New York, USA, e-mail: im2131@columbia.edu
Jose C. S. de **Miranda** (chapter 34, p219)
University of São Paulo, São Paulo, Brazil, e-mail: simon@ime.usp.br
Hans-Georg **Müller** (chapter 35, p225)
University of California, Davis, USA, e-mail: mueller@wald.ucdavis.edu
Antonio **Muñoz-San-Roque** (chapter 2, p9)
Universidad Pontificia de Comillas, Madrird, Spain, e-mail: anotnio.munoz@iit.icai.upcomillas.es
Salvador **Naya** (chapter 36, p231)
University of A Coruña, Spain, e-mail: salva@udc.es
Alicia **Nieto-Reyes** (chapter 37, p239)
Universidad de Cantabria, Spain, e-mail: alicia.nieto@unican.es
Victor M. **Panaretos** (chapter 38, p245)
Ecole Polytechnique Fédérale de Lausanne, Lausanne, Switzerland, e-mail: victor.panaretos@epfl.ch
Efstathios **Paparoditis** (chapter 39, p251)
University of Cyprus, Nicosia, Cyprus, e-mail: stathisp@ucy.ac.cy
Juhyun **Park** (chapter 17, p111)
Lancaster University, Lancaster, U.K., e-mail: juhyun.park@lancaster.ac.uk
Beatriz **Pateiro-López** (chapter 19, p123)
Universidad de Santiago de Compostela, Spain, e-mail: beatriz.pateiro@usc.es
Davide **Pigoli** (chapter 40, p255)
Politecnico di Milano, Italy e-mail: davide.pigoli@mail.polimi.it
Philippe **Preux** (chapter 29, p189)
INRIA Lille - Nord Europe/Université de Lille, Villeneuve d'Ascq, France, e-mail: philippe.preux@inria.fr
Min **Qian** (chapter 33, p213)
The University of Michigan, USA, e-mail: minqian@umich.edu

Alejandro **Quintela-del-Río** (chapter 41, p263)
Universidade da Coruña, Spain, e-mail: aquintela@udc.es
Mustapha **Rachdi** (chapter 13, p85)
Université P. Mendès France, Grenoble, France, e-mail: Mustapha.Rachdi@upmf-grenoble.fr
James O. **Ramsay** (chapter 42, p269)
McGill University, Montreal, Canada, e-mail: ramsay@psych.mcgill.ca
Tim **Ramsay** (chapter 42, p269)
Ottawa Health Research Institute, Canada, e-mail: tramsay@ohri.ca
Timothy W. **Randolph** (chapter 25, p161)
Fred Hutchinson Cancer Research Center, Seattle, USA, e-mail: trandolp@fhcrc.org
Daniel **Reich** (chapter 23, 149)
National Institutes of Health, Bethesda, USA, e-mail: daniel.reich@nih.gov
Daniela **Rodriguez** (chapter 8, p49)
Universidad de Buenos Aires and CONICET, Argentina, e-mail: drodrig@dm.uba.ar
Elvira **Romano** (chapter 43, p277)
Seconda Universitá degli Studi di Napoli, Italy, e-mail: elvira.romano@unina2.it
Juan **Romo** (chapter 20, p131)
Universidad Carlos III de Madrid, Spain e-mail: juan.romo@uc3m.es
Nuria **Ruiz-Fuentes** (chapter 9, p55)
University of Jaén, Spain, e-mail: nfuentes@ujaen.es
María-Dolores **Ruiz-Medina** (chapter 10, p63)
University of Granada, Spain, e-mail: mruiz@ugr.es
Rainer von **Sachs** (chapter 48, p307)
Université Catholique de Louvain, Belgium, e-mail: rainer.vonsachs@uclouvain.be
Ernesto **Salinelli** (chapters 15, p97)
Università del Piemonte Orientale, Novara, e-mail: ernesto.salinelli@eco.unipmn.it
Laura M. **Sangalli** (chapter 40, p255; 42, p269)
Politecnico di Milano, Italy e-mail: laura.sangalli@polimi.it
Pascal **Sarda** (chapter 30, p197)
Institut de Mathématiques de Toulouse, France, e-mail: sarda@math.univ-toulouse.fr
Piercesare **Secchi** (chapter 44, p283)
Politecnico di Milano, Italy, e-mail: piercesare.secchi@polimi.it
Damla **Şentürk** (chapter 35, p225)
Penn State University, University Park, USA, e-mail: dsenturk@stat.psu.edu
Russell **Shinohara** (chapter 45, p291)
Johns Hopkins University, Baltimore, USA, e-mail: rshinoha@jhsph.edu
Leen **Slaets** (chapter 46, p297)
Katholieke Universiteit Leuven, Belgium, e-mail: leen.slaets@econ.kuleuven.be
Ahmad R. **Soltani** (chapter 47, p301)
Kuwait University, Safat, Kuwait, e-mail: soltani@kuc01.kuniv.edu.kw
Mariela **Sued** (chapter 8, p49)
Universidad de Buenos Aires and CONICET, Argentina, e-mail: msued@dm.uba.ar

Javier **Tarrío-Saavedra** (chapter 36, p231)
University of A Coruña, Spain, e-mail: jtarrio@udc.es
Catherine **Timmermans** (chapter 48, p307)
Université Catholique de Louvain, Belgium, e-mail: catherine.timmermans@uclouvain.be
Mariano **Valderrama** (chapter 1, p1)
University of Granada, Spain, e-mail: valderra@ugr.es
Simone **Vantini** (chapter 44, p283)
Politecnico di Milano, Italy, e-mail: simone.vantini@polimi.it
Philippe **Vieu** (chapter 3, p17; 12, p77; 15, p97; 16, p103; 17, p111; 41, p263)
Université de Toulouse, France, e-mail: vieu@math.univ-toulouse.fr
Juan **Vilar-Frenández** (chapter 2, p9)
Universidade de Coruña, Spain, e-mail: eijvilar@udc.es
Valeria **Vitelli** (chapter 44, p283)
Politecnico di Milano, Italy, e-mail: valeria.vitelli@mail.polimi.it
Kamila **Vopatová** (chapter 27, p177)
Masaryk University, Brno, Czech Republic, e-mail: vopatova@mail.muni.cz
Qiwei **Yao** (chapter 31, p203)
London School of Economics, London, UK, e-mail: Q.Yao@lse.ac.uk
Ying **Wei** (chapter 32, p209)
Columbia University, New York, USA, e-mail: yw2148@columbia.edu

Kolibri 360 GmbH
Ferdinand-Jühlke-Straße 7
99095 Erfurt, Deutschland
produktsicherheit@kolibri360.de